P. Narayanasamy, PhD

Immunology in Plant Health and Its Impact on Food Safety

Pre-publication
REVIEWS,
COMMENTARIES,
EVALUATIONS . . .

"**P**rofessor Narayanasamy's book comprehensively highlights the importance of immunological techniques in a broad range of investigative and industrial applications in plant health and food safety. The author has achieved an easily readable balance between generic information and specific detail that will stimulate the interest of both the student and more established scientist. Making a clear link in the mind of the reader between plant health and food safety is one of the book's most topical and significant features."

Caroline Mohammed, PhD
Senior Research Scientist,
CSIRO Forestry and Forest Products;
Senior Lecturer,
University of Tasmania

"**T**his outstanding book reflects tremendous progress in immunology techniques in a fascinating and scientifically valid way that differs from most other texts on plant immunology. It is easy to read, providing essential information on immunological techniques for the detection, qualification, and characterization of microbial pathogens and plant defense-related proteins in plants exposed to biotic and abiotic stress. It also covers their applications in plant management and in the assessment of contaminants in food and feeds. This book is relevant for a wide audience: prospective and practicing plant pathologists, students and researchers, and food/environment protectionists at large."

Dr. Anna Maria D'Onghia
Head of Integrated Pest Management
Department, Mediterranean Agronomic
Institute of Bari, Italy, International Centre
for Advanced Mediterranean
Agronomic Studies (ICAMAS),
Paris, France

NOTES FOR PROFESSIONAL LIBRARIANS
AND LIBRARY USERS

This is an original book title published by Food Products Press®, an imprint of The Haworth Press, Inc. Unless otherwise noted in specific chapters with attribution, materials in this book have not been previously published elsewhere in any format or language.

CONSERVATION AND PRESERVATION NOTES

All books published by The Haworth Press, Inc. and its imprints are printed on certified pH neutral, acid-free book grade paper. This paper meets the minimum requirements of American National Standard for Information Sciences-Permanence of Paper for Printed Material, ANSI Z39.48-1984.

Immunology in Plant Health and Its Impact on Food Safety

FOOD PRODUCTS PRESS®
Crop Science
Amarjit S. Basra, PhD
Senior Editor

Handbook of Formulas and Software for Plant Geneticists and Breeders edited by Manjit S. Kang

Postharvest Oxidative Stress in Horticultural Crops edited by D. M. Hodges

Encyclopedic Dictionary of Plant Breeding and Related Subjects by Rolf H. G. Schlegel

Handbook of Processes and Modeling in the Soil-Plant System edited by D. K. Benbi and R. Nieder

The Lowland Maya Area: Three Millennia at the Human-Wildland Interface edited by A. Gómez-Pompa, M. F. Allen, S. Fedick, and J. J. Jiménez-Osornio

Biodiversity and Pest Management in Agroecosystems, Second Edition by Miguel A. Altieri and Clara I. Nicholls

Plant-Derived Antimycotics: Current Trends and Future Prospects edited by Mahendra Rai and Donatella Mares

Concise Encyclopedia of Temperate Tree Fruit edited by Tara Auxt Baugher and Suman Singha

Landscape Agroecology by Paul A. Wojkowski

Concise Encyclopedia of Plant Pathology by P. Vidhyasekaran

Molecular Genetics and Breeding of Forest Trees edited by Sandeep Kumar and Matthias Fladung

Testing of Genetically Modified Organisms in Foods edited by Farid E. Ahmed

Agrometeorology: Principles and Applications of Climate Studies in Agriculture by Harpal S. Mavi and Graeme J. Tupper

Concise Encyclopedia of Bioresource Technology edited by Ashok Pandey

Genetically Modified Crops: Their Development, Uses, and Risks edited by G. H. Liang and D. Z. Skinner

Plant Functional Genomics edited by Dario Leister

Immunology in Plant Health and Its Impact on Food Safety by P. Narayanasamy

Abiotic Stresses: Plant Resistance Through Breeding and Molecular Approaches edited by Muhammad Ashraf and Philip John Charles Harris

Multinational Agribusinesses edited by Ruth Rama

Crops and Environmental Change: An Introduction to Effects of Global Warming, Increasing Atmospheric CO_2 and O_3 Concentrations, and Soil Salinization on Crop Physiology and Yield by Seth G. Pritchard and Jeffrey S. Amthor

Teaching in the Sciences: Learner-Centered Approaches edited by Catherine McLoughlin and Acram Taji

Durum Wheat Breeding: Current Approaches and Future Strategies edited by Conxita Royo, M. M. Nachit, N. Di Fonzo, J. L. Araus, W. H. Pfeiffer, and G. A. Slafer

Immunology in Plant Health and Its Impact on Food Safety

P. Narayanasamy, PhD

Food Products Press®
An Imprint of The Haworth Press, Inc.
New York • London • Oxford

Published by

Food Products Press®, an imprint of The Haworth Press, Inc., 10 Alice Street, Binghamton, NY 13904-1580.

© 2005 by The Haworth Press, Inc. All rights reserved. No part of this work may be reproduced or utilized in any form or by any means, electronic or mechanical, including photocopying, microfilm, and recording, or by any information storage and retrieval system, without permission in writing from the publisher. Printed in the United States of America.

Material from *Phytoparasitica; Journal of Plant Diseases and Protection;* and *Journal of Agricultural and Food Chemistry* reprinted with permission from American Chemical Society.

Cover design by Marylouise E. Doyle.

Library of Congress Cataloging-in-Publication Data

Narayanasamy, P., 1937-
   Immunology in plant health and its impact on food safety / P. Narayanasamy.
      p. cm.
   Includes bibliographical references and index.
   ISBN 1-56022-286-7 (hard : alk. paper)—ISBN 1-56022-287-5 (soft : alk. paper)
   1. Plant immunology. 2. Plant diseases. 3. Plant-pathogen relationships. 4. Food—Toxicology—Health aspects. I. Title.
SB750.N35 2004
632'.3—dc22
                                                                                    2004001749

To my parents,
for their love and affection

ABOUT THE AUTHOR

P. Narayanasamy, PhD, is former Professor and Head of the Department of Plant Pathology at Tamil Nadu Agricultural University in Coimbatore, India. Prior to teaching, he worked as a Virus Pathologist at the Indian Agricultural Research Institute in New Delhi. He was the principal investigator for research projects funded by national and international organizations such as the Department of Science and Technology, the Department of Biotechnology, the Government of India, the International Rice Research Institute in the Philippines, and the International Crops Research Institute for Semi-Arid Tropics.

Dr. Narayanasamy is the author/co-author of over 200 scientific papers and seven research bulletins reflecting his vast experience and knowledge in investigations on various diseases affecting cereals, legumes, and vegetable crops. He has also authored two books, *Plant Pathogen Detection and Disease Diagnosis, Second Edition* (2001), and *Microbial Plant Pathogens and Crop Disease Management* (2002). He is the former Editor of the *Madras Agricultural Journal,* published from Tamil Nadu Agricultural University.

Dr. Narayanasamy was awarded a postdoctoral fellowship by the Rockefeller Foundation to pursue research on rice virus diseases at the International Rice Research Institute (1966-1967). He has been elected Fellow of the Indian Phytopathological Society in New Delhi. He has developed immunological techniques for the detection of viruses infecting rice and legumes and the pathogenesis-related proteins induced following the application of antiviral principles.

CONTENTS

PART V: ASSESSMENT OF FOOD SAFETY

Foreword

Thanks to spectacular progress in preventive and curative medicine, the average life span of human beings is increasing. Although medical science is helping to add years to life, it is now important to add life to years. As a result, diets based on safe and hygienic foods are assuming great importance. Because of health consciousness, industrialized countries want to raise food safety standards through the *Codex Alimentarius Commission*. The World Trade Organization Agreement on Agriculture also prescribes stringent sanitary and phytosanitary measures. This book, designed to assist in developing effective systems of assessing the levels of mycotoxins and chemicals in foods and feeds by employing suitable immunological techniques, is timely. The book provides remedies to the common maladies relating to quality and safety of dietary material.

Professor Narayanasamy has compiled and presented with great clarity the latest information on all aspects relating to immunology in plant health and food safety. I hope this book will be widely read by scientists, technologists, and research scholars. It will also be useful for those in charge of assessing food safety at the field level. The presentation of protocols for a large number of immunoassays is an admirable feature of this book.

We owe Professor Narayanasamy a deep debt of gratitude for this labor of love in the cause of improving food and feed quality and safety.

M. S. Swaminathan
Chairman, M. S. Swaminathan Research Foundation,
Chennai, India;
Holder of the UNESCO Chair of Ecotechnology;
Chairman of PUGWASH Movement;
and Former Director General
of the International Rice Research Institute, Philippines

Preface

The primary aim of cultivation of crops is to obtain produce of high quality in required quantity for the use of the grower and for sale of the surplus. Frequently, it has not been possible to achieve this goal because the crops are exposed to both abiotic and biotic stresses, leading to marked reduction in the quantity and quality of agricultural produce. This book underscores the need for realization of the importance of protecting plant health and monitoring the levels of food safety, so that a supply of acceptable foods and feeds and preservation of the environment free of pollution are possible.

Immunology has been instrumental in the rapid development of several disciplines of plant sciences. Immunological techniques have been demonstrated to be cost-effective, simple, sensitive, rapid, and reliable. Their potential to replace conventional analytical methods that are time-consuming and cumbersome has been indicated by several studies. Immunoassays have been extensively applied for the detection and quantification of microbial plant pathogens and plant constituents both in healthy plants and also in plants exposed to abiotic and biotic stresses. The usefulness of immunological techniques for effective application of various approaches of plant health management, such as development of resistant cultivars through genetic manipulation, induction of resistance to diseases, and production of disease-free crops, has been vividly demonstrated. Furthermore, immunodetection of contamination of foods and feeds with dreadful mycotoxins capable of causing cancer and hazardous chemical residues in foodstuffs and feeds has provided reliable data for the development of effective systems for monitoring and checking the movement of undesirable agricultural commodities.

This book provides the essential information, based on an extensive literature search, that will help researchers, faculty, graduate students, and food technologists interested in planning and developing programs intended to obtain agricultural produce of high quality acceptable for human and animal consumption. Protocols for various immunological assays are presented to enable researchers and technologists to select suitable techniques for their investigations. It is hoped that this book will stimulate further research, resulting in the development of systems that may improve the quality of foods and feeds in addition to preserving the environment.

Acknowledgments

My heartfelt thanks and sincere appreciation go to Professor M. S. Swaminathan, who has provided dynamic leadership for the preservation of natural resources and the environment, and for writing the foreword for this book.

My thanks to all my colleagues for their useful suggestions, especially Dr. T. Ganapathy for his appreciable technical help in the preparation of figures and to Mrs. K. Mangayarkarasi for her efficient secretarial assistance.

I wish to express my appreciation to my family members—Mrs. N. Rajakumari, Mr. N. Kumaraperumal, Mrs. Nirmala Suresh, Mr. T. R. Suresh, and Mr. S. Varun Karthik—for their sustained encouragement and the wonderful atmosphere that allowed me to devote my time to the preparation of this book.

List of Abbreviations

ABA	abscisic acid
AFA	antifungal activity
ALP	alkaline phosphatase
BSA	bovine serum albumin
BYDV	barley yellow dwarf virus
CMV	cucumber mosaic virus
CP	coat protein
CTV	citrus tristeza virus
CyMV	cymbidium mosaic virus
DAS-ELISA	double antibody sandwich ELISA
DIBA	dot immunobinding assay
DON	deoxynivalenol
ELFA	enzyme-linked fluorescent assay
ELISA	enzyme-linked immunosorbent assay
FCS	fetal calf serum
FITC	fluorescein isothiocynate
GLC	gas-liquid chromatography
HPLC	high performance liquid chromatography
HRGP	hydroxyproline-rich glycoprotein
HRP	horseradish peroxidase
IEM	immunoelectron microscopy
IF	immunofluorescence assay
IgG	immunoglobulin
ISEM	immunosorbent electron microscopy
MAB	monoclonal antibody
MW	molecular weight
PAB	polyclonal antibody
PBS	phosphate-buffered saline
PCD	programmed cell death
PCR	polymerase chain reaction
PEG	polyethylene glycol
PG	polygalacturonase
PLRV	potato leaf roll virus

PPV	plum pox virus
PR	pathogenesis related
PVX	potato virus X
PVY	potato virus Y
RBC	red blood cell
RISA	radioimmunosorbent assay
SA	salicylic acid
SAR	systemic acquired resistance
SBWMV	soil-borne wheat mosaic virus
SDS-PAGE	sodium dodecylsulfate-polyacrylamide gel electrophoresis
TBIA	tissue blot immunoassay
TLC	thin layer chromatography
TMV	tobacco mosaic virus
TomRSV	tomato ringspot virus
ToMV	tomato mosaic virus
ToRSV	tobacco ringspot virus
TSWV	tomato spotted wilt virus
ZYMV	zucchini yellow mosaic virus

PART I:
PRINCIPLES AND TECHNIQUES

Chapter 1

Introduction

AGROECOSYSTEMS AND MICROBES

Agroecosystems consisting of diverse crops occupy about 30 percent of Earth's surface. The productivity and quality of the agroecosystems depend on several factors including microbial diversity. Life on Earth may not be possible without microorganisms. Microbial diversity directly influences plant productivity and diversity by its differential effects on plant growth and development, plant competition, and nutrient and water uptake. In addition, microbes play an important role in the conservation and restoration of agroecosystems. Estimates on the number microbial species present on Earth vary widely, from 10,000 to 10 million (Hawksworth, 1991; Anonymous, 1994). Only a fraction of the microbial species present on this planet have been precisely identified. Virtually nothing is known about the activities of the large number of unidentified ones, and their role in the functioning of agroecosystems is yet to be understood.

Microbes impact agroecosystems in several ways that may be either beneficial or harmful for the development of plants. The beneficial functions of microbes include soil humus formation, cycling of nutrients, and building of soil tilth and structure, whereas the harmful effects include causation of plant diseases, formation of plant-suppressive compounds, and loss of nutrients required for plant growth and reproduction. Plants, in their turn, may act as important selective factors for the diversity of populations of microbes existing on the plant surface and the rhizosphere in the soil through their influence on the availability of nutrients. The rhizosphere soil is the volume of soil adjacent to and influenced by plant roots. Intense microbial activity occurs in the rhizosphere, due to its proximity to plant root exudates. Rhizosphere microbial communities are distinct from those in other soils.

Cropping systems greatly influence the microbial communities existing on and around plants. The type and sequence of crops (crop rotation) grown have marked influence on the nature of plant pathogens, whereas the levels of resistance determine the extent of damage that may be inflicted by pathogens on different crops. In addition, the type and frequency of irrigation, tillage, and levels of fertilization are the other factors that may influence the incidence and spread of different diseases (Narayanasamy, 2002).

Plants are exposed to abiotic stresses such as adverse environmental conditions, toxic materials, unavailability of nutrients, and biotic stresses such as microbial pathogens, pests, and weeds that adversely affect their health and consequently their productivity. Some of the microbial pathogens not only debilitate the plants but also cause several diseases in humans and animals when they consume foods and feeds contaminated with mycotoxins. In an attempt to protect crops from infection by microbial pathogens, several chemicals are applied indiscriminately and frequently in excess. Such acts lead to avoidable problems such as environmental pollution, development of resistance in pathogens to chemicals, and persistence of residues of chemicals in foods and feeds.

PLANT HEALTH MANAGEMENT

Plant health management is "a science and practice of understanding and overcoming the succession of biotic and abiotic factors that limit plants from achieving their full genetic potential as crops" (Cook, 2000, p. 97). Plant health management aims at achieving the genetic yield potential of crops by various approaches. The yield from a crop may be referred to as absolute, attainable, affordable, and actual. The absolute yield is of scientific interest and represents the genetic potential, but it is not practical or realistic to achieve. The attainable yield is achievable, if all inputs and favorable environments (natural and physical) are provided. However, it is rarely possible to provide such conditions. The affordable yield can be realized by taking into consideration the costs of inputs and other practices to reduce the effects of stresses in order to preserve the health of the crops. The actual yield represents the yield of produce actually harvested and available for use by the growers and for sale to others. Small or large gaps may occur between the affordable yield and actual yield because of failure to provide proper or timely nutrition and inadequate protection against stresses. It is therefore essential to have an adequate knowledge of the nature of the cause(s) that adversely affect plant health and the practices that are to be followed for effectively protecting crop plants against different stresses so that cultivation of crops remains a profitable proposition.

Among the factors that affect plant health, biotic and abiotic stresses are important because they induce marked changes in the metabolic activities of stressed plants, leading to adverse effects on the growth and reproduction of plants. Basic knowledge of structure and functions of healthy plants is necessary to recognize the changes that are induced by biotic and abiotic stresses (see Chapter 5). Plants are capable of synthesizing various compounds that are useful for humans and animals as foods, feeds, and fiber. In addition, they produce several protective substances that confer resistance against microbial plant pathogens (Chapter 6). The interactions between plants and abiotic stresses (Chapter 7), fungal pathogens (Chapter 8), bacterial pathogens (Chapter 9), and viral pathogens (Chapter 10) have a dramatic influence on the metabolic functions of plants, resulting in drastic declines in yield levels.

Various approaches have been made to prevent or reduce the adverse effects of microbial pathogens. Enhancement of levels of resistance through genetic manipulation and engineering resistance (Chapter 11) and induction of disease resistance and production of disease-free plants (Chapter 12) are considered the principal strategies that can be applied to effectively contain microbial pathogens. Furthermore, these strategies are ecologically safe and economically feasible and form the important components of sustainable agriculture. The mycotoxins produced by microbial pathogens have been responsible for several dreadful mycotoxicoses ultimately resulting in the death of exposed humans and animals. Some of these mycotoxins pose potential risks as carcinogenic agents (see Chapter 13). Adverse effects of fungicides and pesticides in foods and feeds have also been revealed by several studies (see Chapter 14).

APPLICATIONS OF IMMUNOLOGY

The science of immunology has developed as a multipurpose technology with numerous applications in biological sciences in general and in medicine and agriculture in particular. Immunological techniques have been demonstrated to be highly specific, sensitive, simple, rapid, and cost-effective and can be automated for large-scale applications (Narayanasamy, 2001). Because of these distinct advantages, immunoassays have largely replaced conventional analytical methods, which are time-consuming, cumbersome, and expensive. Chapter 2 explains the principles of immunological reactions that form the basis for all immunoassays. Methods of preparing antisera containing polyclonal and monoclonal antibodies specific to the desired immunogens are described in detail in Chapter 3. The recent development of producing recombinant antibodies and methods of purification of immunoglobulins and antibody fragments may be useful for quantification of antigens. The protocols for basic immunodiagnostic techniques are presented in Chapter 4. Some techniques have been developed for certain specific investigations and are included in other chapters as appendixes. They may have to be used in tandem or in combination with the procedures mentioned in Chapter 4, if required.

This book focuses on the usefulness of immunoassays for gathering precise information on the nature of plant constituents, and their synthesis and distribution in different plant tissues and organs. Immunodetection in situ of the proteinaceous compounds in plant cells has provided vital information on the localization of bioreactive compounds, which was not possible to gather using the conventional techniques. The presence and distribution of several protective substances that are related to development of resistance to microbial pathogens have been studied using appropriate immunological techniques. Furthermore, detection and identification of microbial pathogens and alterations induced by them in compatible (susceptible) and incompatible (resistant) interactions, as determined by immunological techniques, have provided the basis for formulating strategies for developing plant health management systems.

Immunoassays have provided reliable and sensitive methods for assessment of levels of resistance of crop cultivars and also wild relatives, which often form the reservoirs of disease-resistance genes. The extent of expression of resistance genes can be evaluated and their products, primarily proteins and enzymes, can be detected and analyzed. The interaction between the products of resistance (*R*) genes of host plants and avirulence (*avr*) genes of microbial pathogens has been shown using immunological techniques. Production of disease-free seeds and plant-propagative materials such as tubers, setts, and corms has been possible because of the reliable and rapid indexing methods based on immunodiagnostic tests. Food safety can be confidently monitored by employing suitable, sensitive immuno-assays. Researchers; students of plant pathology, microbiology, biochemistry, plant physiology, environmental sciences, and food technology; and food policy planners will find the information presented in this book useful for the production of healthy, acceptable foods, feeds, and fiber, and a pollution-free environment.

SUMMARY

Agricultural crops are exposed to abiotic and biotic stresses. Microbial plant pathogens cause several economically detrimental diseases resulting in significant quantitative and qualitative losses in foods, feeds, and other plant products. Immunological techniques are advantageous for assessing the adverse effects of microbial plant pathogens and for determining the residues of contaminating mycotoxins.

Chapter 2

Principles of Immunological Reactions

ANTIGENS

Animals, following introduction of an antigen, which may be fairly large molecules or particles containing protein or polyaccharides, exhibit a characteristic immune response. Animals possessing lymphoid cells endowed with receptors capable of combining specifically with the antigen molecules are useful for the production of specific antibodies. Antigens may be defined as substances (macromolecules, cells, or cellular constituents) capable of inducing production of specific antibody proteins (immunoglobulins) capable of reacting with the antigen. This property of the antigen is known as the immunogenicity. The second characteristic of the antigen is its ability to specifically react with the antibody that was produced in the animal immunized with the particular antigen. This property is called antigenicity. Van Regenmortel (1982) suggested the use of the term *antigenic reactivity* in place of *antigenicity* or *antigenic specificity,* since they have several other connotations. Thus antigens are both immunogenic and antigenic. Some small molecules with specific structures are unable to stimulate the production of antibodies. However, they are able to react with antibodies produced by other antigens containing the small molecule as part of their structure. Such small molecules are called haptens. Haptens may be coupled with carrier molecules of sufficient size. Then, antisera-containing antibodies specific to haptens can be produced. Hapten technology has been useful for the production of antisera specific to several pesticides and mycotoxins.

Antigens with molecular weights (MWs) of 10,000 or more are more effective than smaller molecules (such as peptides), which can also elicit antibody production. The receptors present at the surface of lymphoid cells recognize the antigen which leads to a proliferation of plasma cells that secrete antibodies specific to the antigen. This type of response is referred to as humoral immunity. Another type of response of the immunized animal, known as cell-mediated immunity, consists of the proliferation of immune lymphocytes possessing antigen-specific receptors without any concomitant liberation of circulating antibodies. This type of response may result in protection against infection by viruses (Sissons and Oldstone, 1980).

The antigenic reactivity of a substance depends on its ability to bind with lymphoid cell receptors or antibodies. This ability (reactivity) is due to the presence of specific parts of the antigen known as antigenic determinants or epitopes. An epitope has a three-dimensional structure that is complementary to the binding site present in the antibody molecule. Each epitope in a protein may have five to seven amino acid residues. Two types of epitopes have been generally recognized. A continuous determinant (epitope) has a contiguous sequence of amino acid residues exposed at the surface of a native protein and distinctive conformational features. On the other hand, a discontinuous determinant contains residues that are not contiguous in the primary structure of the native protein, but the distant residue may be placed contiguously through the folding of the polypeptide chain or by juxtaposing two separate peptide chains. The presence of discontinuous determinants has been conclusively demonstrated only in the case of lysozymes (Atassi and Lee, 1978). Another type of epitope, known as cryptotopes (Jerne, 1960), may become antigenically active only after breakage, depolymerization, or denaturation of the antigen. Viral capsid proteins may have cryptotopes on the surfaces of protein monomers that are turned inward and become buried after polymerization. Such cryptotopes are exposed when the capsid protein is denatured. In the case of viruses, polymerized proteins may have specific epitopes formed either due to conformational changes of the protein induced by intersubunit bonds or placement of juxtaposing residues from neighboring subunits. Such epitopes, which are not present in the individual protein subunits, are called neotopes (Van Regenmortel, 1966). In addition, two more kinds of epitopes have been recognized. Metatopes are present in both polymerized and dissociated forms of plant viral proteins as in the case of tobacco mosaic virus capsid. Neutralization epitopes are recognized by their ability to neutralize the infectivity of plant viruses in the neutralization test. These epitopes cannot be analyzed outside of the operational context of neutralization of infectivity of the viral antigen (Van Regenmortel and Dubs, 1993).

Serological properties may be affected by changes in the amino acid sequences of proteins. Mutants of plant viruses may show single amino acid substitution in coat proteins resulting in alterations in biological properties such as vector transmission. The role of a specific epitope in the transmission of potyviruses (represented by potato virus Y) by the aphid vectors was studied by Jayaram et al., 1998. If the amino acid substitution results in variation of antigenic reactivity, the particular amino acid is considered to contribute to the structure of the epitope. The antigenic reactivity may be altered by fragmentation of the protein. The serological properties of intact viruses may show significant differences when compared with those of denatured viral coat protein. The serological properties of antigens have to be compared under identical conditions because the antibodies to intact virus particles (proteins) are directed against the unique conformation of the native protein, while antibodies to the protein fragments are directed against epitopes of unfolded protein molecules.

If the epitope(s) present in a protein is (are) characterized, it is possible to synthesize a series of short peptides corresponding to the sequences of the epitopes. The antisera prepared using these synthetic peptides are useful to differentiate variants or strains of viruses, based on the differences in the antigenic structure. The antigenic structure is considered an antigenic signature representing the viruses or isolates, providing a useful basis for virus identification. A system of antigenic signature analysis for the rapid differentiation of soybean mosaic virus isolates was proposed by Hill et al. (1994).

ANTIBODIES

Antibodies produced in response to the introduction of a specific antigen belong to a group of proteins known as immunoglobulins (IgG). They are capable of binding specifically to the antigen concerned. They are produced by lymphoid (B) cells or β-lymphocytes and are found in the serum of immunized animals. The B-cells may be present in the spleen, lymph nodes, Peyer's patches of the digestive tract, and peripheral (circulating) blood. One antibody specific for one epitope (antigenic site) on the antigen is produced by each B-cell.

All IgGs have a similar basic structure (Figure 2.1). The IgG molecule in rabbits consists of one pair of identical light (L) chain (MW 25,000) and heavy (H) chain (MW 50,000) which are linked together by noncovalent forces and disulfide bonds. The L chains may be of two types, namely, κ (kappa) and λ (lambda) chains, but the light chains of any one IgG molecule will be of the same type. Each light and heavy chain has two regions, a variable and constant region. The amino acid sequences of a variable region (V_L) located in the N-terminal half of the different antibodies exhibit extensive variations, whereas the amino acid sequences of the constant region (C-terminal half) may show variation to a much lesser degree. The sequence variability of the variable region (V_L) is not uniformly distributed along the V_L region, but confined to three regions totaling about 25 residues known as hypervariable regions. These hypervariable residues deter-

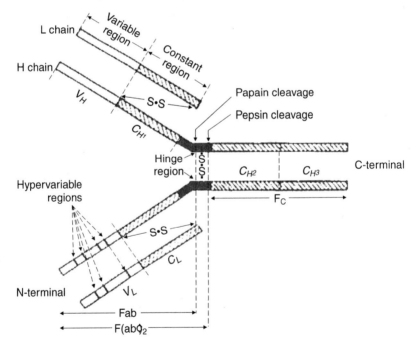

FIGURE 2.1. Structure of rabbit IgG molecule and the fragments formed after cleavage by papain and pepsin

mine the nature of antibody combining sites coming in contact with the epitopes (antigenic determinants) present on the antigens. The N-terminal quarter of the H chains is also variable (V_H).

The IgG molecule has a Y shape and the variable regions (V_H and V_L) are located at the extremity of the two arms of the molecule. The L chains are found externally in the arms and have one variable and constant region. On the other hand, the H chains have two parts. One variable and one constant (CH 1) region in the arms and two parts (CH 2 and CH 3) toward the C-terminal end. The CH 1 and CH 2 domains are separated by a short region called the hinge region which is sensitive to proteolytic enzymes. The enzyme papain attacks the hinge region, cleaving the IgG molecule into two parts. The fragment of the arms of the IgG molecule formed after cleavage with papain is known as Fab, whereas the fragment of the IgG molecule obtained after cleavage with pepsin is called F(ab')$_2$. The fragment of the IgG molecule in the C-terminal end is designated Fc fragment. Different antibodies exhibit a unique antigen binding site, as they retain a common three-dimensional structure since the framework of the Fv region is not affected by the hypervariable residues present in loops that extend into the solution.

The constant region located in the C-terminus of the antibody molecule defines the class of the antibody such as IgG, IgM, IgA, IgD, and IgE or subclass such as IgG 1, IgG 2a, IgG 2b, IgG 3, IgG 4, IgM 1, IgM 2, IgA 1, and IgA 2. The class and subclass of an antibody can be classified using commercially available isotyping reagents. Generally, IgG and IgM are formed in response to antigenic stimulation. Hence most of the monoclonal antibodies generated may be of IgG or IgM isotype. Plant viruses may induce predominantly IgG, which is resistant to mercaptoethanol, whereas IgM is sensitive to mercaptoethanol, and IgA has a sensitivity level in between IgG and IgM. The IgG, IgA, and IgM fractions have sedimentation coefficients (S_{20}, W) of 7, 7 to 19, and 19, respectively. The amino acid sequences of the hypervariable region V_L and V_H domains primarily determine the specificity of each antibody species. However, different antibodies may be able to bind a single epitope, but with different degrees of affinity. The reverse situation, in which one antibody species reacts with different epitopes, may also exist. It is difficult to assess the polyfunctional nature of the combining sites. The Fc fragment of IgG has a high affinity for the protein A molecule (MW 42,000) isolated from the cell walls of the bacterium *Staphylococcus aureus*. This property has been exploited in studies relating to immuno- chemistry, precipitation of antigen-antibody complexes, solid phase immunoassays, and affinity chromatography for virus purification.

The serum containing antibodies (immunoglobulins) is known as the antiserum. Serological tests have been shown to be more sensitive when purified immunoglobulins are used, instead of whole anti- serum. During the antigen-antibody interaction, the noncovalent in- termolecular forces that hold together the antibody-combining sites (or paratopes) and antigenic determinants (or epitopes) are similar to those involved in the stabilization and specific configuration of pro- teins. When paratopes and epitopes come in close contact, these forces become operative. The antigen-antibody bond is stronger with closer contacts established between the reactants. For more details re- garding features of antigen-antibody interaction, see Van Regen- mortel (1982).

SUMMARY

The charactreristics of the principal reactants of immunological re- actions, namely antigens and antibodies have been described. The structure of immunoglobulins induced by antigens, distinguishing properties of different globulins, and the specificity of serological re- actions are discussed.

Chapter 3

Preparation of Antisera

The availability of an antiserum containing the antibodies specific to an antigen is the basic requirement for performing different kinds of serological techniques for various purposes. The immunogenicity of antigens differs considerably depending on their size and structure. Polyclonal and monoclonal antibodies have been prepared and used for the detection, quantification, and determination of serological relationships between antigens.

POLYCLONAL ANTISERA

Plant viruses are simpler than other microbial pathogens, fungi, and bacteria that affect plant health adversely. Hence, preparation of antisera against plant viruses is described as a model. Animals such as rabbits, mice, fowl, and horses, have been used for the production of anitsera. However, rabbits are the most commonly used test animals. Partially or highly purified virus preparations may be injected intramuscularly, intraperitoneally, or intravenously (Matthews, 1957; Van Regenmortel, 1982). Intravenous injections are administered into the marginal veins of the ear. With intravenous injections, small doses of virus are administered at first and then the dose is gradually increased in subsequent injections.

With unstable viruses, the antigen is emulsified with an equal volume of Freund's complete or incomplete adjuvant. The complete adjuvant contains killed and dried *Mycobacterium butyricum* cells in addition to a mixture of mannide monooleate (1.5 parts) and paraffin oil (8.5 parts). The incomplete adjuvant does not contain the bacterial cells. The adjuvant stabilizes the antigen and also prevents rapid translocation of the immunogen. In addition, the adjuvant induces an inflammatory reaction, which has favorable influence on the formation of antibodies. The number of injections, quantity of immunogen required, interval between injections, and interval between last injection and bleeding vary considerably with immunogens.

The immunized animals are usually allowed a rest period of two weeks after the last injection of the immunogen. The marginal ear veins are cut with a sterile razor blade and the blood is collected in sterile tubes. After allowing sufficient time for coagulation of red blood cells, the supernatant serum containing antibodies specific for the immunogen(s) is centrifuged to remove the residual blood cells. This antiserum is stored in small vials after addition of preservatives such as glycerol, phenol, sodium azide, merthiolate, or cialit at 5°C in a deep freeze or as lyophilized powder. Although serological tests may be performed using unfractionated whole serum, purified globulins should be used to enhance the sensitivity and specificity of the serological tests.

A novel method of producing polyclonal antibodies (PABs) uses bacterially expressed viral coat proteins as immunogen. When purified viruses or viral proteins are not available, the coat protein (CP) gene of the virus is cloned and expressed in the cells of the bacterium *Escherichia coli*. The polypeptides produced in the bacterial cells are fractionated and employed as immunogens to produce PABs in rabbits. These PABs are highly specific, and the sensitivity and reliability of the serological tests are enhanced by using them. Furthermore, undesirable reactions in test animals following the injection of complex viruses purified from infected plants can be avoided. For further details, see Narayanasamy (2001).

MONOCLONAL ANTISERA

Methods of producing antisera that contain monoclonal antibodies (MABs) have been developed for different immunogens, including microbial plant pathogens. Because plant viruses are good antigens with comparatively simple structures, the standard protocol is described in this section. Many modifications have been made to suit the needs and purpose of experiments. The hybridoma technology introduced by Köhler and Milstein (1975) has provided a revolutionary advancement in the process of antibody production that avoids many problems associated with the use of polyclonal antibodies. This technology offers the principal advantage of obtaining an unlimited supply of MABs. The MABs are ideal serological reactants for taxonomic, diagnostic, structural, and biochemical analyses of plant viruses and other microbial pathogens. Since each MAB specifically reacts with a single antigenic determinant group or epitope, analyses can be made on that basis. Hybridomas are somatic cell hybrids formed by the fusion of β-lymphocytes (antibody-producing cells) with myeloma cells (capable of indefinite multiplication). The hybrids derive the ability to produce a specific antibody from the β-lymphocyte and to be cultured indefinitely in vitro from the myeloma cells. Each hybridoma clone produces identical antibodies that are specific for a single epitope present in the immunogen.

The principal steps in producing the hybridoma clones are as follows:

1. Immunization
2. Culturing of a mouse myeloma cell line
3. Fusion of spleen cells with myeloma cells
4. Cloning antibody—producing hybridomas
5. Screening hybridomas for MAB production
6. Production of ascitic fluid containing selected MABs

Immunization

Generally, BALB/c (six weeks old) mice are used for production of MABs by giving two to four intraperitoneal injections (50 to 100 μg of purified virus in phosphate-buffered saline) emulsified with an equal volume of complete Freund's adjuvant (for initial immunization) or incomplete Freund's adjuvant for subsequent injections at an interval of seven days. Other strains of mice such as DBA/2, B 10 G, CBA/J, SJL, or C57 B 1/6 have also been used (Mernaugh and Mernaugh, 1995). After a period of rest (up to three weeks), the mice giving the highest specific response (as determined by enzyme-linked immunosorbent assay [ELISA]) are given two intravenous booster injections (10 μg of virus in saline). The spleens are excised at three days after the second booster injection and spleen cells (β-lymphocytes) are used for fusion.

Culturing of Mouse Myeloma Cell Lines

Myeloma cell lines that are deficient in the ability to synthesize hypoxanthine (or adenosine) phosphoribosyl-transferase (HPRT or APRT) or thymidine kinase (TK) are used as cell fusion partners for β-lymphocytes. The desired myeloma cell lines such as P3/NS 1/1-Ag 4-1 (NS 1/1), P3 x 63 Ag 8.653 and SP 2/0 Ag 14 may be obtained from the American Type Culture Collection (ATCC). Dulbecco's modified Eagle's Medium (DMEM) is a good medium for growing myeloma or antibody-producing hybridoma cells. DMEM contains glutamine-pyruvate antibiotic β-mercaptoethanol (GPAM) plus 10 percent fetal calf serum (FCS). The mouse peritoneal macrophages from peritoneal cavities and thymocytes from thymus of mice are suspended in DMEM-GPAM-HAT 10 percent FCS medium and used as feeder cells for the hybridoma clones.

Cell Fusion and Cloning Hybridoma

Cell fusion procedures using polyethylene glycol (PEG) have been commonly used. Before cell fusion is taken up, the immunized mouse is sacrificed and the spleen is surgically removed. The spleen cells are dislodged by repeatedly injecting intact spleen with serum-free DMEM medium (5 to 10 ml). The spleen cells are mixed with the myeloma cells at a ratio of 1:1 or 2:1 (spleen cells: myeloma cells). Then PEG (4,000 MW at 1 mg/ml) is added to the cell mixture and gently agitated for one minute. Then DMEM-GPAM (10 ml) is added slowly, followed by centrifugation for ten minutes at 50 g_n and resuspension of the pellet in DMEM-GPAMC FCS-HAT medium.

The cell suspension is then dispensed at 50 μl/well in the 96-well Falcon plates in which the feeder cells (100 μl/well) have already been added and incubated in a 7 percent CO_2 incubator at 37°C for 10 to 15 days for the completion of fusion of spleen cells and myeloma cells to form hybridomas. The supernatant solution in each well is tested for the extent of secretion of specific antibody by performing an ELISA test. The desired hybridomas are cloned by limiting dilution for which the cells are distributed into the wells containing thymocytes or macrophages as feeder cells, at a cell density of 0.5 cell/well. Alternatively, the hybridoma may be cloned by growing cells in soft agar (agarose) on which individual clones may appear as white spots after one to two weeks. Cloning by limiting dilution has been found to be easier and less time-consuming than cloning in soft agar (Mernaugh and Mernaugh 1995). Each hybridoma clone produces a specific MAB against the immunogen concerned.

Large quantities of MABs can be obtained from mouse ascites which is found in the fluid accumulating in the peritoneal cavity (abdomen) injected with hybridoma clones in DMEM (0.5 ml). The immunosystem of the mouse is suppressed by injecting pristane (0.5 ml) prior to injection of hybridoma cell gowth and antibody production. The concentration of MABs in ascites may vary from 0.5 to 20 ng/ml. The ascites fluid is obtained at ten days after injection of hybridoma clones using a 19-gauge syringe needle and centrifuged at 300 g_n for ten minutes to remove cells and debris (Van Regenmortel and Dubs, 1993).

Monoclonal antibodies have been produced against a large number of plant viruses and some fungal, bacterial, and phytoplasmal pathogens infecting economically important crops. They have been employed for rapid detection, differentiation, and establishing relatedness among microbial pathogens. The practical utility of MABs in plant quarantine, disease survey, and breeding for resistance programs has been widely realized.

PHAGE-DISPLAYED RECOMBINANT ANTIBODIES

Bacteria are infected by viruses (bacteriophages) that replicate in bacterial cells and they are released following the lysis of the bacterial cell wall. *Escherichia coli* is infected by M13 phage which has ss-DNA as the genome and a flexuous (flexible) filamentous shape. M13 phage components such as phage DNA, gene 8 coat proteins, gene 3 attachment proteins, and other proteins that may be fused with the phage components, are continuously produced in the bacterial cells without undergoing lysis. The phage-displayed recombinant antibodies can be produced by genetically linking the DNA from antibody-producing β-lymphocytes or hybridomas with the phage gene 3 DNA. When the *E. coli* cells are infected by M13 phage carrying the gene 3 DNA-antibody DNA fusion, coexpression of proteins encoded by antibody DNA and gene 3 DNA occurs. Any antibody DNA linked to phage DNA and antibody proteins fused to phage proteins will be assembled and secreted just like phage proteins.

Phage-displayed recombinant antibody technology has been found to be useful for the detection of several plant viruses such as potato leafroll virus (Harper et al., 1997; Toth et al., 1999), potato virus Y (Boonham and Barker, 1998), African cassava mosaic virus and tomato leafcurl virus (Ziegler et al., 1998), cucumber mosaic virus (He et al., 1998; Gough et al., 1999), potato virus X (Franconi et al., 1999) and beet necrotic yellow vein virus (Griep et al., 1998). Production of antibodies displayed on phage can be made rapidly and at a lower cost than production of MABs. These recombinant antibodies are equally specific as MABs and can be used for the rapid identification and differentiation of the plant viruses. The protocol for production of phage-displayed antibodies is available elsewhere (Narayanasamy, 2001).

A single-chain Fv antibody fragment (scFv) may be prepared from a synthetic phage antibody after several rounds of selection against a purified preparation of antigens (viruses). A scFv fragment may be engineered to recognize a specific epitope of antigenic protein and expressed in plants by employing an appropriate vector such as an epichromosomal expression vector derived from potato virus X (PVX). The gene encoding the secretory scFv is then used to transform the plant. The engineered scFvs may become inexpensive tools for the detection of the antigens. These "plantibodies" may find wide applications including detection of human viruses. A bispecific scFv product in transgenic tobacco plants was able to recognize certain neotopes and cryptotopes of the tobacco mosaic virus coat protein (TMV-CP) (Fischer et al., 1999). Functional expression of a scFv both in *E. coli* and plant cytoplasm was reported by Tavladoraki et al. (1999). Likewise, the expression of genes encoding antihuman IgG in perennial transgenic alfalfa plants expressing light- and heavy-chain-encoding mRNAs were obtained (Khoudi et al., 1999). These studies indicate the usefulness of plant systems for the production of large amounts of monoclonal antibodies in a reliable, stable, and cost-effective manner.

PURIFICATION OF IMMUNOGLOBULINS AND ANTIBODY FRAGMENTS

Purified immunoglobulins should be used in place of unfractionated whole antiserum to enhance the sensitivity and specificity of serological tests. Some of the procedures employed are briefly described.

Precipitation Methods

Ammonium Sulfate Precipitation

Ammonium sulfate is known to be a good protein precipitant. When added to the antiserum, it reaches a salt concentration of one-third to one-half saturation for the precipitation of immunoglobulins. The albumin and other serum proteins will remain in the supernatant. One ml of 4 M ammonium sulfate solution is added to 1 ml of the antiserum with constant stirring. The pH of the mixture is adjusted to pH 7.8. Use 1 N NaOH and allow the mixture to stand for one hour for the completion of precipitation of immunoglobulins which are removed after centrifugation for ten minutes at 8,000 g_n. The precipitate is then dissolved in half volume of the original serum taken for precipitation. The precipitation and dissolution cycle is repeated twice. The solution containing the final precipitate is dialyzed against phosphate-buffered saline, pH 7.8, and then centrifuged to remove the suspended impurities. The immunoglobulins in solution are preserved under frozen conditions (Van Regenmortel, 1982).

Rivanol Precipitation

Immunoglobulins (IgG) of very high purity can be obtained by using rivanol (2-ethoxy-6, 9-diaminoacridine lactate) which precipitates albumin and other serum proteins, while IgG remains in solution. By using 0.1 N NaOH, the pH of the antiserum is raised to 8 to 8.5. Rivanol (0.4 percent) is added in drops to the antiserum with constant stirring, at the rate of 3.5 ml to 1 ml of antiserum. After the completion of precipitation, the mixture is centrifuged and the precipitate is reextracted with water. The supernatant containing IgG and the water used for extracting the precipitate are combined and treated with excess of saturated aqueous potassium bromide to remove the rivanol by conversion into the insoluble bromide form (rivanol bromide forms yellow precipitate). The precipitate is removed after centrifugation for 30 minutes at 15,000 rpm.

The supernatant containing IgG is treated with equal volumes of 4 M ammonium sulfate to precipitate IgG and centrifuged at 10,000 rpm for 20 minutes. Further steps are the same as in ammonium sulfate precipitation method (Hardie and Van Regenmortel, 1977).

Caprylic Acid Precipitation

Serum proteins, except IgG, are precipitated by caprylic acid under acidic conditions. Acetate buffer (60 mM, pH 4) is used to obtain fourfold dilutions of serum or ascitic fluid containing IgG and the pH is adjusted to 4.5 using 0.1 N NaOH. To this solution, caprylic acid (25 µg/ml) is slowly added in drops with constant stirring for 30 minutes. The precipitate formed is removed by centrifugation at 10,000 g_n for 30 minutes. The pH of the supernatant containing IgG is adjusted to 7.4 using 1 M NaOH. The IgG may be further purified by following the ammonium sulfate precipitation procedure, if required (McKinney and Parkinson, 1987; Weiss and Van Regenmortel, 1989).

Chromatography Methods

DEAE Cellulose Chromatography

This method can be used for further purification of IgG obtained after ammonium sulfate precipitation, by employing the anion exchanger diethylaminoethyl (DEAE) cellulose at a pH of 6.5. The contaminants bind to the DEAE cellulose column. The cellulose column is first treated with ten volumes of 5 mM sodium phosphate, pH 6.5 followed by addition of IgG in the same buffer. Alternatively, by raising the pH to 8.5 using 10 mM Tris buffer, the IgGs may be bound to the cellulose column and then they are eluted with Tris buffer containing increasing concentration of NaCl (50 to 200 mM) (Harlow and Lane, 1988).

Protein A Chromatography

Protein A from the cell wall of *S. aureus* exhibits a strong affinity for antibodies. The immunoglobulins may be purified by using this property of protein A. Depending on the capacity of globulin to bind with protein A, low or high salt concentration is used before addition to the column, the salt concentration being higher for globulins with low affinity for protein A (Ey et al., 1978; Harlow and Lane, 1988).

The globulin with high affinity is purified by adjusting the pH of the crude preparation to 8 using one-tenth volume of Tris (1.0 M) buffer, pH 8.0. The globulin solution is then passed through a column of protein A-Sepharose beads (at the rate of 10 to 20 mg globulin per ml of wet beads). The beads are washed in succession with ten column volumes of 100 mM Tris, pH 8.0, and with ten column volumes of 10 mM Tris, pH 8.0. The column is eluted with 100 mM glycine, pH 3.0. The presence of globulins is identified using absorbance at 280 nm as the basis and the fractions with globulins are neutralized with 1 M Tris, pH 8.0.

Purification of immunoglobulin with low affinity has to be done at high salt concentration. Borate buffer containing glycine (1.5 M) and NaCl (3.0 M), pH 8.9 is used to equilibrate the protein A-Sepharose column and immunoglobulin solution is added. The column is then washed with borate-glycine-NaCl buffer, pH 8.9. In the case of mouse IgG1, a citrate buffer (0.1 M), pH 6.0 is used to elute the immunoglobulin and it is neutralized with Tris buffer (1.0 M), pH 8. Other proteins may be eluted with citrate buffer (0.1 M), pH 3.5.

Antibody Fragment Preparation

Immunoglobulins may be broken into two major fragments (designated Fab and F(ab')2 and Fc due to the action of papain and pepsin (Figure 2.1). These fragments of IgG have been employed in some serological techniques such as immunoelectron microscopy (IEM) and ELISA and also for the determination of virus-antibody interactions. By using a column of immobilized papain, IgG molecules are digested to obtain Fab fragments. Papain is mixed in excess with Affi-gel 10 (BioRad) and the column is treated with phosphate buffer (150m M), pH 6.5, containing β-mercaptoethanol (10 mM) and ethylene diaminetetraacetate (EDTA) (2 mM) to activate papain. The IgG is passed through the column after removing the mercaptoethanol from the column using the phosphate buffer containing EDTA (2m M). Digestion is continued till whole IgG is not detected by sodium dodecylsulfate-polyacrylamide gel electrophoresis (SDS-PAGE). The digest is then cycled through a Sepharose 4B protein A column in phosphate buffer, pH 8.0, for purification of Fab fragment.

SUMMARY

   The methods of preparation of the polyclonal and monoclonal antisera and purification of immunoglobulins are described. The relative specificity and sensitivity of polyclonal, monoclonal and recombinant antibodies are highlighted. The advantages and limitations of immunological techniques employing these antibodies are discussed.

Chapter 4

Basic Immunodetection Techniques

Plant proteins have a pivotal role in the maintenance of normal physiological functions that are required for growth and reproduction. Basic immunological techniques for the detection, localization, and quantification of various proteinaceous compounds in healthy and stressed plant cells/tissues and microbial plant pathogens are described in this chapter. Specialized techniques pertaining to detection of antigens in certain tissues/conditions are described in Chapters 6, 8, 9, 10, 11, and 13.

PRECIPITATION AND AGGLUTINATION REACTIONS

The size of the reacting antigen forms the basis of differentiating precipitin or precipitation and agglutination tests. The term precipitation is applied for the tests based on the insolubilization of macromolecules (proteins) or plant virus particles. Agglutination involves clumping of plant cells or red blood cells or bacterial cells or particles of similar sizes.

Precipitin or Precipitation Tests in Liquid Medium

Precipitin Test in Tubes

Precipitin tests are usually performed in small glass tubes. Two-fold dilutions of antigen and antiserum containing specific antibodies are prepared in 0.85 percent NaCl or phosphate-buffered saline (PBS) solution. About 0.5 ml of suitably diluted antigen and antiserum are mixed in clean glass tubes and incubated in a water bath at 25 to 40°C. The formation of visible precipitate is recorded at 20-minute intervals for one to two hours of incubation. The highest dilution of the antiserum that forms visible precipitate, known as the titre of the serum, is recorded. Likewise, the highest dilution of antigen forming visible precipitate, is termed antigen (virus) end point, and is also noted.

Ring Test or Ring Interface Test

This test, a variation of the precipitation test, is conducted in narrow glass tubes. The antiserum diluted with glycerin (10 to 30 percent) in saline (0.1 to 0.2 ml) is placed in the tube. An equal volume of the antigen preparation is gently layered over the antiserum solution. Care should be taken to release the antigen solution from the pipette so that the antigen solution slowly flows through the sides of the glass tube containing the antiserum. A ring of precipitate appears at the interface between antiserum and antigen, indicating a positive relationship between the antibody and the antigen.

Slide Test

The serological relationship between the antigen and antibody can be studied by this simple test. Drops of the antigen and antiserum, suitably diluted, are mixed on grease-free glass slides. The slides are incubated in a moist chamber, after covering the drops with a layer of mineral oil to prevent drying of the reactants. The formation of precipitate can be viewed under a dark field microscope (×50 to ×100 magnification).

Microprecipitin Test

This test, an improvement over the slide test, is carried out in the bottom plate of a petri dish coated with hydrophobic film of polyvinyl formaldehyde (Formvar) (0.1 percent) in chloroform or silicone. Drops of antigen and antiserum (1:5 v/v) mixture are placed on the precoated dishes in an orderly array and they are then mixed well with a thin glass rod. The drops are covered with a thin layer of mineral oil to prevent drying and incubated at room temperature for three to six hours or at 10 to 12°C for 24 hours. Long incubation periods at low temperatures enhance the sensitivity of the test by four times. This test, developed by Van Slogteren (1955), saves time and space. Forty to 60 tests can be performed in one petri dish, making economical use of antiserum, when compared with tube precipitin test.

Agglutination Tests

The sheep red blood cells (RBC) or carrier particles of similar size (such as latex, bentonite, or barium sulfate) are sensitized by attaching either the antigen or antibody to the surface of RBC or carrier.

RBC Agglutination Test

The test using RBCs is performed in plastic microtiter plates with rows of wells as follows:

1. Wash fresh sheep RBCs with saline several times.
2. Suspend RBCs (25 ml) in PBS, pH 7.2.
3. Dialyze the RBC suspension with 50 ml of formaldehyde (35 percent) for three hours.
4. Wash the RBCs several times with saline; adjust the final concentration of RBCs to 25 percent and store at 4°C.
5. Add 1 ml of tannic acid (0.1 mg/ml) to a suspension of RBCs (2.5 percent) in PBS, pH 7.2 for coupling and incubate for ten minutes.
6. Wash the RBCs with PBS; resuspended in 1 ml of PBS; mix with an equal volume of purified antigen (virus) preparation (0.5 mg/ml) or with 1 ml of specific antibody (0.1 mg/ml of IgG).
7. Add 50 µl of antiserum dilution and 1 percent suspension of antigen-coated RBCs to each well in the microtiter plate and incubate for one to three hours.

Agglutinated RBCs forming a network of precipitate (clumps) and covering a large area of well surface indicate a positive reaction. Presence of unagglutinated RBCs at the deepest point of the well shows a negative reaction (Figure 4.1).

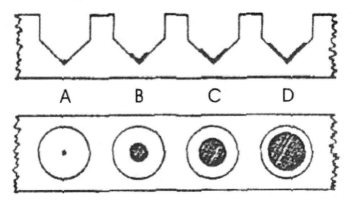

FIGURE 4.1. Formation of different patterns of hemagglutination of erythrocytes: (A) negative reaction—erythrocyte sediment as a small button at the center of the well; (B, C, and D)—positive reactions—increasing levels of aggregation of erythrocytes

Latex Agglutination Test

In the latex test, either the antigen or antibody is adsorbed onto polystyrene latex particles as follows:

1. Prepare a suspension of latex particles (0.8 μm diameter, Difco Laboratories, Detroit, Michigan) to have a dilution of 1:15 with saline.
2. Dilute the antibody (1:50 to 1:2000) with 0.05 M Tris-HCl buffer, pH 7.2.
3. Mix equal volumes of latex and antibody suspensions, incubate for one hour; recover antibody-coated latex particles by low speed centrifugation and wash twice with 0.05 M Tris-HCl buffer, pH 7.2, containing 0.02 percent polyvinyl-pyrrolidone as a stabilizer.
4. Suspend the final sediment in Tris-HCl buffer containing 0.02 percent sodium azide; store in refrigeration until required (maximum of six months).
5. Mix 50 μl of antigen solution with 25 μl of sensitized latex (coated with antibody) suspension in the wells of a microtiter plate; prepare antigen solution using Tris-HCl buffer, pH 7.2,

containing 0.2 percent polyvinyl pyrrolidone and place the plates in a rotary shaker at 200 rpm for 15 minutes.
6. Read the results after about one hour.

The latex agglutination test can be performed in capillary tubes or in small glass tubes (Khan and Slack 1978). The agglutination tests have been found to be more sensitive than precipitin tests.

Virobacterial Agglutination (VBA) Test

The virobacterial agglutination test (VBA) has been shown to be effective in differentiating potyviruses and comparable in sensitivity to the ISEM technique, but less sensitive than direct ELISA format. However, it is more sensitive and simpler than the latex particle agglutination test (Walkey et al., 1992) (Figure 4.2).

1. Use the Cowan 1 strain of *S. aureus* cells cultured in nutrient agar medium for 24 to 48 hours at 37°C.
2. Dilute the antiserum (specific to the antigen)-glycerol mixture (1:1) with 24 volumes of PBS containing Na_2HPO_4. $12H_2O$ (2.9 g), KH_2PO_4 (0.8 g), NaCl (8.0 g) and KCl (0.2 g) in distilled water (1 l), pH 7.2 and add sodium azide (NaN_3) at a rate of 0.2 g/l.
3. Mix the diluted antiserum (five volumes) with *S. aureus* (one volume) in the ratio of 5:1 for conjugation between the reactants; add ethanol-saturated basic fuchsin as a coloring agent for easy recognition of bacterial agglutination; store, if necessary for a maximum of six months at 4°C; if stored, shake the conjugates well on Vortex stirrer and check for autoagglutination using a hand lens.
4. Prepare the viral antigen by crushing the infected leaf tissues within a small polyethylene bag using a pestle or by triturating in a mortar with pestle using a small amount of K_2HPO_4 (10 g/l) solution and prepare the extract from healthy leaves in a similar manner.
5. Dispense aliquots of bacterial-antiserum conjugate (4 µl) and test samples (2µl) on multitest slide (Flow Laboratories Ltd., United Kingdom); maintain identical controls with healthy material and mix the reactants carefully.

6. Observe the formation of agglutination using a hand lens, holding the slide over a black background lit by diffuse light.

Positive reaction is indicated by agglutination of reactants within 30 seconds to 5 minutes.

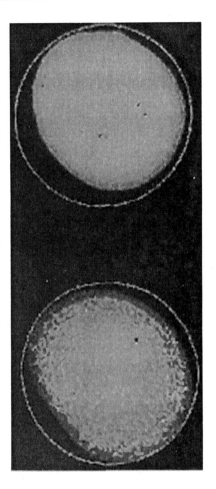

FIGURE 4.2. Virobacterial agglutination (VBA) test—detection of turnip mosaic virus in infected plant extract: (top well) negative reaction (control); (bottom well) postive reaction, indicated by clumping of bacterial cells (*Source:* Walkey et al., 1992; Reprinted with permission by Blackwell Scientific Publications, Oxford, United Kingdom.)

IMMUNODIFFUSION TESTS

Immunodiffusion tests are carried out in gels, instead of free liquid. The antigen and antibody diffuse through the gel and visible precipitation lines are formed at the position where both reactants reach optimal concentrations. These tests are more sensitive than precipitation tests conducted in liquids.

Single Diffusion Technique

Single Diffusion in Tubes

In this method, narrow glass tubes (2 to 7 mm × 80 mm) are precoated with 0.1 percent agar by filling the tube with melted agar and then draining the agar and drying the tubes. The internal reactant is the antigen mixed with 0.4 percent agar. When the mixture forms a gel, 0.2 ml of antiserum is added as the external reactant and the tube is then sealed. The position of the leading edge of the precipitate that is formed in the antigen-agar column is proportional to the square root of time. This method is rarely followed for establishing the antibody-antigen relationship (Commoner and Rodenberg, 1955).

Radial Diffusion Method

The radical diffusion method is performed in petri dishes, using the dissociated proteins of rod-shaped viruses that cannot diffuse easily in the agar, as follows:

1. Mix the antiserum containing 1 percent sodium dibutyl napthalene sulfonate and 0.4 percent sodium azide with an equal volume of 1 percent liquid Ionagar No. 2 buffered with 0.1 M Tris-HCl (pH 7.2) containing 1.7 percent NaCl.
2. Transfer aliquots of 15 ml of the antiserum-agar mixture to each petri dish (90mm).
3. Place the (barley) seeds in moist chamber for 20 days at 20°C for germination and excise the primary leaves produced from the germinating seeds.

4. Prepare leaf segments (1 mm) and embed them in the solidified agar in the petri dishes by gently pressing the leaf segments using forceps. The leaf segments are spaced in rows approximately 1 mm within rows and 2 mm between rows.

About 500 to 600 samples can be assayed in one petri dish. Visible precipitin lines are formed around leaf segments from infected seeds. This technique has greater sensitivity (tenfold) compared with conventional double diffusion tests (Slack and Shepherd, 1975).

Double Diffusion Technique

Double diffusion is the most frequently used technique to demonstrate the relationship between antigen and antibody. Melted Bacto-agar (0.7 to 1.5 percent) or agarose in a suitable buffer is poured into silicone-coated glass petri dishes or plastic petri dishes to a depth of 2 to 3 mm. Wells are formed in the agar, after solidification, in a pattern with a central well of 4 mm diameter, surrounded by eight peripheral wells of 4 mm diameter at a distance of 3 mm from the edge of the central well. A corkborer or suitably designed cutter is used to form a well-defined geometric design. The antigen and antiserum are pipetted into the wells carefully in drops so that the reactants do not overflow and come in contact with each other directly. The plates are then incubated in a moist environment for several days at a constant temperature. The formation of precipitin lines is observed by placing the plates against a light source. Permanent records of the results can be made by simple contact printing of precipitin lines onto photographic paper or the lines can be stained (Almeida et al., 1965; Crowle, 1973; Simmonds and Cumming, 1979). The double diffusion technique is advantageous because direct comparisons can be made between several antigens by placing a reference antiserum in the central well and the antigens to be compared in the peripheral wells. The identity and relationship of the unknown (uncharacterized) antigen can be established by placing the antigen in the central well and the antisera of the known antigens in the peripheral wells (Figure 4.3).

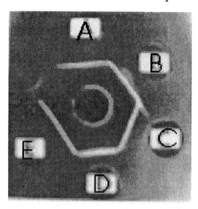

FIGURE 4.3. Serological relationship of fraction I proteins from different plant species: (A) *Chenopodium quinoa;* (B) *Cucurbita pepo;* (C) *Nicotiana tabacum;* (D) *Zea mays;* (E) *Petunia hybrida* (*Source:* Van Regenmortal, 1982; Reprinted with permission from Elsevier.)

IMMUNOELECTROPHORESIS

Using immunoelectrophoresis, complex mixtures of antigens can be differentiated, based on two independent characteristics—electrophoretic mobility and antigenic specificity (Van Loon and Van Kramaner, 1970).

Immuno-Osmophoresis (Ragetli and Weintraub, 1964)

This method is based on the migration of negatively charged antigens toward the anode and simultaneous electroosmotic flow of the weakly charged antibodies toward the cathode. The protocol of the technique is detailed as follows:

1. Place glass slides (2.5 × 7.5 cm) in close-fitting aluminum foil boats; pour Bactoagar (1 percent) in 0.01 M Tris-succinate buffer, pH 6.75 to a depth of 4 to 5 mm.
2. Form two rows of wells positioned exactly opposite each other at a distance of 5 mm; fill the wells with 0.5 μl of antigen at the cathode side and with 0.5 μl of antiserum at the anode side.

3. Place the slides on two open petri dishes each containing about 30 ml of Tris buffer; connect the slides and the buffer through a filter paper wick.
4. Place the platinum electrodes in the petri dishes and connect them to a power pack and pass a current of 40 to 50 m ampere at 300 V.
5. Observe the formation of visible precipitin lines within a few minutes.

LABELED ANTIBODY TECHNIQUES

Immunodetection of antigens may be made rapidly and efficiently by attaching a label that can be detected even in minute quantities. The antibodies may be labeled using enzymes, fluorescent dyes, or radioactive materials. The use of radioactive materials is not preferred due to the risk involved.

Enzyme-Linked Immunosorbent Assay (ELISA)

In this method, reactants are immobilized on a plastic surface and the reaction between antigen and antibody is detected by enzyme-labeled antibodies. Different formats of ELISA are used to study different aspects of relationships between the reactants, localization of antigenic compounds in plant tissues, and the detection and differentiation of different microbial plant pathogens in infected plants, fruits, and vegetables, and in soil. The presence of various agrochemicals in foods and feeds, dairy products, as well as in water can be detected using different ELISA formats. Because of the greater sensitivity and economic use of reagents, ELISA has become the preferred procedure replacing precipitin and immunodiffusion assays.

Preparation of Antibody-Enzyme Conjugate
(Clark and Bar-Joseph, 1984)

The enzyme alkaline phosphatase (ALP) is preferred to horse radish peroxidase (HRP), because it is stable and can be conveniently linked to the antibody (protein) by a gluturaldehyde bridge. However, it is limited by its high cost relative to HRP and its inefficiency of conjugation to protein A. The protocol for one-step conjugation of the purified antibody to ALP or HRP is described as follows:

1. Prepare a solution of purified antibody (IgG) (1 mg/ml) and transfer 2 ml solution to a glass tube.
2. Dissolve the enzyme (5 mg) directly in the IgG solution; dialyze at least three times against PBS (1 l), if the antibody is purified by ammonium sulfate precipitation and centrifugation.
3. Add freshly prepared 2.5 percent glutaraldehyde solution (50 µl) to the IgG-enzyme mixture; mix the contents gently and incubate for four hours at 30°C or at room temperature.
4. Observe for the development of a pale yellow-brown color; transfer the mixture to a dialysis tube and dialyze three times against PBS (1 l) to remove excess glutaraldehyde.
5. If necessary, centrifuge at low speed to remove the precipitate, if any.
6. Add bovine serum albumin (BSA) at 5 mg/ml and store at 4°C with 0.02 percent sodium azide for ALP conjugates (up to six months); for long-term storage, store at –20°C after adding an equal volume of glycerol.
7. For HRP conjugates do not use sodium azide; store with glycerol (50 percent) at 4°C or at –20°C or feeze-dry and store in glass vials under vacuum.

Double Antibody Sandwich (DAS)-ELISA
(Clark and Adams, 1977; Clark and Bar-Joseph, 1984)

Among the different ELISA formats, double antibody sandwich (DAS)-ELISA is more frequently employed for the detection and identification of viruses and proteinaceous compounds.

1. To each well of the microtiter plate, transfer aliquots of 200 µl of the purified immunoglobulins (antibodies) diluted to have 1 to 10 µg/ml in coating carbonate buffer containing Na_2CO_3 (15 mM), $NaHCO_3$ (35 mM), and NaN_3 (0.2 g/l), pH 9.6 and cover the plates to prevent evaporation.
2. Incubate at 30°C for two to four hours or at 4°C overnight.
3. Wash the wells, after emptying them, by flooding with PBS-T buffer, pH 7.4 containing NaCl (8.0 g/l), $Na_2HPO_4 \cdot 2H_2O$ (1.44 g/l), KH_2PO4 (0.2 g/l), KCl (0.2 g/l) and Tween-20 (0.5 ml/l of PBS) and soak for three minutes.

4. Repeat washing and soaking twice; empty the plates and remove the residual moisture by shaking.
5. Block the plastic surface with 1 to 2 percent bovine serum albumin (BSA) in PBST, if necessary.
6. Add aliquots of 200 μl of samples containing antigen diluted in PBST to duplicate wells; incubate for three to six hours at 30°C or two hours at 37°C or overnight at 4 to 6°C.
7. Wash the plates three times as in step 3.
8. Into each well, dispense aliquots of 200 μl of specific alkaline phosphatase enzyme-antibody conjugate appropriately diluted in PBST (1/200 to 1/2000 immunoglobulin); cover the plates and incubate at 30°C for three to six hours.
9. Wash the plates as in step 3; add enzyme substrate *p*-nitrophenyl phosphate (1 mg/ml) in diethanolamine buffer, pH 9.8.
10. Incubate for one hour at room temperature or until color develops to a desired level; terminate the reaction between the enzyme-substrate by adding 3M NaOH at 50μl/well.
11. Record the color intensity (absorbance) at 405 nm in a spectrophotometer or ELISA reader.

Fluorescent Antibody Techniques

The intracellular location and distribution of antigens within the tissues of plants and vectors of plant viruses may be studied using fluorescent antibody techniques.

Fluorescence Test with Leaf Tissues

The presence of plant viruses in plant tissues can be visualized by employing the fluorescence test.

1. Prepare small aluminum foil boats containing gelatin (20 percent) in glycerol (5 percent); place leaf segments in gelatin; freeze the contents with dry ice at –20°C and remove the aluminum foil.
2. Make cross sections (10 to 20 μm) at room temperature; mount them on clean glass slides smeared with gelatin-glycerol adhesive.

3. Pipette drops of antibody conjugated with 0.01 percent fluorescein isothiocyanate (FITC) on sections and incubate in a moist chamber for 45 to 60 minutes in darkness.
4. Wash the sections with PBS, pH 7.2 for 10 to 15 minutes to remove excess stain.
5. Mount the sections in a glycerol-phosphate mixture (acid-free glycerol [1 ml] + phosphate buffer [9 ml], pH 7.2).
6. Observe fluorescence in the sections using a fluorescent microscope.

Enzyme-Linked Fluorescent Assay (ELFA)

This test is considered to be more sensitive than DAS-ELISA in detecting plant viruses (Dolares-Talens et al., 1989).

1. Coat the wells of microtiter plates with purified IgG diluted in 0.05 M sodium carbonate buffer, pH 9.6.
2. Block well surface with 2 percent ovalbumin and 2 percent polyvinylpyrrolidone in 0.02 percent M sodium phosphate buffer, pH 7.2, containing 0.05 percent Tween 20 and 0.85 percent NaCl (PBST).
3. Empty the wells and wash three times with PBST.
4. Add the antigen at appropriate dilution and incubate for twelve hours at 4°C.
5. Add the second antibody labeled with biotin and diluted with PBST and wash with PBST.
6. Add 4-methylumbelliferyl phosphate (MUP) (0.1 mM in 1.0 mM diethanolamine, pH 9.8 containing 0.01 mM $MgCl_2$).
7. Add NaOH (to have a final concentration of 0.27 M) to stop reaction for colorimetric assay or $Na_2 HFO_4$ adjusted to pH 10.4 with KOH (to have a final concentration of 0.2 M) for fluorescent assays; record the color intensity/fluorescence at 405 nm in spectrophotometer or fluorometer.

Tissue Blot Immunoassay (TBIA)

The tissue blot immunoassay (TBIA) involves direct blotting of plant tissue onto nitrocellulose membranes for the detection of microbial pathogens and other antigens (Hsu et al., 1995). TBIA is simple, rapid, and large numbers of samples can be conveniently assayed. The presence and translocation of fungal protein that elicits plant defense response in tobacco could be demonstrated (Bailey et al., 1991).

Direct TBIA (Hsu et al., 1995)

1. Prepare strips of nitrocellulose membrane (0.2 or 0.45 μm).
2. Excise plant tissues (leaves, petioles, stems, or flower buds); roll thin tissues such as leaves into a tight cylinder; cut tissues with a new razor blade to have a single plane-cut surface.
3. Press the newly cut surface for about one or two seconds onto a nitrocellulose membrane strip to obtain a tissue blot; do not squeeze the tissue.
4. Immerse the tissue blots in a blocking solution containing 0.5 bovine serum albumin, 2 percent nonfat dry milk in PBS for 30 to 60 minutes at room temperature and shake occasionally.
5. Wash the blots for one minute with washing solution containing PBS and Tween-20 at 0.5 ml/l of PBS and shake gently (one rotation per second on a mechanical shaker).
6. Transfer the tissue blots to a glass petri dish (12 cm diameter) containing alkaline phosphatase-labeled specific antibody; allow reaction between antigen (in tissue blots) and antibody for 60 minutes at room temperature and ensure the antibody solution covers the blot fully.
7. Wash the blots in washing solution for 30 minutes; shake gently and repeat washing three times.
8. Prepare the substrate solution containing nitroblue tetrazolium (14 mg) (NBT) and 5-bromo-4-chloro-3-indolyl phosphate (7 mg) (BCIP) in 40 ml of substrate buffer consisting of 0.1 M Tris, 0.1 M NaCl, 5m M $MgCl_2$, pH 9.5. (Dissolve NBT in 300 μl methanol and dissolve BCIP in 50 μl dimethyl sulfoxide; add NBT solution to substrate buffer and then add BCIP solution.)
9. Immerse the blots in the substrate solution for two to five minutes at room temperature and rinse the blots in distilled water for a few seconds.
10. After ten minuntes, add stopping solution containing 0.01 M Tris-HCl, 0.05 M EDTA, pH 7.5 to stop the reaction and repeat with one or two changes of stopping solution for ten minutes each.

11. Dry the blots using two or three layers of tissue paper in a dust-free environment.
12. Observe the development of purple color on the blots indicating positive reaction and if necessary use a dissecting microscope with ×10 to ×20 magnification.

IMMUNOELECTRON MICROSCOPY

Electron microscopy techniques (using ultrathin sections of embedded plant tissues) have been employed to study the ultrastructure of plant cells and organelles. Details of various techniques are available in several publications (Hall and Hawes, 1991; Harris and Oparka, 1994; Narayanasamy, 2001). Immunoelectron microscopy has been demonstrated to be a very useful tool to visualize immunological reactions on electron microscope grids and to reveal the localization and distribution of various proteins in different plant tissues. Enzyme-labeled antibodies may be applied after embedding thin sectioning of plant tissues or they may be allowed to diffuse inside fixed cells and to interact with antigenic sites prior to thin sectioning. Some of the immuno-cytological techniques frequently used for the localization of desired antigens in plant cells are described as follows.

Immunosorbent Electron Microscopy (ISEM)

The serologically specific electron microcopy (SSEM) technique developed by Derrick (1973) was renamed as immunosorbent electron microscopy (ISEM) by Roberts and Harrison (1979). This test involves attachment of the antigen to the electron microscope grids coated with antibody raised against the antigen.

1. Coat the grids with polyvinyl formaldehyde (Formvar) strengthened with a layer of evaporated carbon.
2. Prepare suitable dilution of antiserum (1:1000 to 1:5000), using 0.05 M Tris-HCl buffer, pH 7.2; float the grids, film side down on drops (10 to 50 µl) of diluted antiserum for about 30 minutes at room temperature to facilitate adsorption of antibodies forming a layer on the film; or incubate at 37°C.

3. Remove the excess (unabsorbed) antibodies by floating the grids on buffer solution and place the grids on filter paper to drain the buffer solution.
4. Float the grids on drops of antigen solution or extract containing antigen diluted to appropriate level using Tris-HCl buffer.
5. Stain the antigen (virus particles) with uranyl acetate and wash excess stain with buffer.
6. Observe under electron microscope for the presence of antigen (virus particles).

Immunocytochemistry

Immunocytochemical techniques are useful to gather information on

1. chemical composition of cell structures;
2. changes in protein distribution;
3. accumulation of callose, lignin, and hydroxyproline-rich glyco-proteins (HRGPs) for reinforcement of cell walls during biotic and abiotic stresses;
4. accumulation sites of newly synthesized gene products;
5. antimicrobial activities of specific molecules; and
6. expression of transgenes and detection of their products in transgenic plants.

Immunocytochemistry has been widely employed because of the possibility of using the gold conjugates as electron-opaque markers of cell constituents. The principal advantage is that colloidal gold particles, due to their electron-opaque nature and uniformity in shape, make them distinct from biological structures and highly visible, even at low magnification. The gold-labeling technique has wide applications, since gold particles can be conjugated to a wide variety of probe macromolecules. Enzyme gold localization of substrates, lectin-gold recognition of specific sugar residues, and immunogold identification of antigens (microbial plant pathogens) are important applications of this technique. Immunogold labeling is simple because specific primary antibodies can be used to detect the antigens exposed at the cut surface of a section of plant tissue. When observed under the electron microscope the colloidal gold particles, conjugated to secondary antibodies or to protein A, attach to the bound primary antibodies serving as electron-opaque markers of the antigen. Using appropriate antibodies, proteins, carbohydrates, and glycoproteins may be localized and identified rapidly, because of the versatility, sensitivity, and exquisite specificity of the reaction between the antibody and antigen molecules. Both polyclonal and monoclonal antibodies have been employed for these studies.

Detection and Localization of Plant Macromolecules

Preparation of gold marker (Beesley, 1992; Harris, 1994):

1. Add 1 ml of aqueous gold chloride to 79 ml of distilled water.
2. Prepare the reducing mixture containing 1 percent trisodium citrate, $2H_2O$ (4 ml), 1 percent tannic acid (2 ml), 25 mM potassium carbonate (2 ml), and water to make 20 ml; warm the solution to 60°C.
3. Add the reducing mixture to the gold solution slowly with constant stirring; red color of the mixture indicates sol formation.
4. Heat the mixture to boiling and cool the solution.
5. Determine the amount of protein required to stabilize the sol by mixing increasing amounts of protein with 0.25 ml of sol; determine the concentration of protein necessary to prevent a color change to blue; this concentration of protein is required for stabilization of sol.
6. Add immunoglobulin at 1 mg/ml in TBS containing 0.01 M Tris and 0.15 M NaCl, pH 8.2, to 30 ml gold sol to exceed by 10 percent the concentration needed for stabilization; add 0.3 ml of 10 percent BSA for complete stabilization and centrifuge to remove aggregates, if necessary.
7. Dialyze against 20 mM Tris, 20 mM sodium azide, 1 percent BSA, 20 percent glycerol, pH 8.2 and store the gold sol in aliquots at 4°C.

Immunogold labeling (Vandenbosch, 1991):

1. Place the ultrathin sections on electron microscope grid into 20 μl drop of blocking buffer containing 1 percent BSA, 0.02 percent sodium azide, and 0.05 percent Tween 20 in Tris-buffered saline (TBS-10 mM Tris-HCl, pH 7.4, plus 150 mM NaCl);

incubate in a drop of primary antibody (raised against desired antigen) diluted in blocking buffer for one hour at room temperature or overnight at 4°C.

2. Transfer the grid successively through a series of drops of TBS; drain the excess buffer from the grid using filter paper.
3. Incubate the grid in 20 µl drop of secondary antibody-gold or protein A-gold, diluted in blocking buffer for one hour at room temperature.
4. Rinse the grid in TBS, followed by a short wash in glass-distilled water and stain the sections with 2 percent aqueous uranyl acetate for five minutes.
5. Observe under electron microscope.

Detection of Microbial Pathogens

Immunogold labeling of viruses (Milne, 1993):

1. Float filmed grids on drops of purified preparation of extracts of infected plant tissues for the adsorption of virus particles and rinse the grid with 0.1 M PBS-T.
2. Incubate the grids with 0.1 to 1.0 percent BSA for 15 minutes to block nonspecific adsorption site; rinse the grids with PBS-T.
3. Incubate the grids with drops of antiserum containing antibodies (against the virus to be detected) diluted with 0.1 M phosphate buffer, pH 7.0, and rinse the grids with PBS-T.
4. Incubate the grids with protein A gold (PAG) in PBS for 60 minutes; rinse the grids with PBST and then with glass-distilled water.
5. Stain with uranyl acetate (1 percent) for contrast enhancement.

Immunogold labeling of fungal antigens (Svircev et al., 1986):

1. Add 1 percent aqueous sodium citrate (4 ml) to a boiling solution of 0.01 percent chloroauric acid (100 ml); cool for five minutes; watch for the development of wine red color and store this colloidal gold suspension in darkness.
2. Using potassium carbonate, adjust the pH of colloidal gold suspension to 6.9; add protein A suspension (0.3 mg in 0.2 ml of distilled water) and centrifuge at 48,000 g_n at 4°C.

3. Resuspend the pellet containing dark red protein A-gold complex in 0.01 M PBS (10 ml), pH 7.4, and store at 4°C.
4. Place ultrathin sections of infected plant tissues on drops of solution of sodium periodate kept on filmed grids for two to three minutes and rinse the grids with distilled water three times.
5. Block the nonspecific binding sites with 1 percent BSA for five minutes and rinse the grids with distilled water.
6. Float the grids with sections on specific antiserum (against the virus concerned) for 30 minutes at room temperature and wash the grids by floating on several drops of water successively.
7. Treat the sections with protein A-gold conjugate for 30 minutes; wash the sections as done earlier; stain with 3 percent uranyl acetate for 20 minutes and observe under electron microscope.

Lectin Cytochemistry

Lectins form a group of carbohydrate-binding proteins (usually glycoproteins) of nonimmune orgin. They are present in several plant species. Because they bind specifically with carbohydrates, they have been used as probes for recognition of and attachment to specific molecules present on cell walls. They are also used for the in situ identification of oligosaccharides that are released from cell walls of plants or pathogens and capable of functioning as elicitors of plant defense responses. Lectins with molecular weights >15 kDa can be complexed with colloidal gold and applied directly to ultrathin sections of plant tissues. In smaller lectins (<15 kDa), an indirect method involving the use of a secondary reagent that has affinity for the lectin has to be followed (Roland and Vian, 1991).

Preparation of gold particles (15nm):

1. Take 100 ml of 0.01 percent tetrachloroauric acid (in double-distilled water) in ultraclean or siliconized Erlenmeyer flask and heat to boiling.
2. Rapidly add 4 ml of 1 percent sodium citrate with gentle heating; color becomes violet purple initially and after five minutes it turns reddish orange; store at 4°C as long the color remains the same without the formation of aggregates.

Preparation of gold-lectin (or enzyme) probes:

1. Maintain the pH level close to the isoelectricpoint of the protein (enzyme or lectin) by adding 0.2 M K_2CO_3 to raise the pH or 0.1 M acetic acid to lower the pH; determine the pH of the colloidal gold after adding two drops of 1 percent PEG (20,000 Da).
2. Estimate the minimal amount of protein (probe) required to stabilize the colloidal gold by setting up a series of test tubes, each containing 500 μl of colloidal gold; add to each tube 100 μl of ten serial dilutions of probe protein in water; shake the tubes well and, after five minutes, add 100 μl of 10 percent NaCl and determine the minimum amount of probe proteins required to prevent the color change from blue (flocculated gold) to red (stabilized gold).
3. Prepare a suspension of the protein at the concentration determined in step 2 in 200 μl of distilled water; add 10 ml of colloidal gold (at optimal pH); homogenize the solutions; after three minutes add PEG (20,000 Da) to saturate uncoated gold sites, if required.
4. Centrifuge at 4°C for 30 minutes at 60,000 g_n (higher speed required for smaller gold particles); use the bottom fraction containing red mobile pool of gold probe complex, rejecting the top clear supernatant and the middle fraction containing flocculated uncoated gold particles.
5. Repeat centrifugation twice; resuspend the mobile sediment in 2 ml of buffer containing 0.2 mg/ml of PEG (20,000 Da) after adjusting the pH to the required level (step 1) and store at 4°C.

Labeling with lectin (or enzyme)-gold complex:

1. Place ultrathin sections of plant tissues on coated grids; float the grid on buffer 0.01 M PBS buffer for ten minutes in a moist chamber.
2. Float the grid on a drop of lectin-gold complex and incubate for 30 to 40 minutes in a moist chamber; rinse the grids with PBS buffer several times followed by rinsing with distilled water and air dry the grids.
3. Stain the grids with uranyl acetate or lead citrate, if necessary.
4. Observe under electron microscope.

IMMUNOCAPTURE-REVERSE
TRANSCRIPTION-POLYMERASE CHAIN REACTION
(IC-RT-PCR) TECHNIQUE

The IC-RT-PCR technique, in principle, involves trapping the antigen (virus) on the wall of a specific antibody-coated tube, followed by washing to remove inhibitory substances of plant origin. Using viral RNA as a template, the cDNA is synthesized using reverse transcriptase and then the cDNA is amplified in PCR with virus-specific primers. The PCR product is analyzed by electrophoresis in agrose gel. The IC-RT-PCR technique has been reported to be 100 to 1,000 times more sensitive than ELISA in detecting plant viruses in extracts of infected leaves (Nolasco et al., 1993; van der Vlugt et al., 1997; Mumford and Seal, 1997).

Immunocapture of Antigen (Virus) from Plant Tissues

1. Grind infected plant tissues (1:10 w/v) in 500 mM Tris-HCl, pH 8.2, containing 2 percent polyvinyl pyrrolidone, 1 percent PEG (6,000 Da), 140 mM NaCl, 0.05 percent Tween-20, and 3 mM NaN_3; centrifuge for five minutes at 5,000 g_n and collect the supernatant.
2. Transfer 50 µl of the supernatant to each well of the microtiter plate already coated with antigen (virus)-specific antibody (as described in the earlier section on ELISA) and incubate overnight at 4°C.
3. Wash the wells three times carefully by flooding with PBS-T ensuring that no cross contamination between wells occurs.

RT-PCR Amplification

1. To each well in the microtiter plate, add 20 µl of the reverse transcription mixture consisting of 50 mM Tris-HCl, pH 8.3, 75 mM KCl, 3 mM $MgCl_2$, 1 mM each dNTP, 25 units of ribonuclease inhibitor, 1 µM downstream primer, 200 units of M-MLV reverse transcriptase (BRL), and incubate for one hour at 37°C.
2. Perform amplification of cDNA in the same wells (using a Techno PHC-3 thermocycler with a microtiter plate adaptor) by adding 80 µl of amplification mixture consisting of 60 mM Tris-

HCl, pH 9.0, 15 mM KCl, 2.1 mM $MgCl_2$, 20 mM $(NH_4)_2 SO_4$, 0.2 mM each dNTP, 0.2 mM each 0.2 μM each primer, 0.005 percent BSA and overlay with a drop of mineral oil.

3. Heat the mixture at 94°C for two minutes; cool to 72°C and add 1.6 units of thermostable DNA polymerase.

4. Proceed through 30 to 35 cycles; each cycle consists of an annealing step of one minute at 52°C, an elongation step of one minute at 72°C, a denaturation step of 30 seconds at 93°C, and an elongation step of five minutes for the final cycle.

Detection of Amplified DNA Products by Southern Hybridization

Transfer the DNAs to nylon membranes (Hybond N+; Amersham, Inc.) using the alkaline transfer procedure; soak the membranes in sodium citrate (SSC) (20 × SSC = 3 M NaCl, 0.3 M sodium citrate, pH 7.2) and hybridize to probes labeled with digoxigenin.

IMMUNOCAPILLARY ZONE ELECTROPHORESIS (I-CZE)

The I-CZE technique has the specificity of immunological methods combined with the sensitivity, rapidity, and automation of capillary zone electrophoresis (CZE) for the detection of antigen (viruses) in purified preparations and extracts of infected plant tissues. CZE-based immunoassays provide the distinct advantage of analyzing antigen-antibody reaction in free solution without the need for immobilization of antigen or antibody on a solid support as required by conventional labeled antibody techniques. The I-CZE assay can detect as little as 10fg of viral antigen in purified preparations and leaf extracts (Eun and Wong, 1999).

Preparation of Plant Extracts

1. Grind fresh plant tissue samples (0.5 g) in 0.01 M sodium borate buffer (1 ml), pH 7.5 percent; centrifuge the suspension at 8,000 g_n for 30 minutes and store the supernatant at 4°C.

2. Dilute the antibodies and antigen (virus) using 50 mM sodium borate buffer, pH 9.7, to produce two sets of serial dilutions.

3. Mix antigen and antibody solutions at appropriate dilutions; incubate for 10, 20, 30, and 50 minutes and maintain healthy plant extract sample as negative control and purified antigen sample as positive control.

Capillary Zone Electrophoresis Analysis

1. Carry out CZE analyses with the Beckman P/ACE system 2100 under normal polarity conditions, the anode and cathode being positioned at the capillary inlet and outlet, respectively.
2. Add the test samples into separate holders and place them in the inlet tray; capillary rinsing, sample injection, sample separation and detection, and data processing and generation of electropherograms are performed by the fully automated analytical process system.
3. Use an uncoated 57 cm fused capillary with an internal diameter of 75 μm housed in a temperature-regulated cartridge; maintain a temperature of 25°C within the capillary during the CZE analysis.
4. Condition the capillary by successfully rinsing with 0.1 M HCl for five minutes, 0.1 M NaOH for ten minutes, and deionized water for five minutes before commencement of analysis for the day.
5. Perform rerun rinsing before each run, with 50 mM sodium borate buffer, pH 9.7, for two minutes; postrun rinse successively with 0.1 M NaOH for two minutes and deionized water for two minutes.
6. Inject the samples pneumatically at 3.45 kPa for five seconds and use 50 mM sodium borate buffer, pH 9.7, as running buffer for all runs.
7. The separated components, as they pass a glass window situated close to the outlet of the capillary, are detected by a wavelength-selectable UV detector system consisting of a deuterium lamp, mirrors, wavelength-selectable UV filters, and a photomultiplier tube.
8. Set the detection absorbance at 280 nm for all runs; link the UV detector system to a computer for plotting the signal graphically in the form of an electropherogram (relative absorbance against elution time).

9. Identify the major peaks in the electropherogram and record elution times after each CZE separation. Carry out 50 successive CZE runs to elute each identified peak and store all pooled fractions from each peak at 4°C for examination under electron microscope to identify the virus in the fractions.

SUMMARY

Various immunodetection techniques ranging from simple precipitation tests, which depend on the direct reaction between antigen and antibody, to labeled antibody techniques that depend on the development of clear visible color reactions, are described. The labeled antibody techniques provide highly specific, sensitive, and reliable results within short periods, compared with conventional analytical techniques. Immunoelectron microscopy has been demonstrated to be useful for in situ detection of microbial pathogens, mycotoxins produced by the fungal pathogens, and plant defense-related protein in plant tissues. The comparative efficacy of immunodiagnostic techniques and their application for detection, identification, and differentiation of related pathogen species, strains, and pathotypes are indicated.

PART II:
STRUCTURE AND CONSTITUENTS
OF HEALTHY PLANTS

Chapter 5

Structure and Functions
of Plant Components

Plants have different kinds of organs/tissues capable of carrying out various functions required for normal growth and reproduction. Under optimal conditions of adequate supply of nutrients and availability of favorable environmental conditions, plants are able to synthesize various compounds that are essential for their growth and reproduction. The form and structure of plants are altered when the conditions become unfavorable. Plants have several organs such as roots, stems, leaves, flowers, seeds, and fruits. Each one performs specific function(s) resulting in normal development. Each organ, in turn, has many types of tissues that are made up of millions of cells. The cell, thus, is the ultimate unit from which all living organisms—both plants and animals—are built. The cells of all organisms, however much they may be diversified, possess many common features and show similarity in the manner in which many biochemical processes are carried out. The basic features of the cellular units are briefly described.

Cells are classified into two types—prokaryotic and eukaryotic—based on the complexity of internal structures. Prokaryotic cells, such as bacteria, do not have internal membrane-bound structures and well-defined nuclei. However, they possess all metabolic machinery required for normal growth and reproduction. On the other hand, the eukaryotic cells of plants have a well-defined nucleus surrounded by a membrane (nuclear envelope). In addition, they have a clearly developed internal system of membranes that separates the cell into distinct areas known as organelles. These include ribosomes, chloroplasts, and mitochondria, which are involved in specific biochemical functions required for various metabolic processes. Several specialized functions, such as photosynthesis and storage of compounds, are associated with clearly identifiable tissues, such as leaves and seeds, respectively.

COMPONENTS OF PLANT CELLS

The cell boundary is demarcated by distinct cell walls that determine the form of the cells in addition to providing a physical barrier to the entry and spread of microbial pathogens. The presence of the cell wall differentiates plant cells from animal cells. Plant cell walls are made up of cellulose and other complex sugars that are responsible for the rigidity/flexibility of different cells. Since plants do not have skeletons, rigidity is provided by the cell walls which assume the distinct shape and size of specific plant species. Cell walls also contain lignin, suberin, waxes, proteins, enzymes, calcium, boron, and water (Cassab, 1998). The plasma membrane forms the internal boundary of the cell, isolating internal contents of the cell from its environment. It also acts as a selectively permeable membrane, controlling the movement of molecules into and out of the cell. The important feature of the plasma membrane is the lipid bilayer consisting primarily of phospholipids and proteins. A hydrophobic barrier regulates the passage of polar molecules such as inorganic ions, sugars, and proteins.

Cytoplasm is enclosed by the plasma membrane and is composed of cytosol and several structures such as the nucleus, ribosomes, chloroplasts, mitochondria, cytoskeleton, endoplasmic reticulum, and Golgi bodies. Cytosol contains many organic compounds such as sugars, amino acids, proteins, and inorganic compounds. The nucleus is enveloped by a double membrane (nuclear envelope) in which holes or pores permit the exchange of materials between the cytoplasm and the nucleus. Among the functions of the nucleus, the synthesis of nucleic acids (deoxyribonucleic acid [DNA]) and ribonucleic acid [RNA]) is the most important. Messenger RNA (mRNA) and ribosomal RNA (rRNA) synthesized in the nucleus are transferred to the cytoplasm for protein synthesis. The DNA present in the nucleus determines the overall genetic constitution of the plants. Nuclear DNA forms a complex with proteins resulting in the formation of chromatin. The nucleolus, a dense area within the nucleus, contains the DNA required for the production of ribosomes.

The endoplasmic reticulum (ER) contains a system of membranes continuous with the nuclear envelope. Primarily it contains a network of membrane-bound tubes and may be rough or smooth. Synthesis of proteins that are secreted from the cell occurs on the rough ER. Proteins targeted to the nucleus are also synthesized in the rough ER. The ribosomes attached to the rough ER give the rough appearance. The smooth ER, which is continuous with rough ER, is free of ribosomes. The smooth ER is involved in the production of enzymes required for the lipid synthesis and detoxification of potentially harmful compounds. Disruption of cells leads to the breakup of the membranous structure of the ER resulting in the formation of many small fragments. These small fragments, known as microsomes, enclose enzymes that are associated with the ER. Such enzymes are called microsomal enzymes.

Golgi bodies, also known as the Golgi apparatus or dictyosomes, are specialized areas of the ER. Microscopic observations of smooth ER reveal the presence of several layers stacked one above the other. At the ends of each layer known as cisternum, small vesicles capable of transporting the products of biosynthetic processes, are present. They seem to be the major site of packaging, modification, and sorting of cellular products. The addition of sugar residues to form glycoproteins is one of the modifications carried out by Golgi bodies. Golgi bodies produce secretory vesicles that move to cell membranes. The membrane enclosing the vesicle merges with the cell membrane and then the contents of the vesicles are discharged outside the cell. This process is known as excytosis or reverse pinocytosis.

Young plant cells contain many vacuoles bounded by a single membrane called a tonoplast. As the cells become older, the vacuoles merge to form a larger one. The turgidity of the cells varies due to the movement of water to and from vacuoles. They act as storage houses for compounds that are to be separated from the primary biochemical processes occurring in the cells. Pigments and excretory materials may be stored in vacuoles.

Plant cells also contain peroxisomes, another type of vesicle, which are the sites of specialized amino acid and fatty acid degradation. During the process of degradation, hydrogen peroxide and free radicals are produced. Peroxisomes regulate antioxidant mechanisms and prevent possible damage to cells following the release of highly reactive chemicals during degradation. The enzyme present in peroxisomes degrades hydrogen peroxide to water and oxygen. Likewise, the antioxidant vitamin E present in the peroxisome membrane reacts with the free radicals, converting them into stable and less reactive compounds.

Mitochondria, considered the powerhouse of cells, are the major sites of adenosine triphosphate (ATP) synthesis and oxidative metabolism. Their similarity in shape, structure, and size to bacteria supports the view that mitochondria might have evolved from bacteria which had a symbiotic relationship with early eukaryotic cells. Mitochondrial DNA, RNA, and ribosomes also have striking similarity to bacteria. Some proteins are synthesized in situ in the mitochondrial matrix, while others are formed from nuclear DNA in the cytoplasm. Mitochondria are 75 percent protein and 25 percent lipid and have a double membrane structure. The outer membrane has a simple structure and remains unfolded, with limited biochemical activity. The inner membrane is highly folded and forms cristae that project into the central area of the mitochondria. The inner membrane regulates important biochemical processes and forms the site of the mitochondrial electron transport chain and synthesis of ATP by oxidative phosphorylation. The central compartment of the mitochondria, known as matrix, has a concentrated aqueous solution made up of many enzymes such as those involved in tricarboxylic acid (TCA) cycle and fatty acid oxidation.

Other important organelles are chloroplasts, which have some features similar to mitochondria. Chloroplasts have a double membrane structure and the ability to trap energy in the form of ATP. However, the sources from which energy is trapped by chloroplasts and mitochondria are different. Chloroplasts convert energy in sunlight into ATP whereas mitochondria utilize the chemical energy released during the oxidation of organic compounds. The inner membranes of chloroplasts have numerous folds known as thylakoid discs with several units that are involved in photosynthesis. The specialized pigments, chlorophylls (a and b), present in the inner membrane are responsible for trapping the solar energy. Chloroplasts also have aqueous contents in the stroma.

Plant cells have a system of protein filaments and tubules known as cytoskeleton. The cytoskeleton system regulates the mechanism of cell movement; is a framework for the organization, positioning, and relocation of subcellular organelles, and strengthens the cell, resulting in the formation of the characteristic cell shape. Cytoskeletons have a complex structure. They have several components of which three have been characterized: actin filaments, microtubules, and intermediate fibers. Actin filaments have thousands of units of the protein actin. Depending on the requirements of the cell, actin may polymerize or depolymerize. Myosin, another protein, can bind with actin and also to subcellular organelles. In the presence of ATP, myosin may move the organelles around the cell along the framework of actin filaments, thus providing a possible mechanism for cytoplasmic streaming. Microtubules have hollow tubular structures containing α- and β-tubulin in addition to two more proteins, kinesin and dynein, which seem to have similar functions as myosin. Microtubules have an important role during mitosis, when the separation of replicated DNA into two cells occurs. Intermediate filaments appear to have a role in providing a framework under the cytoplasmic membrane for the stability of cell shape and positioning of subcellular organelles. Prokaryotic cells carry out various biochemical functions independently. On the other hand, in eukaryotic cells and multicellular organisms such as plants, the functions of cells have to be restricted and coordinated to have the right sequence of well orchestrated metabolic processes required for the normal growth and reproduction.

ORGANIZATION OF TISSUES OF THE PLANT BODY

Every plant cell is totipotent with the capacity to regenerate into a whole plant. A whole plant can be regenerated by culturing the cells in appropriate media. Plants have an array of specialized cell types that are organized into specialized tissues and organs. The process of organization of specialized cells is known as differentiation. The functions of specialized tissues and organs such as roots, stems, leaves, fruits, seeds, and tubers are specific and distinct. Stems, roots, and leaves are involved in vegetative growth of plants, whereas flowers are involved in sexual reproduction resulting in the formation of fruits and seeds.

Plant growth is the result of repeated division of localized groups of embryonic or meristematic cells present at the very tip of stems. The apical meristem is responsible for linear growth, as it continuously produces cells that differentiate into various tissues. Similar cellular division occurs also at the tips of roots. Whereas the cells of the apical meristem are undifferentiated, some differentiation can be seen in the primary meristem, which is just below the apical meristem. The protoderm forms a single cell layer at the surface, while the procambium has discrete strands in the inner portion of the stem embedded in the ground meristem. The protoderm matures into the epidermis in due course, providing protection for the internal tissues. The epidermis has a vital role in preventing pathogen attack, external injury, and desiccation. From the epidermis, several types of appendages such as glandular and nonglandular hairs and scales are formed. The epidermal cell walls are impregnated with the fatty substance cutin and a common waxy layer cuticle. The epidermis has many intercellular fissures (openings) surrounded by pairs of guard cells. The openings and guard cells constitute the stoma through which gaseous exchange and entry of pathogens may occur.

The vascular tissues made up of phloem and xylem are formed when the cells of procambium are differentiated. The vascular tissues are primarily involved in transport of water, nutrients, and photosynthates and provide support. They are present as discrete strands known as vascular bundles. The ground meristem gives rise to a generalized tissue known as parenchyma, a supportive tissue called collenchyma, and another supportive and protective tissue termed sclerenchyma. These tissues are organized into pith, present in the center of the stem, cortex present in between the epidermis and vascular bundles, and pith rays found between vascular bundles (Bold et al., 1980; Scagel et al., 1984; Chrispeels, 2003).

Parenchyma Tissue

Parenchyma tissue matures from ground meristem and is made up of unspecialized, isodiametric, and thin-walled cells. Parenchyma tissue is involved in photosynthesis, storage, and maintenance of cell turgidity. The tissue containing a large number of chloroplasts actively takes part in photosynthesis. This tissue, known as chlorenchyma, is abundant in leaves. The mesophyll containing chlorenchyma has two types of cells: (1) The palisade mesophyll is present just below the epidermis and has columnar cells whose longitudinal axes are at right angles to the leaf surface. (2) The spongy mesophyll is composed of cells with different shapes that are loosely packed, thus possessing considerable intercellular spaces.

Collenchyma Tissue

Collenchyma tissue also forms from the cells of ground meristem. The primary function of collenchyma is support to the plant body. The cell walls are flexible and can stretch to some extent to provide support for primary tissues that are undergoing elongation.

Sclerenchyma Tissue

Sclerenchyma tissue forms after maturation of cells of the ground meristem. This tissue is somewhat heterogenous when compared to parenchyma and collenchyma. It has thick cell walls impregnated with lignin. Although functional, these cells are dead and contain empty lumen as the protoplast becomes nonfunctional. The sclerenchyma provides support and protection to plant organs.

Vascular Tissues

The term stele is applied to the vascular tissues, plus some associated parenchyma, that occupy the central portion of a plant's axis. The outermost layer of the stele is known as pericycle and the parenchyma present in the middle of some steles is called pith. The cortex around the stele has many layers of cells, the innermost layer of which is referred to as endodermis. The vascular tissue consists of phloem, which is involved in the transport of carbohydrates, water and ions, and xylem through which water and ions are transported. Most of the cells in the xylem have lignified secondary walls and many are functional though dead. Xylem is made up of four types of cells: tracheids, vessel elements, fibers, and parenchyma. The term *tracheary element* refers to the tracheids associated with vessel elements. Tracheids have imperforate elongated cells with numerous pits in the secondary walls. Passage of water through tracheids occurs through pit membranes placed between tracheid cells. Vessel elements lacking pit membranes have perforation plates through which water can pass. The cells of vessel elements are joined in tandem to form a vessel. Fibers are present in the xylem and they are considered to have evolved from tracheids through pit reduction followed by increases in thickness of cell walls. Xylem parenchyma cells are living, but differ from other parenchyma cells in that they have thick, lignified secondary walls. Their primary function is storage.

The transport of products of photosynthesis, carbohydrates, water, and some ions, is the primary function of phloem. It may serve an additional support function because of the presence of fibers. Parenchyma and sieve elements are also present in phloem. Sieve cells and sieve tube elements are collectively known as sieve elements. The sieve cells have many pores in the lateral and end walls constituting the sieve areas. The protoplasts and adjacent sieve tube elements are connected through the pores of a common sieve area. Each pore is lined with callose. Deposition of callose increases with age of phloem, resulting in gradual reduction in pore size. Ultimately, the pores become entirely occluded with callose and the phloem becomes nonfunctional. The sieve tube elements appear to function for a short period. The protoplasts of mature sieve tube elements do not contain nuclei and vacuoles, but mitochondria, endoplasmic reticulum, and plastids may be present. Phloem parenchyma has thin-walled cells that perform a storage function. Specialized companion cells and albuminous cells are also found in the phloem parenchyma of some plants. Protoplasmic connections between companion cells may carry out some cytoplasmic functions for sieve elements, since the cells of sieve elements are enucleate.

SUMMARY

Plants are made up of a vast number of cells which form the basic unit in all kinds of plant tissues/organs. The cells are transformed, differentiated, and grouped to form various tissues or organs that have specific function(s). Although the components of the cells in different tissues may be the same, the functions are different. The functions of various types of tissues and their contributions to the health of plants are discussed.

Chapter 6

Biochemical Constituents of Plants

Plants synthesize a variety of compounds to meet the demands for growth and reproduction. The substances produced in excess of such demands are stored in organs such as seeds, tubers, and fruits. Among the various kinds of substances produced by plants, carbohydrates, proteins, enzymes, nucleic acids, fatty acids, alkaloids, and phenolics are important because they are not only required for the health of the plants but also essential sources of food, feed, and fiber for all organisms. Important features of these compounds are described.

CARBOHYDRATES

Carbohydrates, including sugars and related compounds, are used as source of energy by all cell types in plants as well as for raw materials for many chemical syntheses. Sugars are water soluble and hence they can be easily transported through phloem to other tissues. The chemical names of sugars and more complex carbohydrates use the suffix "-ose" (e.g., pentoses, hexoses), and are based on the number of carbon atoms present in the compound. Nonstarch polysaccharides are the basic materials from which the structural components are synthesized. Structural integrity depends on the carbohydrates that make up plant cell walls. The cell walls are composed of polysaccharides, which account for about 90 percent of the structural materials, while proteins and phenolic compounds (lignins) are the other constituents of cell walls. Cellulose, hemicellulose, and pectin are the major polysaccharides present in cell walls.

Cellulose

Plant cell wall structure is primarily determined by cellulose fibers which provide much of its strength. The cell wall is composed of cellulose rods or microfibrils embedded in an amorphous matrix of noncellulose polysaccharides such as hemicelluloses and pectins. Cellulose contains long linear chains of glucose residues covalently linked by β-(1,4)-glucosidic bonds. Glucans made up of long linear chains of glucose molecules (2,000 to 6,000) residues form the primary wall. Estimates report that each cellulose fiber (about 4 nm diameter) may have 30 to 40 glucan chains held together by a very large number of hydrogen bonds.

Hemicellulose

The principal hemicelluloses present in dicotyledonous plants are xyloglucans containing primarily glucose and xylose. In some plants, xyloglucan may also contain fucose, galactose, and small amounts of arabinose. The hemicelluloses commonly found in moncotyledons and legumes, are arabinoxylans containing β-(1,4)-linked xylose residues. The polymeric xylan backbone can attach itself to cellulose through hydrogen bonds as the xyloglucan present in dicotyledons.

Pectins

The presence of linear chains of galacturonic acid residues (polygalacturonans) is the most important feature of pectins. Some of the galacturonic acid residues may be present in the form of methyl esters. Among the cell wall components, pectins are the most soluble and can be easily extracted with hot water. Pathogens produce several enzymes capable of degrading pectins present in the primary and secondary cell walls and middle lamellae resulting in the loss of structural integrity. The pectin in primary and secondary cell walls is called protopectin and has more COOH groups esterified than in the middle lamella. The COO-groups of the polygalacturonan in the middle lamella are held together by Ca^{2+} cross links. The presence of cell wall pectins can be visualized by using appropriate monoclonal antibodies (Sutherland et al., 1999).

Glucosinolates

Glucosinolates are glycosides of β-D-thioglucose with aglycones. Under the influence of glucosinolase (myrosinase), toxic isothiocyanate, thiocyanates, nitriles, or oxazolidone, derivatives such as

goitrin are formed from β-D-thioglucose. Goiter, a condition characterized by the swelling of the thyroid, is a predominant symptom of
glucosinolate poisoning. Varieties of oilseed rape (canola) with low
levels of glucosinolates have been produced to overcome the problem
of toxicity to animals.

Cyanogenic Glucosides

Cyanogenic glucosides are glycosides with a sugar and cyanide-
containing aglycone. Cyanogens such as amygdalin and prunasin are
present in almonds, apricots, apple seeds, and wild cherries, while
linamarin is found in white clover, cassava, linseed, and lima beans.
The presence of dhurrrin in sorghum is also known. Cyanogens are
broken down to release hydrogen cyanide (HCN) in sufficient concentrations to become toxic by the action of glucosidases and hydrox-
ynitrile lyases. HCN is known to inhibit cytochrome oxidase which
catalyzes the final step in the electron transport chain. The glucosides
present in the vacuoles of plant cells and degrading enzymes found in
the cytoplasm are brought together when the cells collapse or mechanical action occurring during feeding thus initiating the process of
formation of HCN. Some cassava cultivars may contain cyanogens
harmful to human beings. Cooking may destroy the enzymes, preventing cyanide production.

PROTEINS

Plant proteins have several vital functions including acting as
enzymes, as structural components of the cells, and in molecular recognition of signals resulting in susceptibility/resistance to biotic
stresses. Cellular activities are primarily controlled by the production
of various enzymes/specific proteins for which the genetic information is carried in the plant DNA. The synthesis of proteins, which are
composed of amino acids, is an essential process occurring in all
kinds of cells. Hence, an adequate supply of various amino acids is
required. Amino acids are small molecules containing both $-NH_2$
(amino) and $-COOH$ (carboxylic acid) groups. Appropriate amino
acids are linked together to form proteins. Some amino acids are
found free in plant cells, unlinked with others. Generally, the free

amino acids are present in low concentrations. However, when protein synthesis is reduced due to certain abiotic stresses such as water, salt, and low light/darkness, the levels of some free amino acids such as proline may be considerably high.

Amino Acids

About 20 amino acids are known. By combining appropriate amino acids different proteins are formed. Proteins contain L-isomers of amino acids. Proline and hydroxyproline, though generally called amino acids, do not have a true amino group but contain a nitrogen atom which forms part of a five-membered ring. The amino acids are classified as aliphatic, hydroxy, sulphur-containing, aromatic, basic, acidic, and imino, based on the nature of the R group. Hydrophobic or hydrophilic properties, as well as other properties, are determined by the nature of the side chain (R) and the charges it carries.

Nonprotein amino acids exist in the free form in plants, especially in legumes. Some of them have been found to be toxic, causing adverse physiological effects on animals. Jack bean *(Canavalia ensiformis)* seeds contain canavanine, a nonprotein amino acid. It resembles arginine in structure and is capable of interfering with the metabolism of arginine and incorporation of arginine into proteins in animals that consume the seeds. Selenium-containing amino acids are present in high concentrations in plants grown in selenium-rich soils. Plants containing Se-methylselenomethionine or Se-methylselenocysteine are toxic to animals and cause "alkali disease" or "blind staggers." These amino acids get incorporated into proteins, substituting normal sulphur-containing amino acids. Another nonprotein amino acid is mimosine, which is present in the leaves and seeds of the tropical legume *Leucaena leucocephala*. Because of the presence of toxic mimosine, the use of this legume for feeding animals is limited, despite its high nutritional value. The toxic effects of mimosine and its breakdown product, dihydroxypyridine (DHP), have been overcome. In one study of Australian animals Hawaiian rumen bacteria was introduced which protected them against mimosine poisoning. Human beings and animals can contract neurolathyrism caused by lathyrogens present in seeds of *Lathyrus sativus* (white pea) and *L. odoratus* (annual sweet pea). The compound β-N-oxalyl α-β-diaminopropionic acid found in white pea attacks nerve cells, leading to weakness and ultimately paralysis of the legs (Chesworth et al., 1998).

Peptides and Proteins

Amino acids are joined together by condensation reactions to form chains (polymers) of amino acids, known as peptides. The bonds that link the amino acids are called peptide bonds. Proteins are made up of several peptides (polypeptides) and consist of about 100 to 3,000 amino acid residues, with a molecular weight ranging from 10,000 to 300,000. Proteins have a variety of functions. They may function as enzymes, storehouses of nitrogen in a biologically accessible form, structural components of cells, and as products of genes of both host plants and pathogens. The interaction between the gene products may determine the levels of host plant resistance or susceptibility. Proteins have a unique ability to fold into well-defined three-dimensional shapes or conformations so that they can bind very strongly to other molecules and recognize them specifically. This ability is responsible for the antibody-antigen and enzyme-substrate reactions. The specific recognition by proteins may be drastically altered or abolished by even very small changes in the shape of proteins.

Protein Structure

The primary structure of a protein is determined by the sequence (order) in which the amino acids are linked by the peptide bonds. This amino acid sequence is dictated by the genetic constitution of the cell. A free amino group (N-terminal) is present at one end of a protein, while the other end of the chain has a carboxyl group (C-terminal). The N- and C-terminals normally carry positive and negative charges, respectively. The side chains (R groups) may also carry positive or negative charges. Distinct arrangements of the polypeptide chains (secondary structure) may occur because of hydrogen bonding between groups which are close together in the same polypeptide chain. This may lead to the formation of a helix (α-helix), as the chain gets twisted. Such an arrangement is observed in coat proteins of rod-shaped plant viruses. In the β-pleated sheet (another type of secondary structure), several individual peptide chains, laid side by side, are held together by hydrogen bonds between peptide groups. The globular proteins may have regions composed of α-helix and β-pleated sheet structures.

Most of the proteins in biological systems may be in the form of solutions or suspension in water. The presence of water affects the shape of the proteins appreciably. The protein chain may arrange itself so that hydrophobic groups point toward the middle of the protein structure, while the hydrophilic groups point toward the surrounding water. The tertiary structure of protein thus shows the changes in the polypeptide backbone that result in bends and twists. The quaternary structure of protein reveals the presence of prosthetic groups attached to proteins. The prosthetic groups do not contain amino acids, but they are required for the activity of the protein to which they are linked. Complex carbohydrates, metal ions, or complex polycyclic compounds such as haem in haemoglobin, myoglobin, or cytochromes are the prosthetic groups linked to proteins.

Protein structure is maintained by the interactions between protein subunits. Biological activity may be lost if the subunits dissociate. Some proteins are further processed by the action of enzymes, as in the case of collagen. Proline, present originally, is subsequently hydroxylated to form hydroxyproline by the action of proline hydroxylase which is activated by ascorbic acid. Collagen is involved in the formation of crosslinks between adjacent molecules.

Many proteins, by covalent attachment, have carbohydrate molecules that are important in determining the functions of proteins. Proteins have some properties in common. Based on the solubility of plant proteins in water, salt solutions, and organic solvents, a system of classification has been developed. Albumins, globulins, glutelins, and prolamins can be differentiated based on their solubility (Chesworth et al., 1998). The solubility of most proteins may be increased by the addition of small amounts of inorganic salts. However, addition of larger amounts of these salts will decrease the solubility of proteins, indicating that the ionic strength of the solution, rather than the concentration of the salt determines the solubility of the protein. "Salting out" is a common procedure used to precipitate proteins from solution without denaturing them. Ammonium sulphate is often used to salt out proteins and plant viruses during the process of purification. Trichloroacetic acid (TCA) has been used to precipitate proteins at acid pH levels. Adjustment of pH of the solution to reach the isoelectric point at which both positive and negative charges are at the same level results in the precipitation of proteins. The biological activity of proteins is drastically affected by higher temperatures. The secondary, tertiary, and quaternary structures are lost when solutions of proteins are heated.

Molecular analysis has established that three classes of compounds are capable of storing information on variations in a large scale. The nucleic acids, proteins, and complex carbohydrates including glycoproteins can store different types of biological variations. Proteins can conserve most of the information contained in nucleic acids. Proteins are strategically placed in the center of the sequence of synthesis. Most of the information on variation is received by proteins from nucleic acids. Proteins are more versatile and reactive than nucleic acids in utilizing the information. On the other hand, proteins can filter out the information on qualitative variations, making their products relatively poor storehouses of genetic variation.

Localization/Distribution of Protein in Plant Cells

The presence of cell walls not only differentiates plant cells from animal cells but also forms a physical barrier to the entry and spread of microbial pathogens. Cell walls are vital in maintaining the health of plants. Among the several constituents of cell walls such as polysaccharides, lignin, suberin, waxes, calcium, boron, and water, proteins and enzymes have an important role during growth and development of plants, environmental stresses, and infection. The cell wall is constantly modified by enzyme action (Cassab, 1998). Proteins localized in cell walls are relatively abundant in plants and green algae (see Chapter 4). Generally, they are rich in one or two amino acids and are highly or poorly glycosylated. Hydroxyproline-rich glycoproteins (HRGPs) or extensins, arabinogalactan proteins (AGPs), glycine-rich proteins (GRPs), proline-rich proteins (PRPs), and chimeric proteins that contain extensin-like domains are important. Among cell wall proteins, extensins are the most well studied. They are preferentially localized in sclerenchyma and cambium cells and are also associated with secondary xylem and phloem tissues. Localization of extensin in several tissues and plants has been observed by Ye et al. (1991) and Ye and Varner (1991) using polyclonal antibodies raised against extensin from soybean seed coat in the immunocytochemistry tissue printing technique (see Chapter 4). Extensins may have a role during pollen germination, fruit ripening, cell elongation, and wound healing when cells are required to be loosened or reinforced. Moreover, extensins may contribute to defense against pathogens. Immunochemical studies have revealed the accumulation of extensins in cell walls close to sites where pathogen growth has been arrested (Esquerré-Tugayé et al., 1985).

AGPs are present in cell membranes, the extracellular matrix (ECM), and gum exudates. Particularly in *Acacia* spp. Gum arabic AGPs are secreted in large quantities upon wounding and they may act as a physical barrier by forming a gel plug. AGPs may contain about 90 percent (w/w) of carbohydrates and about 10 percent of proteins. Localization of AGPs in carrot protoplasts was studied by using a specific monoclonal antibody, JIM4. This antibody was able to recognize AGP epitopes (specific reactive surfaces) present in a specific set of cells during the development of carrot root apex (Stacey et al., 1990). AGPs are considered to have important functions in pollination such as pollen recognition and adhesion on the stigma, in addition to serving as nutrient molecules, since they have high sugar content and stickiness (Fincher et al., 1983; Lord and Sanders, 1992). Glycine-rich protein (GRP), another cell wall protein, may play an important role in the development of vascular tissues, nodules, and flowers. Because GRP from soybean aleurone layers can be extracted with hot water suggests that GRP may be associated with cell wall polysaccharides by nonionic bonds (Matsui et al., 1995).

Enzymes

Enzymes act on the specific substrates and convert them into different products that are required for the growth and reproduction of plants. Enzymes are true catalysts, bringing about changes in substrates, while they themselves do not undergo any change. They can increase the rates of reactions by many thousand times, under optimal conditions, converting substrates into products with very high efficiency. Enzymes generally function by decreasing the activation energy of reactions. An unstable intermediate (transition state) with higher energy content than the original reactant is formed when molecular reactants are brought together. The intermediate represents an energy barrier through which molecules must pass and the rate of reaction is determined by the number of molecules with sufficient energy to pass over the barrier. This activation energy barrier is lowered by the action of enzymes. However, the overall free energy change for the reaction, i.e., difference between the energy of the reactants and products, remains unchanged by the enzyme activity.

The formation of an enzyme-substrate (ES) complex is the first step in any enzyme-catalyzed reaction. Groups of atoms present on the specific surfaces of both enzyme and substrate interact with one another. Enzymes have active sites that bind with the substrates through a combination of many forces that are responsible for the maintenance of the protein structure. The binding of substrates with enzymes is a reversible process and dissociation of substrates may occur freely. The substrate has to fit the active site precisely (as in the lock-and-key model) indicating that this property determines the specificity of enzyme activity. Any condition affecting the three-dimensional structure of an enzyme may drastically reduce or inactivate the reaction. Among the several factors influencing the rates of enzyme-catalyzed reactions, enzyme concentration, substrate concentration, temperature, pH, and presence of inhibitors and cofactors are important. Enzyme inhibitors may have considerable adverse effects on plants and animals. Insecticides containing organophosporus and carbamate compounds are toxic to all types of animals. Organophosphorus compounds have been used as nerve gases in chemical warfare and many herbicides act as enzyme inhibitors.

The presence of small molecules known as coenzymes is required for the activity of some enzymes. Biological oxidation and reduction reactions are dependent on redox coenzymes. Removal of pairs of hydrogen atoms results in oxidation of the molecules, as in the oxidation of malate in the TCA cycle. Adenosine triphosphate (ATP) and adenosine diphosphate (ADP) are important coenzymes involved in the conversion and transfer of energy in many reactions in biochemical pathways, particularly those involving sugars. The conversion of ATP into ADP is accompanied by the release of an appreciable amount of energy required for the progress of biochemical reactions in plants.

Molecular Detection and Recognition

The ability to bind and recognize other molecules is one of the most important properties of proteins. The interaction between host plants and microbial pathogens has been studied intensively. Using the techniques that assess this property has provided a better under-standing of the nature of the plant-pathogen relationship. Many proteins, such as receptors and antibodies, also have similar specific binding properties. A receptor molecule binds specifically with another molecule. This is analogous to the binding of substrates to enzymes. Because the receptor has high affinity and specificity, the presence of a certain other molecule(s) can be detected. Antibodies are proteins produced in response to the introduction of antigens into animal systems. Various plant proteins can be detected by producing antibodies against them. These antibodies can be used to detect and quantify the plant proteins present in different cells/tissues. The methods of producing antisera containing specific antibodies and various methods for immunodetection of different biochemical constituents of plants are described in Chapters 3 and 4, respectively.

Characterization and subcellular localization of some plant enzymes have been achieved by immunological techniques. The antibodies raised against intact exo- and endoglucanases were employed to characterize these enzymes (cellulases) present in maize coleoptile cell walls, which are capable of mediating the hydrolysis of noncellulosic β-(1,3) (1,4)-glucan in situ (Inoushe et al., 1999). In soybean, lipoxygenase (LOX) is considered to be involved in the reserve lipid mobilization during germination. By using the immunogold labeling technique, the presence of LOX protein in the cytoplasm was demonstrated. LOX appeared in vacuoles of epidermal cells of cotyledons of germinated seeds (Wang et al., 1999).

Protective Plant Proteins

Plant proteins have a distinct and pivotal role in the maintenance of plant health. They are important in many aspects of host defense, both as constitutive resistance factors and as part of the complex cascade of active resistance responses. Based on the expression patterns of protective proteins, they can be broadly divided into two groups. The first group consists of constitutive or tissue-specific proteins whose expression is not dependent on infection or damage. They may be present in specific organs, tissues, or cell types. The presence of such proteins in seeds and other storage tissues protects them against infection by microbial pathogens or damage (Mauch et al., 1988). The second group includes proteins that are induced in response to infection, damage, or predation. Pathogenesis-related (PR) proteins are the most intensively studied proteins. The protease inhibitor PI-II of potato also is included in this group.

Constitutive or Tissue-Specific Proteins

The inhibitors of hydrolytic enzymes such as proteinases and amylases are the most widely recognized constitutive proteins. They are found in high concentrations as storage proteins in seeds and other storage organs. These enzyme inhibitors are classified based on their amino acid sequences and other specificities. They specifically inhibit proteinases, especially serine proteinases such as trypsin, chymotrypsin, and subtilisin. Cystatins, inhibitors of cysteine proteinases from rice, and inhibitors of α-amylases from plants, fungi, and bacteria have also been characterized (Richardson, 1991; García-Olmedo, Carmona, et al., 1992). Tubers (potatoes), bulbs, gum secretions from *Acacia arabica,* and latex from rubber contain protein inhibitors.

Many kinds of proteins, other than enzyme inhibitors, are also present in seeds. In addition to enzyme inhibitors, barley seeds have been shown to contain proteins with protective functions such as thionins, endochitinases, ribosome-inactivating proteins (RIPs), β-glucanase, phospholipid transfer protein (LTP), lectin, and thaumatin-like proteins (TLPs) (Hejgaard et al., 1991; Shewry, 1993, 1995). The nature and antimicrobial properties of the important protective proteins are discussed in the following.

Thionins

The presence of thionins in plants and their antimicrobial properties were first reported in wheat (flour). A thionin named purothionin was shown to be toxic to yeast (Balls et al., 1942). Purothionins are basic and cysteine-rich polypeptides with an MW of about 5 kDa. Thionins from the endosperm of wheat, rye, oat, and barley were later isolated and sequenced (Hernandez-Lucas et al., 1978; Wada, 1982; García-Olmedo et al., 1989). All of these cereal thionins are localized in the endosperm of seeds and can be considered a single class (type I). In addition to seed thionins (seed-specific hordothionins), leaf-specific thionins that can be induced by pathogens have also been isolated from barley (Bohlmann et al., 1988). These inducible thionins are placed in another class (type II). The type I thionins are located intracellularly at the periphery of the protein body, while the location of type II thionins is not clearly established (Shewry and Lucas, 1997).

The thionins of wheat (purothionins) and barley (hordothionins) have been reported to be inhibitory to several bacterial pathogens such as *Clavibacter michiganensis, Pseudomonas solanacearum, Phytophthora infestans, Colletotrichum lagenarium,* and *Fusarium solani* (García-Olmedo, Carmona, et al., 1992; Bohlmann, 1994; Florack and Stiekema, 1994). Transgenic tobacco plants expressing type I α-hordothionin gene under the control of the CaMV 35S promoter exhibited greater level of resistance to two pathovars of *Pseudomonas syringae* (Carmona et al., 1993). However, no enhancement in resistance of transgenic tobacco plants was observed by Florack et al. (1994). Expression of a viscotoxin (from mistletoe *[Viscum album]*) and endogenous *Arabidopsis* thionins at high levels led to significant enhancement of resistance to *Plasmodiophora brassicae* (clubroot disease) and *Fusarium oxysporum* f. sp. *matthiolae* (wilt disease), respectively (Epple et al., 1997; Bohlmann, 1999). Thionins may inhibit cell-free protein synthesis under in vitro conditions. Some thionin genes may be induced by methyl jasmonate which is involved in the induction of proteinase inhibitors in tomato (Andersen et al., 1992; Brümmer, et al., 1994). The usefulness of thionin gene(s) for engineering disease resistance is yet to be clearly demonstrated.

Ribosome-Inhibiting Proteins (RIPs)

Ribosome-inhibiting proteins (RIPs) are widely distributed in the plant kingdom. They are toxic N-glycosides, capable of modifying ribosomes by cleaving a specific adenine residue on a highly conserved sequence of 28 S rRNA. The RIPs depurinate the universally conserved α-sarcin loop of large rRNAs, resulting in blocking of translation and consequently protein synthesis. The possible involvement of RIPs in plant defense against pathogens was first indicated by the demonstration of inhibition of infection by tobacco mosaic virus (TMV) by the extracts of pokeweed *(Phytolacca americana)* (Irvin et al., 1980). RIPs have received considerable attention because of their possible use in the development of chimeric toxins and therapeutic agents against viruses. α-trichosanthin (α-TCS) has been found to be effective in blocking the replication of human immunodeficiency virus 1 (HIV-1) in cells and macrophages in vitro (Piatak and Habuka, 1992).

RIPs may be present in all plant parts and may accumulate to reach amounts varying from 0.1 percent to more than 1 percent of total tissue weight. Generally, RIPs do not inactivate the ribosomes of the same plant or closely related species and hence, they are presumed to have a role in resistance to microbial pathogens. RIPs have been grouped into two types. Type I RIPs have a single peptide chain that may not be glycosylated and they have MWs of 23 to 25 kDa and alkaline pI of 8.0 to 10.0. Type II RIPs contain an A chain, which is basically equivalent to type I RIP, disulfide linked to a lectin-like B chain that binds the RIP to cell surfaces and facilitates entry of the A chain into the cytosol. Another type of related protein, agglutinin, has also been recognized. Agglutinins may be bifunctional lectins and agglutinate cells very efficiently because of their structural features. Type III RIPs are synthesized as inactive precursors (proRIPs) and they require proteolytic processing events to occur between two amino acids involved in the formation of the active site. This type of RIP has been characterized from maize and barley. Immunological cross-reactions with antibodies to proRIP of maize indicates that sorghum may also contain a related proRIP (Nielsen and Boston, 2001)

Among plant RIPs, pokeweed antiviral protein (PAP) and *Mirabilis* antiviral protein (MAP) have been studied intensively. PAP is a mixture of three proteins present in spring leaves, summer leaves, and seeds (Barbieri et al., 1982). PAP localized in the cell walls of leaf mesophyll cells can be purified. Exogenous application of PAP entirely prevents mechanical transmission of viruses such as TMV to many different host plants (Chen et al., 1991). In transgenic tobacco and tomato plants expressing PAP, the protein was enriched in the intercellular fluid and these plants showed resistance to potato virus X (PVX), potato virus Y (PVY), and cucumber mosaic virus (CMV). PAP may either enter the host cells along with the virus and block translation of viral RNA or bind to the virus or the cell wall preventing the entry of the virus into the cell (Lodge et al., 1993).

To elucidate the mechanism of action of PAP, Zoubenko et al. (2000) showed that PAP is associated with ribosomes and depurinates tobacco ribosomes in vivo by removing more than one adenine and guanine. PAPn, a mutant of PAP, with a single amino acid substitution (G75D), did not bind ribosomes efficiently and was not toxic when expressed in transgenic tobacco plants (as the PAP). In PAPn-expressing transgenic plants, basic PR-proteins, the wound-inducible protein kinase, and protein inhibitor II were induced and these plants were resistant to viral and fungal infection. The results indicated that PAPn activates a salicylic acid (SA)-independent, stress-associated signal transduction pathway conferring resistance to pathogens in the absence of ribosome binding, rRNA depurination, and acidic PR-protein production. Simultaneous expression of PAP and enzymatically inactive mutant resulted in overproduction of PR-protein in transgenic tobacco plants which were resistant to fungal pathogens. Such a coordinate expression of RIP and PR-protein seems to be specific for PAP-expressing tobacco (Zoubenko et al., 1997).

MAP isolated from *Mirabilis jalapa* was sequenced and expressed in *E. coli* whose ribosomal activity was inhibited by MAP (Habuka et al., 1989). The MAP gene was expressed as a fusion protein with β-galactosidase. By using an ELISA assay, the amount of MAP produced was estimated to be about 150 µg/l (Pitak and Habuka, 1992). Two type I RIPs were isolated from the roots of *Mirabilis expansa*. These RIPs were toxic to several soil-borne bacterial species, *Fusarium* spp. and nonpathogenic *Trichoderma* spp. even at µg levels. Differential sensitivity of microbial pathogens to the RIPs from *M. expansa* was observed. The fungal pathogen *Pythium irregulare* was sensitive, while *Pythium ultimum* showed resistance to the RIPs (Vivanco et al., 1999). The RIP from *Dianthus sinensis* is capable of inhibiting the infectivity of TMV and this RIP gene has 59 percent homology with PAP and MAP genes (Cho et al., 2000).

Another type I RIP with antifungal activity was isolated from barley grains. The growth of *Trichoderma reesi, Botrytis cinerea,* and *Rhizoctonia solani* was inhibited by this RIP in vitro (Roberts and Selitrennikoff, 1986). The activity of barley RIP was dramatically increased when it was combined with two other barley seed proteins viz., β-(1,3)-glucanase and an endochitinase (Leah et al., 1991). Transgenic rose plants expressing a type I RIP from barley showed reduced susceptibility to black spot disease *(Diplocarpon rosae)* by secreting the RIP in the extracellular space (Dohm et al., 2001). The maize b-32 protein is a functional RIP, inhibiting in vitro translation in the cell-free reticulocyte-derived system and exhibits specific Nglycosidase activity on 28S rRNA. The *opaque-2 (O2)* mutant kernels lacking b-32 were susceptible to fungal infection and insect feeding. The transgenic tobacco lines expressing *b-32* gene showed greater tolerance to *R. solani* (Maddaloni et al., 1999). Sativin isolated from sugar snap (*Pisum sativum* var. *macrocarpin*) showed similarity to pisavin, a RIP from *P. sativum* var. *arvense* in the N-terminal amino acid sequence. But sativin showed little ribonuclese activity and inhibited translation in a rabbit reticulocyte lysate system. Sativin exhibited antifungal activity against *F. oxysporum* but not against *R. solani* (Ye et al., 2000).

A novel protein designated Crip-31 with properties similar to RIPs, was isolated from *Clerodendrum inerme*. Crip-31 induced both localized and systemic resistance against CMV, PVY, and tomato mosaic virus (ToMV) in tomato. This inducer protein is basic in nature and has hydrophobic residues and a molecular mass of 31 kDa. Treatment with proteinase K (endopeptidase K) does not alter the ability of Crip-31 to induce resistance in tomato (Shelly et al., 2001).

Polygalacturonase-Inhibiting Proteins (PGIPs)

Microbial plant pathogens causing necrosis (death of cells or tissues) are known to produce a variety of extracellular hydrolytic enzymes that act on host cell wall components. The French bean *(Phaseolus vulgaris)* contains a cell wall protein capable of inhibiting the endopolygalacturonases produced by fungal pathogens. Albersheim and Anderson (1971) isolated the polygalacturonase-inhibiting protein (PGIP) from bean *(P. vulgaris)*. The PGIPs isolated from four bean cultivars increased the lifetimes of phytoalexin elicitor-active oligogalacturonides generated by endo-PGs produced by three races of *Colletotrichum lindemuthianum* causing anthracnose disease (de Lorenzo et al., 1990). Purification of the PGIP from bean and cloning of the gene encoding the PGIP have been achieved (Cervone et al., 1987; Toubart et al., 1992). The presence of PGIPs in soybean seedlings (Favaron et al., 1994), tomato fruits (Stotz et al., 1994), pears (Sharrock and Labavitch, 1994), apples (Yao et al., 1995), potato leaves (Machinandiarena et al., 2001), and roots and stems of cotton (James and Dubery, 2001) has been reported.

The PGIP from bean is a secreted protein located in the extracellular matrix. It is synthesized as a precursor protein which is later processed after translation, finally becoming a polypeptide with a MW of 34 kDa (Toubart et al., 1992). The PGIPs seem to have inhibitory activity against endopolygalacturonases (endo PGs) produced by fungi, but not against fungal exoPGs, pectate lyases, or pectolytic enzymes from bacteria (Johnston et al., 1993). The inhibitory activity of PGIPs varies with endoPGs produced by different fungal pathogens, so also with the isoforms (isozymes) of the same endoPG. PGIPs capable of binding fungal PGs, exhibit varying inhibition specificities and kinetics. Purified bean PGIP inhibited several fungal PGs, whereas pear PGIP was effective only against *B. cinerea* (Stotz et al., 2000) [see the appendix at the end of this chapter]). The presence and distribution of PGIPs in various plants and tissue types indicate that they are constitutive proteins. However, PGIPs can also be induced by wounding, stress, or infection by pathogens. PGIPs are a family of relatively heat-stable glycoproteins in the cell walls of a variety of dicotyledonous plants. They are included in the class of leucine-rich repeat (LRR) proteins that are involved in the recognition of signals derived from protein-protein interactions between host plants and pathogens. Binding of fungal PGs by PGIPs may contribute to constitutive resistance to soft rot pathogens, whereas activation of PGIP genes may form a component of inducible (active) plant defense (Shewry and Lucas, 1997; Machinandiarena et al., 2001).

Defensins

Defensins are basic antimicrobial peptides that have been implicated in plant defense against microbial pathogens. Initially, they were considered to be similar to thionin and named γ-thionins. However, they were found to be structurally unrelated. Terras et al. (1995) proposed the name plant defensins based on structural and functional similarities with insect defensins. Defensins have been isolated from a wide range of monocot and dicot plants. Using immunological or in situ hybridization techniques, the localization of constitutively expressed plant defensins in their tissue of origin has been studied. They accumulate preferentially in peripheral cell layers. Highest concentrations of defensins were detected in the outer cell wall lining the epidermis, cotyledons, hypocotyl, and endosperm of radish seeds (Terras et al., 1995). In potato tubers, transcription of defensin gene reached maximum levels in the epidermis and leaf primordia (Moreno et al., 1994). In contrast, plant defensins were localized primarily in the xylem, stomatal cells, and cell walls lining substomatal cavities in sugar beet leaves (Kragh et al., 1995).

Plant defensins form a family of small (about 5 kDa), usually basic, peptides that are rich in disulfide-linked cysteine residues. Comparison of amino acid sequences indicates that relatively few residues are conserved in all plant defensins. The conserved residues are confined to eight cysteines and a glycine at position 48 (da Silva Conceicão and Broekaert, 1999). The three-dimensional folding pattern of plant defensins appears to be very similar to that of drosomycin, a pathogen-inducible antifungal peptide from the fruit fly *Drosophila melanogaster* (Landon et al., 1997). The defensin alfalfa antifungal peptide (alfAFP) was isolated from the seeds of alfalfa *(M. sativa)* and exhibited strong activity against *Verticillium dahliae* infecting potato. Expression of the alfAFP peptide in transgenic potato plants provided robust resistance against *V. dahliae* under both greenhouse and field conditions. No earlier demonstration exists of a single transgene imparting a disease-resistance phenotype that offers protection to a level equivalent to the fumigant (Gao et al., 2000). A defensin gene that may have a protective role in the flowers of sunflower was identified. A full-length sunflower cDNA displaying the structural features and consensus sequences of plant defensins was isolated. The predicted protein has a signal peptide of 31 amino acids and a mature peptide of 47 amino acids with a molecular mass of 5,300 Da (Urdangarín et al., 2000).

Many plant defensins inhibit the growth of fungal pathogens either by reducing hyphal elongation and increasing branching (morphogenic) or by reducing hyphal extension without marked morphological distortions (nonmorphogenic). Plant defensins have been isolated from wheat, barley, sorghum, pea, cowpea, potato, radish, sugar beet, and dahlia. Plant defensins exuded from germinating seeds of radish account for about 30 percent of the total released proteins and may provide protection to the germinating seeds. The defensins present in leaves are induced following infection by *Alternaria brassicola* (Terrass et al., 1995). Induction of defensins in potato tubers, flowers, stems, and leaves, effective against *C. michiganensis,* ssp. *sepedon-*

icus, P. solanacearum, and *F. solani* has also been observed (Moreno et al., 1994).

Lectins

Lectins are proteins capable of recognizing and binding to specific sugar sequences. They may agglutinate cells (forming clumps) that possess carbohydrates on the cell surface or precipitate glycoproteins (proteins containing one or more carbohydrate groups bound to them). The presence of a lectin in a plant can be detected by testing whether an extract of the plant tissue agglutinates erythrocytes (mammalian red blood cells) and by demonstrating that the agglutination is sugar specific, as revealed by inhibition of reaction by simple or complex polysaccharides under suitable conditions. Purified lectins agglutinate erythrocytes at concentrations as low as 0.1 to 1.0 μg/ml. Some lectins of plant origin, notably from cereal, legumes, and solanaceous species (Shewry and Lucas, 1997) have been characterized. Concanavalin A from jack bean *(C. ensiformis)* can recognize oligomannosyl N-linked sugars, whereas wheat germ agglutinin binds sialic acid and *N*-acetylglucosamine. Ricin from castor is able to bind galactose. Lectins have an essential role in the phenomenon of self-incompatibility in plants. The germination of pollen grains is arrested in cross-pollinated plants, if the lectins in the pollen recognize the stigma of the same plant. Lectins have been useful for the study of carbohydrate structure and function, by employing them as probes. Furthermore, they are also toxic to animals making it necessary to avoid feeds with toxic constituents.

Chitin-binding lectins are considered to have a role in plant defense against microbial plant pathogens, especially fungal pathogens that contain chitin on their cell walls (Bramble and Gade, 1985). The seeds of many graminaceous species contain lectins that specifically bind the sugar *N*-acetylglucosamine (G/c NAc), its oligomers and chitin, a polymer of G/c NAc residues (Chrispeels and Raikhel, 1991). Some lectins possess antimicrobial properties: wheat germ agglutin (Mirelman et al., 1975), potato lectin (Callow, 1977; Garas and Kuć, 1981; Andreu and Daleo, 1988), pokeweed lectin (Bramble and Gade, 1985) and stinging nettle lectin (Broekaert et al., 1989). The stinging nettle lectin was more effective against *B. cinerea* (gray mold) than chitinase (Broekaert et al., 1989). Likewise, hevein (from rubber) inhibited the growth of several fungal pathogens such as *B. cinerea, F. oxysporum,* and *Pyrenophora tritici-repentis* at concentrations of 300 to 500 µg/ml, while tobacco chitinase was ineffective even at 1 mg/ml (Van Parijs et al., 1990, as cited in Chrispeels and Raikhel, 1991). The use of lectins for engineering resistance to fungal pathogens appears to be a distinct possibility.

Pathogenesis-Related (PR)-Like Proteins (PRLs)

Pathogenesis-related proteins (PRs) may be defined as proteins coded for by the host plant but induced only in pathological or related situations (Antoniw et al., 1980). The classical PRs are considered as a collective set of novel proteins associated with host defense in incompatible interactions. However, in compatible interactions these novel host proteins have been detected only occasionally and have not been adequately characterized. Many plant species are known to contain proteins that share sequence homology with well-characterized PRs or proteins that possess some of the characteristics of PRs. These proteins are termed PR-like proteins (PRLs). Many seed-specific proteins belong to this group. PR-like chitinases and thaumatin-like proteins (TLPs) isolated from seeds of wheat, barley, oats, sorghum, and maize have antifungal activities (Roberts and Selitrennikoff, 1986; Hejgaard et al., 1991; Vigers et al., 1991; Radhajeyalakshmi et al., 2000; Krishnaveni et al., 1999). PRLs may exert synergistic effect, leading to a level of resistance significantly broader than any one of them may provide (Hejgaard et al., 1991).

Localization and accumulation of rye seed chitinases (RSC-a and RSC-c) with antifungal activity were studied by employing immunoblot and ELISA techniques. An antiserum specific to chitin-binding domain (CB-domain) of RSC-a and another antiserum specific to the catalytic region of the RSC-a and RSC-c were used. Immunoblot analysis revealed the presence of both RSC-a and RSC-c in the endosperm of the rye seeds. ELISA and immunoblot techniques showed that both RSC-a and RSC-c accumulated in the seed during a later stage of development (Taira et al., 2001). During seed germination, many seed protein fractions with potent antifungal activity (AFA) were generated. The germination-related increase of AFA was observed in wheat *(Triticum durum).* The increase was more rapid in seeds incubated in vitro, indicating that at least part of the antifungal protein (AFP) generation was not dependent on gene expression. Seven antifungal proteins effective against important pathogens such as *P. infestans* and *F. graminearum (Gibberella zeae)* were isolated. The generation of AFPs is transitional in nature, but it could play an important role in the protection of plants in the early stage of development, when the defense system may not be in operation (Wang et al., 2002).

TLPs have a molecular mass of 22 to 26 kDa and extensive similarities with thaumatins, which are extremely sweet-tasting proteins from the African shrub *Thaumatococcus danielli.* Many plant proteins isolated from seeds have similar amino acid sequences. Osmotin, a TLP, was produced by tobacco cells that were exposed to increasing concentrations of NaCl (Singh et al., 1987) (see Chapter 7). The presence of zeamatin, another antifungal protein in maize flour, caused rupture of tips of fungal hyphae (Huynh et al., 1992). Transgenic potato plants expressing high levels of osmotin-like protein constitutively were more tolerant to *P. infestans,* which causes late blight disease (Zhu et al., 1996). Likewise, overexpression of TLP gene in rice resulted in significant enhancement of resistance to *R. solani,* which causes sheath blight disease and for which no resistance source is available (Datta et al., 1999).

Permatins, another set of TLPs, have shown antifungal properties. Permatin cDNA clones were produced and their expression in developing barley and oat seeds was studied. Developing barley and oat seeds accumulated permatin mRNA in an unusual pattern. Permatin mRNA and protein were abundant at the time of pollination and decreased to nondetectable levels rapidly thereafter. By employing specific antibodies, expression of permatin gene and protein initially in the ovary wall and then in the aleurone and ventral furrow of developing seeds was observed (Skadsen et al., 2000).

Other Antimicrobial Proteins

The search for antimicrobial proteins in plants has revealed the presence of different kinds of antimicrobial peptides of diverse origin. An antifungal protein Ace-AMP1 was isolated from onion. Scented geranium transformed with the gene encoding this antifungal protein showed enhanced level of resistance to *B. cinerea* leaf infection (Bi et al., 1999). Two cysteine-rich antimicrobial peptides

(MWs of 6.8 and 10.8 kDa) were detected in cowpea seeds. Based on the sequence analysis of these peptides, a defensin and lipid-transfer protein (LTP) with a high degree of homology to other antifungal peptides from plants, could be differentiated. They were inhibitory to the development of *F. oxysporum* and *F. solani*. Immunofluorescence assays showed that the LTP was localized in the cell wall and in cytosolic compartments in the cotyledons and embryonic axes of seeds (Carvalho et al., 2001). From the soluble fraction of strawberry leaves, fragarin, a preformed antimicrobial compound (phytoanticipin), was isolated. The growth and oxygen consumption of the bacterial pathogen *C. michiganensis* were rapidly inhibited by fragarin. Fragarin may act at the membrane level and its action is correlated with a decrease in cell viability (Filippone et al., 2001). An amlyase inhibitor from *Lablab purpureus* (AILP) with a molecular mass of 30 kDa was demonstrated to inhibit α-amylase of *Aspergillus flavus* which causes ear rot disease of corn (maize). α-amylase is known to promote aflatoxin production by *A. flavus* in the endosperm of infected corn kernels and aflatoxin is known to be carcinogenic (see Chapter 13). The amino acid sequence of AILP is similar to the lectin arcelin, and belongs to the α-amylase inhibitor family described in common bean (Fakhoury and Woloshuk, 2001).

Harmful Plant Proteins

Plant pollen grains are known to contain allergens responsible for respiratory disorders in human beings. The pollen of castor *(Ricinus communis)* contains a potential allergen. The allergenic protein was isolated by sodium dodecyl sulfate-polyacrylamide gel electrophoresis (SDS-PAGE) technique. The allergenic nature of this protein was demonstrated by direct skin tests and ELISA inhibition by reaction with serum IgE, the general procedure followed to identify allergens (Sanjukta et al., 1999).

NUCLEIC ACIDS

The structure of DNA molecules present in the cells of all organisms determines the sequence (order) in which amino acids are arranged in proteins. The amino acid sequence of proteins, in turn, determines the properties of proteins. Hence the DNA is considered to be the "blueprint" that dictates the nature and functions of proteins. Likewise, all cellular functions are controlled by the information contained in the DNA. This information is transmitted from one generation to another generation faithfully and accurately by the DNA, until the nature of the DNA is altered by phenomena such as mutation and recombination. Although the information for protein synthesis is stored in DNA, RNA is required for the translation of information in the DNA into the amino acid sequences in proteins.

Nature of Nucleic Acids

Both types of nucleic acids (DNA and RNA) have a backbone consisting of alternating sugar and phosphate groups. RNA contains ribose sugar, while DNA has deoxyribose sugar. The sugar molecule is linked to a nitrogenous base which may be either a purine or pyrimidine. The purine and pyrimidine bases are generally attached to sugar groups to form nucleosides or to sugar phosphates to form nucleotides as observed in nucleic acids. Purine and pyrimidine derivatives such as ATP and nictinamide adenine dinucleotide (NADH) function as coenzymes involved in energy metabolism. Each sugar molecule in DNA and RNA carries a base attached to the 1' position. DNA has purine bases adenine (A) and guanine (G) and pyrimidine bases cytosine (C) and thymine (T). RNA, on the other hand, contains the same purine bases as DNA, but the pyrimidine base cytosine remains unaltered, and thymine is substituted by uracil (U). DNA and RNA have long chains of nucleotides, each of which contains a sugar-phosphate-nitrogenous base (A, G, C, T, or U).

DNA in plants is mostly double stranded (ds) with two strands (chains) running alongside but in opposite directions, and are designated antiparallel. The two strands are held together by hydrogen bonds between A and T or between C and G. This bonding, known as base pairing, is highly specific and involves particular combinations of purine-pyrimidine pairs. The two chains of the DNA are considered to be complementary, since the base of one is determined by and complementary to the other. In contrast, some plant viruses such as geminiviruses, contain single-stranded (ss) DNA as their genome.

Structure of DNA

The ds-DNA has a double helix structure in which the two sugar phosphate chains form the backbone of the molecule and are twisted around one another in a spiral. The bases lie perpendicular to the axis of the helix and point inward toward each other, since they are paired together. Each turn of the helix has about ten nucleotides occupying ca. 3.4 nm of the length of the chain of nucleotides. Two types of grooves, narrow and wide, may be recognized along the surface of helix. Interactions between the DNA and other molecules such as proteins may occur at these grooves. The DNA helix may show a type of coiling known as supercoiling.

Prokaryotes such as *E. coli* have circular DNA in the nucleoid region of the cytoplasm. Bacterial DNA is associated with small amounts of proteins and folded to form about 100 supercoiled loops. On the other hand, in eukaryotes, which have well-defined nuclei, the DNA is present in the form of characteristic chromosomes within the nucleus. DNA forms a complex with large quantities of basic proteins including histones. This complex is a constituent of the chromatin of which the chromosomes are made. The nonhistone proteins may have a role in gene expression by switching "on and off" in some parts of chromatin. Plant organelles, chloroplasts, and mitochondria also have DNA, in addition to nuclear DNA. The synthesis of some plant proteins occurs when the information carried by these DNA is translated. The organelle DNA resembles prokaryotic DNA.

Structure and Types of RNA

RNA also has a backbone consisting of alternating sugar and phosphate molecules. An OH group at the 2' position is present in ribose sugar. The pyrimidine uracil replaces thymine present in the DNA. Generally, RNAs are single stranded (ss). However, within the RNA structure, short regions may be coiled to form paired regions. Some plant viruses contain double-stranded RNAs as their genomes. Rice dwarf virus and clover wound tumor virus have ds-RNAs.

Three distinct types of RNAs are present in plant cells and they differ in their size and function. RNA-polymerases are involved in the synthesis of all RNAs that occurs throughout the life of the cell, but not restricted to periods of cell division as in the case of DNA synthesis. The nucleotide sequences of the DNA determine the sequence of bases assembled into RNA. Thus the information in the DNA is transcribed into the RNA, which in turn, is translated by the RNA into amino acids.

Messenger RNA (mRNA) is one of the three types of RNAs formed in plants and other organisms. mRNA accounts for about 2 percent of the total RNA content, and contains sequences of bases required for placing the amino acids in the desired sequence in a protein. For the synthesis of each protein, one mRNA is required. Hence thousands of different mRNAs may be produced by most eukaryotic cells. In contrast, some mRNAs of prokaryotes may be polycistronic, carrying the codes for the synthesis of several proteins. Another type of RNA is transfer RNA (tRNA) which makes up approximately 16 percent of the total RNA contents. The molecules of tRNA are small, uniform in size, and may have 75 to 90 nucleotides. The tRNA is used as an intermediate carrier, as it has a site for attachment of amino acids and another site that is complementary for recognizing specific sections of mRNA molecules. The tRNA brings amino acids to ribosomes where they are assembled into proteins. The ribosomal RNA (rRNA) is the most predominant RNA and accounts for about 80 percent of the total RNA contents of the cells. The rRNA forms the major component of ribosomes on which synthesis of protein occurs. The ribosomes consisting of large and small subunits are constituted by rRNA (65 percent) and protein (35 percent). The large and small subunits are differentiated by their sedimentation constants expressed in Svedberg units (S). The rRNA is synthesized in the nucleolus, a specialized region of the nucleus (see Chapter 5). Initially, the rRNA is synthesized as one long chain and complexes with proteins. This complex is then split into many smaller pieces which constitute the ribosome.

The Genetic Code

DNA is synthesized during the replication of chromosomes prior to cell division, whereas RNA is produced almost complementarily as part of the protein synthesizing machinery of the cell at varying rates, depending on the supply of nucleotides. The enzymes DNA-polymerase and RNA-polymerase are involved in the synthesis of DNA and RNA, respectively. Three bases in DNA or mRNA determine the nature of a single amino acid and this base sequence is known as a triplet code. The flow of information from DNA occurs as follows:

DNA→ (transcription) RNA → (translation) protein

The mRNA molecules carry the information required for protein synthesis in the ribosomes. Three adjacent bases in the mRNA determine the nature of amino acid in a protein. Because there are four bases, adenine (A), cytosine (C), guanine (G), and uracil (U), it is possible to have 64 different combinations of three bases which function as triplet codes (Table 6.1).

TABLE 6.1. Universal genetic code in organisms

First position	Second position				Third position
	U	C	A	G	
U	Phe	Ser	Tyr	Cys	U
	Phe	Ser	Tyr	Cys	C
	Leu	Ser	STOP	STOP	A
	Leu	Ser	STOP	Trp	G
C	Leu	Pro	His	Arg	U
	Leu	Pro	His	Arg	C
	Leu	Pro	Gln	Arg	A
	Leu	Pro	Gln	Arg	G
A	Ile	Thr	Asn	Ser	U
	Ile	Thr	Asn	Ser	C
	Ile	Thr	Lys	Arg	A
	Met	Thr	Lys	Arg	G
G	Val	Ala	Asp	Gly	U
	Val	Ala	Asp	Gly	C
	Val	Ala	Glu	Gly	A
	Val	Ala	Glu	Gly	G

Source: Chesworth et al., 1998.

SUMMARY

Plants synthesize several kinds of substances that are essential for their growth and reproduction. In addition, the substances produced in excess are stored and they form the sources of food, feed, and fiber for all organisms. Carbohydrates, proteins, enzymes, nucleic acids, fatty acids, alkaloids, and phenolics are important as cellular and structural components and are required for various physiological processes in plants. The functions and distribution of proteinaceous substances, as detected by immunological techniques and their role in the development of resistance/tolerance to both biotic and abiotic stresses are discussed.

APPENDIX:
ASSAY OF POLYGALACTURONASE-INHIBITING PROTEIN (PGIP) BY IMMUNOBLOTTING TECHNIQUE (STOTZ ET AL., 2000)

Antiserum Production

1. Deglycosylate purified pear fruit PGIP using trifluoromethane sulfonic acid; gel purify and electroelute.
2. Inject the deglycosylated PGIP preparation (100µg) mixed with Freund's complete adjuvant intrapopliteally into New Zealand white rabbits; at two-week interval inject two boost injections (50 µg each) of glycosylated PGIP mixed with Freund's incomplete adjuvant through popliteal.
3. Collect blood serum after eight weeks and purify the antibodies against PGIP.

Gel Electrophoresis and Immunoblotting

1. Separate proteins by sodium dodecyl sulfate-polyacrylamide gel electrophoresis (SDS-PAGE); electroblot polyacrylamide gels (10 percent) onto polyvinylidenedifuloride membranes (Millipore, Bedford, Massachusetts).
2. Screen the blots with antibodies against deglycosylated pear PGIP at a dilution of 1:2000; treat the membrane with alkaline phosphatase-conjugated goat antirabbit IgG at a dilution of 1:2000.
3. Develop blots with nitroblue tetrazolium and 5-bromo-4-chloro-3-indolylphosphate as substrate.

PART III:
PLANT-STRESS INTERACTIONS

Chapter 7

Plant-Abiotic Stress Interactions

Plants are exposed to abiotic (physiogenic) and biotic (pathogenic) stresses, resulting in varying degrees of alterations in metabolic activities. Such changes may lead to retardation of growth and reproduction of plants ultimately resulting in significant yield loss. Abiotic stresses may be varied and cause a variety of alterations among which protein syntheses are commonly observed. Protein-based responses of plants subjected to abiotic stress include overall changes in protein synthesis, especially alterations in the levels of specific proteins. The changes in the protein levels may depend on the nature, duration, and severity of stress and they may be reflected in an already existing pool of proteins or in the de novo appearance of some protein(s).

Abiotic stresses may be caused by excess or suboptimal levels of temperature and water supply, or salinity, metal toxicity, gaseous pollutants, and UV radiation. Tolerance or sensitivity toward an abiotic stress depends on the genetic and biochemical makeup of the plant species. Abiotic stresses may induce plants to respond with some kind of defense response, as in the case of biotic stresses. However, chronic exposure to abiotic stress may modify the developmental pattern of the stressed plants.

Abiotic stresses may modify gene expressions in plants (Vierling, 1991). As a result of such modifications, accumulation or depletion of certain metabolites, alterations in the activities of several enzymes, and overall changes in protein synthesis are often observed. The synthesis of certain specific proteins may indicate the nature of the stress as in the case of heat stress. Heat shock proteins (HSPs) synthesized in plants exposed to high temperatures, appear to have specific functions such as protein-protein interactions, translocation of proteins across cellular components, and protecting the organism from heat stress (Cordewener et al., 1995). Likewise, plants under salinity stress synthesize new proteins, which, along with amino acids and soluble nitrogenous compounds, may act as components of a salt-tolerance mechanism (Naqvi et al., 1995; Igarashi et al., 1997). Since tolerance to abiotic and biotic stresses depends on the genetic constitution of plants, attempts have been made to differentiate the tolerant/resistant/sensitive genotypes of crops on the basis of profiles or levels of soluble proteins, specific enzymes in germinating seeds, and growing plant parts (Perezmolphebalch et al., 1996; Elsamad and Shaddad, 1997). Among the abiotic stresses, the salt-, water-, and heat/cold-induced stresses have been found to be widespread and the effects of these stresses are discussed as follows.

SALT STRESS

Crop health and productivity are appreciably affected by soil salinity in arid and semiarid tracts in the tropical countries. Agricultural lands extending to several millions of hectares have become unsuitable for cultivation, due to constant buildup of salinity in the soils. Salinity affects seed germination, nutrient uptake, metabolism, and, consequently, plant growth. Synthesis of specific proteins is a characteristic feature of salt stress, while the levels of total and/or soluble proteins may show decreases or increases (Cassab, 1998). Salt-tolerant and salt-sensitive genotypes exhibit distinct differences in levels of proteins when they are grown under salt stress. The contents of soluble proteins and free amino acids, in addition to sugars, sucrose, starch, and phenols decreased in all three genotypes of mulberry as the salinity levels increased (Agastian et al., 2000). The specific proteins produced under salt stress may possibly have a role in either induction of salt tolerance or adaptation to salinity.

In tobacco, cells that adapt to sodium chloride (NaCl) contain a unique protein (26 kDa) in addition to higher concentrations of two proteins (20 and 32 kDa) compared to unadapted cells. Specific alteration in gene expression in salt-adapted cells leads to the synthesis of several novel proteins such as 26 kDa protein, which is involved in the process of cellular adaptation to osmotic pressure. The specific protein osmotin (26 kDa) accumulates in cells undergoing osmotic adjustment to salt or desiccation stress (Singh et al., 1987). Osmotin, a specific protein associated with NaCl-adapted tobacco cells, accounts for about 10 to 12 percent of the total protein in the cells (Singh et al., 1985). Production of specific proteins following exposure to salt stress in other crops such as rice (Naqvi et al., 1995), barley (Hurkman and Tanaka, 1987; Popova et al., 1995), citrus (Ben-Hayyim et al., 1989; Naot et al., 1995), and tomato (Ben-Hayyim et al., 1989; Elenany, 1997) has been reported. The proteins induced in salt-stressed plants are found to be species specific.

Some plant species, in response to salt stress, show de novo synthesis of certain hydrophilic proteins and of their mRNAs, which may also be induced by water deficit or treatment with abscisic acid (ABA) (Artlip and Funkhouser, 1995). Salt stress during seed development results in the synthesis of late embryogenesis-abundant (LEA) proteins in cotton. Similar LEA synthesis in carrot, barley, and maize has also been observed (Ramagopal, 1993). Transgenic rice plants expressing LEA protein gene *HVA1* from barley had significant accumulation of HAV1 protein in roots and in leaves and these plants were more tolerant to salinity. A cDNA clone *oslea 3* encoding a group of three LEA proteins accumulated to greater levels in the salt-tolerant rice variety (Pokkali) compared with sensitive cultivar (TN1) under salt stress (Moons et al., 1997). LEA proteins may have an important role in the development of salinity tolerance in rice plants.

A general decrease in protein synthesis and loss of ribosomes have been observed in plants exposed to salt stress (Artlip and Funkhouser, 1995), resulting in reduction in protein levels. Such a condition may be due to decreased protein synthesis, suboptimal levels of amino acids, and denaturation of enzymes required for the syntheses of amino acids and proteins. Varying effects of salt stress on the activities of different enzymes such as proteolytic, amylolytic, nucleolytic, phosphorolytic, oxidative, antioxidant, photosynthetic, and nitrogen assimilatory enzymes have been observed. The changes in the enzyme activities either increase or decrease depending on the nature of enzymes, sensitivity/tolerance of cultivars (genotypes), plant parts (seeds/leaves), and the intensity of salt stress. The activities of nucleases, proteases, peptidases, phosphatases, and oxidases are increased in rice seedlings due to salt stress (Dubey, 1999). Enhanced activities of β-amylase in barley plants grown in the presence of NaCl (200mM) were detected by Dreier et al. (1995). A differential response of salt-tolerant rice cultivars was evident in the activities of malatedehydrogenases that were inhibited to 16 to 100 percent when exposed to NaCl for a period of 5 to 20 days (Kumar et al., 2000). In salt-tolerant cells of *Citrus sinensis,* lipoxygenase (LOX) was induced very rapidly and in a transient manner. Preferential induction of a 9-lipoxygenase activity occurred very rapidly in salt-tolerant cells leading to reduction of hyperoxides to hydroxy derivatives (Ben-Hayyim et al., 2001). On the other hand, no changes in the total apoplastic concentration of cell wall proteins and in the activities of apoplastic peroxidases or xyloglucan endotransglycosylase could be detected in the salt-stressed maize leaves (Cramer et al., 2001). In general, protein contents of plants exposed to salt stress register significant reduction because of accelerated activities of protein-hydrolyzing enzymes. However, accumulation of proline in all organs of different species of *Lycopersicon* was detected, but no general relationship between proline accumulation and salt tolerance could be established (Rajasekaran et al., 2000).

WATER STRESS

Water stress is the most critical factor causing drastic reduction in crop yields. In severe, prolonged drought the entire crop is lost in the semiarid tracts in tropical countries. Water stress during seed germination, seedling establishment, and flowering may lead to huge losses in yield of crops. Protein synthesis, photosynthesis, respiration, and nucleic acid synthesis are the major physiological processes affected adversely in plants grown under water stress. Whereas inhibition of protein synthesis is associated with water stress, enhancement of synthesis of certain specific proteins also occurs due to the changes in gene expression resulting in both quantitative and qualitative changes in the nature of proteins synthesized. Some of the water stress-induced proteins have been well characterized (Claes et al., 1990; Close, 1997; Pelah et al., 1997).

The efficiency of protein synthesis is reduced by mild to moderate water stress. However, normality in the efficiency may be restored when the stress is reversed by watering the plants. The levels of polyribosomes in plants subjected to water stress are considerably decreased resulting in reduction of protein synthesis, as in the cases of wheat (Scott et al., 1979) and maize (Bewley and Larsen, 1982). Plant species that adapt to drought conditions appear to possess a greater capacity to produce polyribosomes, and hence, protein synthesis in these species is not affected. Although protein synthesis is reduced in general, synthesis of specific protein occurs in plants exposed to water stress. Application of ABA has been shown to induce many of the water stress-specific proteins, suggesting that ABA may act as a signal in the response of plants to water stress. Genes governing the synthesis of water stress-specific proteins have been isolated and characterized. The specific proteins have been identified and grouped into LEAs, RABs (responsible to ABA), dehydrins, and vegetative storage proteins (Baker et al., 1995; Artlip and Funkhouser, 1995; Claes et al., 1990; Close, 1997; Pelah et al., 1997).

Among the water stress-specific proteins, synthesis of dehydrins has been commonly reported in barley, maize, and peas. Dehydrin polypeptides have about 100 to 600 amino acid residues. They are considered to function as surfactants resulting in the prevention of co-agulation of a range of macromolecules which leads to preservation of structural integrity of plant cells (Close, 1997). The endogenous levels of ABA, dehydrins and its mRNA have registered simultaneous increases in dehydrated leaves of tomato and maize, indicating that genes encoding dehydrins are ABA-regulated. Dehydrins have been observed to be localized primarily in the cytoplasm of root and shoot cells (Bray, 1995). Rice plants expressing the barley LEA protein gene *HVA1* exhibit greater tolerance to water stress. In rice cell suspension cultures from transgenic plants, a high level of LEA proteins accumulation was noted. Osmotin (26 kDa) synthesis regulated by ABA and its accumulation are also influenced by the extent of water stress. Osmotin, like dehydrins, is present at higher levels under both water and salt stresses (Ramagopal, 1993). The formation of a protein (65 kDa) was detected in leaves of tomato under water stress using the gold labeling technique. Synthesis of this protein in nuclei, chloroplasts, and cytoplasm could be observed (Tabaeizadeh et al., 1995).

Changes in the proline-, threonine-, and glycine-rich proteins in plants exposed to water stress (drought) have been studied. Reduction in the synthesis of a proline-rich protein (12 kDa) in the cell walls of *Lycopersicon chilense* (wild tomato) was observed by Yu et al. (1996). A cDNA clone encoding a proline -, threonine -, and glycine-rich protein (PTGRP) was isolated from *L. chilense*. The corresponding mRNA levels in leaves and stems of drought-stressed plants were reduced by five- to tenfold, which, however, were restored to normal level when plants were rewatered. The subcellular localization of this protein was visualized by using the antibody raised against a glutathione S-transferase/PTGRP fusion protein. PTGRP was local-ized in the xylem pit membranes and disintegrated primary walls in regularly watered plants. On the other hand, a significant reduction in intensity of immunogold labeling in the water-stressed plants was ev-ident, indicating a decrease in the contents of PTGRP in stressed plants. PTGRP appears to be the first drought-regulated protein whose localization in cell walls has been precisely detected (Harrak et al., 1999). Effects of water stress on glycine-rich proteins (GRPs) and expression of the genes encoding GRPs in alfalfa have also been studied (Bray, 1995).

The accumulation of proline in many plant species exposed to wa-ter stress has been observed. However, the significance of such accu-mulation is not clearly understood. The view that proline accumula-tion is beneficial to plants is not universally accepted, since results indicating an opposite effect have also been reported. The study, us-ing antisense soybean plants with an L-Δ^1-pyrroline-5-carboxylate reductase (P5CR) gene, which controls proline synthesis, indicated that transgenic plants had decreased proline synthesis. A wooden box screening showed that proline played a critical role in the survival of soybean plants under water stress, since the transformants failed to survive a six-day drought stress at 37°C. However, the nontrans-formed plants could survive this drought stress (de Ronde et al., 2000). An accumulation of proline in detached leaves of maize geno-types ranging from 58 to 208 percent following exposure to water stress was observed. Proline concentration did not seem to contribute to tolerance of water stress in maize (Wasson et al., 2000).

The specific proteins produced in plants exposed to water stress have been classified into several families and they are involved in the protection of cells from desiccation leading to ultimate death. In the French bean *(Phaseolus vulgaris),* two antigenically related glyco-proteins, p33 and p36, accumulated in the soluble fraction of the cell wall in response to water stress. These proteins could adhere to leaf protoplasts and bind to plasma membrane (PM) vesicles in a divalent cation-dependent manner. The p33 and p36 proteins, based on the amino acid sequence, are considered to be related to proline-rich pro-teins, capable of specifically binding to a PM protein (80 kDa) (García-Gomez et al., 2000). Some of the stress-induced proteins ap-pear to be tissue specific, while the synthesis of other proteins may occur in different tissues or organs.

The activities of several enzymes involved in nitrogen assimilation and photosynthesis are altered by water stress. Decreases in the activity of nitrate reductase in water-stressed plants have been observed (Hsiao, 1973). Adverse effects of water stress on carboxylating enzymes involved in photosynthesis have been revealed by the studies of Kaiser (1987) and Du et al. (1996). Significant decreases in the activities of ribulose-1, 5-bisphosphate (RuBP) carboxylase, PEP carboxylase, fructose-1, 6-biphosphatase, and orthophosphate dikinase in the leaves of water-stressed sugarcane resulted in concomitant reduction in the photosynthetic rates (Du et al., 1996). Inhibition of the activities of fructose-1,6-biphosphatase, and sugar phosphate synthase in the leaves of sorghum exposed to water stress was reported by Reddy (1996). Production of active oxygen species and reduction in the activities of antioxidant enzymes such as catalase, peroxidase, and superoxide dimutase are the effects of water stress (Gogorcena et al., 1995). Enhancemet of enzymatic activity in water-stressed plants has also been reported. Hydrolytic enzymes such as α-amylase exhibit greater activity resulting in accelerated starch hydrolysis and consequent increase in sugar levels and decrease in starch contents of plants exposed to water stress. The activity of sucrose synthase was markedly increased leading to accumulation of sucrose in water-stressed *Populus* sp. (Pelah et al., 1997).

TEMPERATURE STRESS

Temperatures, above or below optimal levels, are harmful for the development of plants. Crops suffer due to heat (high temperature) and cold (low temperature) stress resulting in significant reduction in yield.

Heat Stress

Among abiotic stresses, the most universal response exhibited by all organisms, both eukaryotes and prokaryotes, is to heat stress. In many field crops, heat injury (due to temperatures above 40°C) is observed as drastic reduction in photosynthesis, altered protein metabolism leading to breakdown/denaturation of protein, and inactivation of several enzymes. Heat stress induces both quantitative and qualitative changes in sensitive plants. A decline in protein synthesis results as a consequence of translational repression of most mRNAs. Enhanced synthesis of a set of specific proteins, termed heat shock proteins (HSPs), indicates the qualitative changes in protein synthesis of plants exposed to high temperatures. Formation of HSPs has been observed in diverse plant species exposed to temperatures 10 to 15°C above the optimal levels (Heikkila et al., 1984; Mansfield and Key, 1988; Singla and Grover, 1994; Cordewener et al., 1995).

HSPs provide an endogenous system of thermotolerance to plants. All organisms appear to exhibit the phenomenon of heat shock response (HSR). HSPs localize in organelles such as nuclei, chloroplasts, mitochondria, and plasma membrane (Vierling, 1991). Formation of HSP is initiated as tissue temperature exceeds 32°C. The accumulation of HSPs is positively correlated with increased thermotolerance, in addition to some protection to abiotic stresses (Artlip and Funkhouser, 1995). As temperatures increase beyond 32°C progressive increase in the synthesis of HSPs occurs in crops such as maize, wheat, cowpea, and soybean (Vierling, 1991). Induction of HSP mRNA leading to the production of HSPs may result following treatment with ABA, water stress, and wounding. Differential responses of various tissues and developmental stages of tissues of field-grown cotton to heat stress were observed. Eight unique polypeptides with MWs of 100, 94, 89, 75, 60, 58, 37, and 21 kDa appeared in cotton exposed to 40°C for a few weeks (Burke et al., 1985). In rice seedlings subjected to a temperature of 45°C for one to two hours, accumulation of a polypeptide (104 kDa) was noted (Singla and Grover, 1994). A heat-tolerant maize line exposed to heat stress synthesized a unique set of heat-shock polypeptides of 45 kDa that may have a role in the development of thermotolerance (Bhadula et al., 2001). A polypeptide with 71.7 Ku (u = unified atomic mass unit) was detected in apples exposed to direct sunlight (for 60 to 120 days beyond commercial maturity) followed by exposure to –4°C. Using a polyclonal antiserum raised against wheat HSP 70, the presence of this specific peptide was revealed in the apple peel and cortex of all five apple cultivars tested. The presence of any of the small HSPs could not be detected in these apples. However, three small HSPs could be induced in these apples after exposure to a temperature of 45°C for four hours. These small HSPs were detected using antibodies raised against pea cytosolic HSP 18.1. The results show that apples may respond rapidly to high temperature stress even at advanced stages of maturity by synthesizing small HSPs. These small HSPs may have an important role in protecting cellular biochemical processes during such periods of heat stress (Ritenour et al., 2001).

Heat shock proteins have varying molecular masses and are localized in different organelles. HSPs have been classified based on size. Large HSPs, such as HSP60, HSP70, and HSP 90, are constitutive proteins, localized in the cytoplasm or mitochondria and chloroplasts. They are present in plants grown under normal temperatures, and they begin to accumulate following heat shock. These HSPs may function as molecular chaperones, assist the self-assembly of novel polypeptides, and prevent the formation of an aggregation of nonfunctional proteins formed following heat denaturation (Artlip and Funkhouser, 1995). Small HSPs with MWs of 17 to 30 kDa may accumulate when plants are exposed to heat stress. They bind partially denatured proteins, and prevent irreversible protein inactivation, leading to thermotolerance (Waters et al., 1998). The possibility of involvement of HSPs in signal transduction was suggested by Harrington et al., 1994. Low molecular weight (LMW) HSPs immunochemically related to α-crystallin were detected in the mitochondria of wheat, rye, and maize seedlings. The MWs of these proteins varied from 19 to 24 kDa. When the plants were shifted to 28°C after heat shock (at 42°C for three hours), the organelle-associated LMW HSPs appeared in the cytosol (Korotaeva et al., 2001). In response to heat stress, other proteins or mRNAs also accumulate. These proteins, not regarded as HSPs, include glycolytic enzymes, protein kinases, and ubiquitin. The glycolytic enzymes and protein kinases may possibly be required with metabolic readjustment, whereas ubiquitin is involved in degradation of aberrant proteins formed in plants exposed to heat stress (Artlip and Funkhouser, 1995).

The effects of heat stress on enzymes were assessed by exposing cotton to temperatures of 40° or 45°C. Ribulose-1,5-bisphosphate (RuBP) carboxylase/oxygenase activity was studied. The levels of RuBP registered increase, while the 3-phosphoglyceric acid pool was depleted, indicating that the inhibition of photosynthesis by moderate heat stress was not due to inhibition of the capacity for RuBP regeneration (Crafts-Brandner and Law, 2000). During heat stress, the chloroplast small heat shock protein (Chlp Hsp 24) has been shown to protect photosystem II (PS2) in different tomato genotypes. Phenotypic variations in the production of Chlp Hsp24 by nine tomato genotypes was positively correlated to the thermotolerance of PS2. Significant positive correlation between production of Chlp Hsp24 and Hsp 60 and net photosynthetic rate (PN) was observed (Preczewski et al., 2000). Among the five unique polypeptides formed in heat-tolerant maize lines, three polypeptides had amino acid sequences similar to that of chloroplast and bacterial protein synthesis elongation factor (EF-Tu), as determined by using an antibody specific to maize EF-Tu. A strong reaction between the 45 kDa HSP and the anti-EF-Tu antibody indicates a close relationship. The 45 kDa polypeptides were localized in the chloroplast and may be of nuclear origin. An increase in EF-Tu transcript levels, during heat stress, was followed by increased levels of EF-Tu protein which may play an important role in the development of thermotolerance in maize (Bhadula et al., 2001).

Cold Stress

Cold stress includes two distinct stresses—chilling (temperatures 4 to 15°C) and freezing. Chilling may affect the plasma membrane resulting in electrolyte leakage from cells. Freezing stress may exert intracellular and extracellular effects. Formation of ice crystals can pierce the plasma membrane may become lethal immediately. Ice present in the intercellular spaces and cell walls, because of low water potential, can remove water from the cells thus causing desiccation. Development of tolerance to chilling appears to be a prerequisite for tolerance to freezing. Chilling injury may result in the loss of plasma membrane integrity and irreversible and proportional loss of proteins from cells leading ultimately to the death of cells. On the other hand, in cells dehydrated by freezing, aggregation of proteins and denaturation of soluble proteins may be seen. (Levitt, 1980). Long-term exposure of wheat seedlings to cold stress may lead to changes in the protein-synthesizing systems resulting in optimal synthesizing capacity under altered conditions (Lásztity et al., 1999).

When many plant species are exposed to low nonfreezing temperatures for a prolonged period (a few hours or a day), certain new proteins are produced that provide tolerance or adaptation to subsequent chilling or freezing temperatures. The development of cold tolerance is known as cold acclimation (CA), and is associated with gene expression that requires calcium influx into the cytosol. Alfalfa *(M. sativa)* plant cells treated with agents that block calcium influx were unable to develop CA (Örvar et al., 2000). Since CA results in altered gene expression, several new specific proteins are synthesized. However, the levels of preexisting proteins decline following exposure to low temperatures. The accumulation of specific cold-regulated (COR) proteins has a role in the hardening process and different amounts of COR proteins appear to be related to different degrees of cold tolerance (Mastrangelo et al., 2000).

Cold-acclimated plants contain new proteins that are not found in plants grown at normal temperatures. In winter rye seedlings, 26 new proteins and 11 proteins at higher concentrations were detected, while about 20 proteins in the plasma membrane disappeared (Uemura and Yoshida, 1984). Synthesis of antifreezing proteins in rice, wheat, and barley has also been observed (Griffith et al., 1997; Antikainen and Griffith, 1997). The mechanism controlling accumulation of the COR protein TMC-AP3, a putative chloroplastic amino acid-selective channel protein in barley, durum wheat, and bread wheat, was studied. The frost-resistant genotypes contained higher concentrations of COR TMC-AP3 after nine days of dehardening, when compared to susceptible genotypes, suggesting that the resistant and susceptible genotypes possess different protein degradation rates and/or mRNA translational efficiency (Mastrangelo et al., 2000).

Cold-shock proteins (CSPs) produced in plants in response to cold stress may function differently in various crops. Apoplastic antifreeze proteins protect cells against damage by ice crystals. Chaperones and dehydrins may protect cell macromolecules against cold injury. The proteins that uncouple oxidation and phosphorylation in mitochondria may assist plants in retaining an above-zero temperature for some time and prepare them for tolerating subsequent freezing temperatures (Kolesnichenko et al., 2000). Cold-specific proteins (CSPs) have been characterized and grouped into families. In cold-acclimated wheat *(Triticum aestivum),* four immunologically related proteins belonging to the annexin family were identified. These proteins are intrinsic membrane proteins and their association with the membrane is calcium dependent (Breton et al., 2000). The immunochemical affinity of cold stress protein CSP 310 present among native cytoplasmic, mitochondrial, and nuclear proteins of winter rye was investigated. CSP 310 protein showed affinity to proteins with molecular weights of 230 and 140 kDa. Proteins with immunochemical affinity to CSP 310 were present in cytoplasm, mitochondria, and nuclei and these proteins increased in concentration following exposure of rye plants to cold stress (Kolesnichenko et al., 2000).

The synthesis of several key enzymes (proteins) significantly decreases when plants are exposed to cold stress (Bruggemann et al., 1994; Matsuba et al., 1997). The structure, conformation, and properties of the key photosynthetic enzyme of C3 plants, Rubisco are altered remarkably by chilling. In one study, the activities of Rubisco and stromal fructose-1,6-phosphatase in tomato and maize were entirely lost, following chilling, in sensitive plants (Bruggemann, et al., 1994). Build-up of reactive oxygen species and oxidation of proteins, in addition to decreased activities of catalase, ascorbate peroxidase, and glutathione reductase were the effects of chilling observed in rice plants. The cold stability of catalase and ascorbate peroxidase appeared to be associated with the development of tolerance of rice cultivars to chilling injury (Saruyama and Tanida, 1995).

The requirement of gene expression that depends on calcium influx into the cytosol for the development of CA was demonstrated in alfalfa *(M. sativa)*. Chemical agents that facilitate calcium influx induced expression of CA-specific *(cas)* gene in alfalfa at 25°C, indicating that cytoskeleton recognition is an integral component in low temperature signal transduction in alfalfa cell suspension cultures (Örvar et al., 2000). Calcium-dependent protein kinases (CDPKs) are considered to play a role in signal transduction in plants exposed to cold stress. In rice seeds exposed to 5°C, the leaf and stem tissues showed 37, 47, and 55 kDa protein kinase activities. The 47 kDa protein kinase activity was strictly Ca^{2+} dependent and it was induced by cold stress in both cytosolic and membrane fractions of the stem. The activity of this CDPK was stronger in the cold-tolerant rice cultivars when compared to sensitive ones (Li and Komatsu, 2000). A 56 kDa membrane-bound CDPK that is activated by cold treatment (at 12°C) of rice cv. Don Juan has been identified. This CDPK did not participate in the initial response to low temperature, but rather in the adaptive process to low temperature condition (Martin and Busconi, 2001).

In wheat *(T. aestivum)*, the expression of the *TaADF* gene was studied and the level of expression of the gene was correlated to the extent of freezing tolerance. This gene codes for a protein that is homologous to members of the actin-depolymerizing factor (ADF)/ cofilin family. TaADF protein, a substrate for wheat 52 kDa kinase, is expressed only in cold-acclimated cereals and the accumulation level is comparatively much higher in tolerant wheat cultivars than in less tolerant varieties. This protein may be required for cytoskeletal rearrangements that may occur following exposure of wheat plants to cold stress (Ouellet et al., 2001). Pineapple fruits develop an internal browning disorder called blackheart when they are exposed to cold stress. Polyphenol-oxidase (PPO) (catecholoxidase) activity was dramatically increased by tenfold in cold-stressed fruits. A direct correlation between PPO activity and the severity of blackheart symptoms was discernible. The expression of two *PPO* genes *PINPPO 1* and *PINPPO 2* was strongly upregulated in response to cold stress or wounding, indicating the de novo synthesis of PPO in response to chilling in pineapple fruits. A role for this enzyme in the development of blackheart in cold-stressed pineapple fruits has been suggested by Stewart et al. (2001).

SUMMARY

Plants exposed to abiotic stresses (environmental conditions) may respond to these stresses in various ways. Changes in protein synthesis, both qualitative and quantitative, have been studied to understand the effects of major abiotic stresses caused by salt, water, and temperature (heat and cold). Production of specific proteins in plants exposed to these major abiotic stresses has been found to be a common feature. The involvement and mechanism(s) of action of the specific proteins in the development of tolerance to salt, water, and temperatures (both high and low) have been elucidated in certain crop-stress combinations, although the mechanism(s) of induction of resistance is unclear. The successful attempts to select genotypes/cultivars of some crops showing tolerance to abiotic stresses are highlighted, indicating the possibility of using such tolerant cultivars to overcome the adverse affects of these limiting environmental conditions.

Chapter 8

Plant-Fungal Pathogen Interactions

Plants, growing under various environmental conditions, are exposed to innumerable microbes of which a small proportion is capable of causing diseases. Plant diseases are caused by different microbial pathogens such as fungi, bacteria, phytoplasmas, viruses, and viroids. Plants are susceptible to certain pathogens carrying genes for virulence, whereas plants possessing genes for resistance are not infected by these pathogens. A plant species may show resistance to some pathogens but it may be susceptible to others.

IMMUNODIAGNOSTIC TECHNIQUES

The nature of interactions between a host plant and microbial pathogens depends on the aggressiveness (virulence) of the pathogens and the level of susceptibility/resistance of plant species. Microbial pathogens show wide variations in their virulence, and within a morphologic species, several races, biotypes, and pathovars exist with differing levels of virulence. Hence, it is essential to precisely establish the identity of the pathogens, races, biotypes, or pathovars. Fungal pathogens can be identified by isolating them using appropriate media. By studying morphological characteristics, generally it may be possible to identify fungal pathogens up to species level. Sets of differential plants consisting of different plant species or cultivars have to be inoculated to differentiate races or biotypes present in a morphologic species. Immunological techniques have become the preferred tools to rapidly identify, differentiate, and quantify closely related fungal pathogens. These techniques require only small quantities of pathogens and do not depend on the development of visual symptoms on infected plants. The results may be obtained within a few hours, whereas several days or even weeks may be required for traditional methods involving isolation of pathogens on culture media and inoculation on their respective host plants.

Rapid advancements in recent years have resulted in the improvement of sensitivity and reliability of immunodiagnostic assays. Characterization of antigens against fungal pathogens is a major problem, as fungi are complex antigens and different spore forms are produced during their life cycles. Hence, the preparation of antisera containing polyclonal antibodies against fungal pathogens has been difficult. However, the sensitivity, specificity, and reliability of the immunological reactions have been remarkably increased due to the development of monoclonal antibody technology. Polyclonal and monoclonal antisera have been developed for several fungal pathogens. Furthermore, immunoassays can be employed to differentiate pathogenic and nonpathogenic isolates of a fungus as in the case of *Venturia inaequalis,* which causes apple scab disease. A clear distinction based on the appearance of the fungal cell wall and distribution of fimbrial epitopes labeled with specific antiserum and immunogold complex could be observed (Svircev et al., 2000). Among the various immunodiagnostic assays, ELISA, immunoblots, and immunofluorescence techniques are currently employed to identify and differentiate fungal pathogens present in seeds, planting materials, and whole plants under field conditions. The immunodiagnostic protocols employed for the detection of different fungal pathogens are presented in this chapter's appendix.

Detection of fungal pathogens in seeds and propagative materials such as tubers, bulbs, setts, and cuttings is of great importance because (1) the infected materials rapidly spread disease(s) to other regions in the same country or other countries where the particular disease(s) may not be prevalent or may be of minor importance, and (2) the infected plant materials introduce the disease(s) early into the fields resulting in heavy losses, especially in the case of soilborne diseases. Both domestic and international plant quarantines employ various immunodiagnostic and other molecular techniques to filter out infected plants and propagative materials so that the entry of new pathogens is effectively checked. Under field conditions, detection of pathogens and infected plants is required to check further spread of diseases and also to determine the timing of application of chemicals to restrict the spread of the diseases. Immunological techniques have been used to detect and quantify fungal mycotoxins in foods and feeds leading to rejection of such materials unfit for consumption (see Chapter 13).

Antisera used against fungal pathogens may be generally produced using rabbits or mice for the generation of polyclonal antibodies (PABs) and monoclonal antibodies (MABs), respectively. However, chicken egg yolk was used to raise antibodies against *Colletotrichum falcatum* and *Fusarium subglutinans* infecting sugarcane. The PABs raised in egg yolk (IgY) were equally efficient as PABs in rabbits (IgG) when assessed by indirect double antibody sandwich (DAS)-ELISA. Furthermore, IgY can be prepared and purified less labori-ously and more rapidly (at least tenfold larger quantity per time unit than IgG) and with the same longevity as IgG. The strains of these two sugarcane pathogens could be differentiated and also quantita-tively estimated by using the specific IgY (Vöhringer and Sander, 2001). Phage-displayed recombinant antibodies (see Chapter 4) spe-cific against surface epitopes of *P. infestans* were produced by Gough et al. (1999).

ENZYME-LINKED IMMUNOSORBENT ASSAY (ELISA)

The ELISA technique has been performed to detect, differentiate, and quantify fungal pathogens in plant materials and soil. Various formats of ELISA have been developed to suit different purposes. The double antibody sandwich (DAS-ELISA) format has been widely employed. Pathogens detected are *Aphanomyces euteiches* (Petersen et al., 1996), *Botrytis allii* (Linfield et al., 1995), *Elsinoe fawcettii* (Tan et al., 1996), *F. oxysporum* f. sp. *cucumerinum* (Kitagawa et al., 1989), *Magnaporthe grisea* (Gergerich et al., 1996), *P. cinnamomi* (Cahill and Hardham, 1994), *P. citrophthora* (Timmer et al., 1993), *P. fragariae* var. *rubi* (Olsson and Heiberg, 1997), *Plasmodiophora brassicae* (Wakeham and White, 1996), *Pythium violae* (Lyons and White, 1992), *R. solani* (Thornton et al., 1993), *Septoria nodorum* (Lagerberg, 1996), *Verticillium dahliae* (Plasencia et al., 1996), *Gan-oderma* spp. (Utomo and Niepold, 2000), *C. falcatum* (Viswanathan et al., 2000), *Polymyxa graminis* (Delfosse et al., 2000), *Plasmopara halstedii* (Bouterige et al., 2000) and *Ustilago scitaminea* (Nal-lathambi et al., 2001).

Protein profiles of fungal pathogens may be analyzed by SDS-PAGE technique, and the protein fraction present in common in all isolates or pathotypes of the fungal pathogen is identified. The antiserum prepared against this specific protein or peptides is used to detect the pathogen in plant materials to be tested. Using the antiserum against the 101 kDa polypeptide present in all pathotypes of *C. falcatum* (sugarcane red rot disease), the pathogen was detected in root eyes, buds, leaf scar, and pith region of sugarcane stalk (Viswanathan et al., 1998). A monoclonal antibody (MAb57D3) that specifically bound with a 16 kDa protein produced by *Pyricularia grisea,* reacted only with proteins present in rice blast isolates of *P. grisea* which represented predominant races occurring in the United States. MAb57D3 reacted with intercellular and cell wall antigens of the pathogen as well as with extracts of blast lesions on rice leaves (Nannapaneni et al., 2000). MABs prepared using partially purified extract of *P. halstedii* (sunflower downy mildew) recognized three fungal antigens (68, 140, and 192 kDa). The presence of the pathogens in sunflower seeds could be detected using specific MABs in ELISA (Bouterige et al., 2000). *Tilletia indica* (Karnal bunt of wheat) was also detected by employing the antiserum against a 64 kDa fungal protein in a microwell sandwich-ELISA and dipstick immunoassay (Kutilek et al., 2001). The PABs and MABs prepared against specific proteins of *P. fragariae* were used for the detection of this pathogen in the roots of strawberry (Pekárová et al., 2001). MABs raised against surface antigens from *Pythium sulcatum* exhibited high specificity to the isolates of *P. sulcatum*. The MABs recognized glycoproteins in the cell walls and they could be employed to detect *P. sulcatum* in naturally infected carrot tissues and soil by using indirect competitive ELISA (Kageyama et al., 2002).

The production of various enzymes by fungal pathogens is well-known. The polyclonal antiserum against the purified exopolygalacturonase (exoPG) produced by *F. oxysporum* f. sp. *radicis lycopersici* (FORL) was employed to detect this pathogen in wilt-disease-affected tomato plants. The level of expression of the exoPG was correlated with the development of root rot symptom during pathogenesis (Plantiño-Álvarez et al., 1999). The production of exo-PG by *Sclerotinia sclerotiorum* in infected soybean seedlings was monitored to study the expression of PG-inhibiting proteins (PGIP) constitutively (Favaron et al., 2000).

Antisera prepared against mycelium and spores of fungal pathogen may vary in their efficacy in detecting the pathogens. The anti-mycelium serum (Smy) was more sensitive than antiascospore serum in detecting *S. sclerotiorum* affecting the petals of rape (canola) and in the mycelial extract, but both antisera were equally sensitive when exposed to the ascospore antigen (Jamaux and Spire, 1999). How-ever, the antiserum prepared against *Polymyxa graminis,* an obligate root pathogen and vector of plant viruses could be used to detect dif-ferent stages in the life cycle of the fungus and to detect infection oc-curring in nature (Delfosse et al., 2000). A commercial detection kit was used for detection of *S. sclerotiorum* on canola petals as part of a disease prediction model by (Bom and Boland, 2000). During the di-agnostic survey designated as "Septoria Watch" carried out in the United Kingdom to assess the incidence of *Septoria* diseases of wheat caused by *S. tritici* and *S. nodorum,* immunodiagnosis of *Septoria* infection formed the basis for the timing of fungicide sprays for the control of these diseases (Kendall et al., 1998). Various other applications of ELISA to study the immunological properties of fun-gal pathogens are available in the literature (Fox, 1998; Martin et al., 2000; Narayanasamy, 2001).

Dot-Immunobinding Assay (DIBA)

By using dot-immunobinding assay (DIBA) the nonspecific inter-ference occurring in ELISA formats can be avoided. The seed im-munoblot assay (SIBA) was developed by Gleason et al. (1987) for detecting *Phomopsis longicolla* in infected soybean seeds. The SIBA technique was applied to detect infection of wheat seeds by Karnal bunt caused by *T. indica*. The mycelium or teliospores of the patho-gens form a colored imprint on the nitrocellulose sheet on which the seeds are placed (Anil Kumar et al., 1998). The monoclonal antibody generated against zoospores of *Phytophthora nicotianae* recognized a 40 kDa glycoprotein present on the surface of zoospores (Robold and Hardham, 1998). One distinct advantage of the DIBA technique over ELISA is the possibility of estimating the viable pathogen units, whereas ELISA does not distinguish the live from the dead fungal spores. The production of endoPG by *F. oxysporum* f. sp. *lycopersici* race 2 in infected tomato could be detected by DIBA (Arie et al., 1998).

Immunofluorescence (IF) Assay

Both PABs and MABs have been produced for use in immunofluorescence (IF) assays. MABs specific for *P. cinnamomi* (Gabor et al., 1993) and *B. cinerea* (Salinas and Schots, 1994) have been employed for the detection of these pathogens. MABs generated against components present on the surface of zoospores and cysts of *P. nicotianae* reacted with species-specific epitopes on the surface of the spores. Three MABs reacted with a polypeptide (>205 kDa) which was distributed on the entire zoospore surface including that of the two flagella (Robold and Hardham, 1998). A PAB specific for ascospores of *Mycosphaerella brassicola* (ringspot disease) in cruciferous vegetables was used to detect the pathogen under field conditions (Kennedy et al., 1999). By using an indirect IF assay and PABs raised against intact teliospores of *T. indica,* the antigenic epitopes of teliospores were characterized. The glycoprotein or peptide nature of the epitopes present on the teliospore surface could be inferred (Vikrant Gupta et al., 2001).

PHASES OF FUNGAL DISEASE DEVELOPMENT

The process of disease development (pathogenesis) has five distinct phases:

1. Attachment of pathogens to plant surface
2. Germination of spores or pathogenic units
3. Penetration of host
4. Colonization
5. Symptom expression

Host resistance and environmental conditions have significant influence on the development of diseases. Fungal pathogens produce different kinds of enzymes, toxins, and growth regulators that may be used to obtain the nutrition required for their growth and reproduction in the infected plants. Generally, the enzymes produced by pathogens are involved in the disintegration of the structural components of the host cells/tissues (see Chapter 5), breakdown of inert food substances in the cells, and interference in the functioning of synthesizing systems by affecting the protoplast directly. Toxins produced by some pathogens alter the permeability of protoplasts, causing leakage of electrolytes and ultimate death of cells. Growth regulators of fungal origin may exert hormonal effects on host cells by inducing enlargement (hypertrophy) and excessive cell division (hyperplasia) resulting in tumors (galls).

The role of proteinaceous compounds produced by fungal pathogens in the establishment of infection has been revealed by various studies on immunological techniques.

Attachment of Fungal Pathogens
to Host Plant Surface

Successful infection by a fungal pathogen occurs when it reaches the specific organ or tissue to initiate the process of pathogenesis. Adhesion of spores or infection units is an essential first step. The pathogens appear to employ various mechanisms to bind themselves to specific organs such as roots, stems, leaves, flowers, and fruits. The extracellular enzymes secreted by fungal pathogens appear to facilitate the adhesion of spores by altering the hydrophobicity of the plant surface, as suggested by (Nicholson and Epstein, 1991).

Different species of *Colletotrichum,* which causes anthracnose diseases, produce a water-soluble mucilage containing several enzymes and substances that preserve the viability of the conidia. The cutinases and esterases present in the mucilage produced by *C. graminicola* have been shown to be involved in the degradation of cutin present in plant cuticle (Kolattukudy, 1985; Pascholati et al., 1993). The requirement of esterases and cutinases for the adhesion of appressoria (infection structure) of *Uromyces viciae-fabae* (broad bean rust disease) was revealed by Deising et al. (1992). Inhibition of these enzymes by inhibitors such as diisopropyl-fluorophosphate has been shown to reduce the adhesion ability of fungal spores (Schäfer, 1994). Association of esterase activity with the conidia of *Uncinula necator* (grapevine powdery mildew disease) was demonstrated. A single extracellular fungal protein was identified as a cutinase by its ability to hydrolyze ^{3}H-cutin. The presence of esterase-cutinase inhibitor in the cuticle, however, did not significantly affect adhesion of conidia (Rumbolz et al., 2000).

Proteins or glycoproteins produced by fungal pathogens may facilitate adhesion of spores to plant surfaces as in *P. cinnamomi* (Gubler et al., 1989), *Pythium aphanidermatum* (Estrada-Garcia et al., 1990), and *Colletotrichum musae* (Sela Burrlage et al., 1991). *Nectria hae-matococca*, a 90 kDa glycoprotein of pathogen origin aided the adhesion of the macroconidia to the susceptible plant surface (Kwon and Epstein, 1993). Likewise, a glycoprotein produced by the uredospores of *U. viciae-fabae* was necessary for their adhesion to plant surface (Clement, Butt, et al., 1993; Clement, Martin, et al., 1993). The adhesive substance produced by *C. graminicola* consisted of a preformed protein and a secreted glycoproteins (Mercure, Kunoh, et al., 1994; Mercure, Leite, et al., 1994).

The conidia of *C. lindemuthianum* have a filbrillar "spore coat" and also a cell wall that they adhere to aerial plant parts for the initiation of pathogenesis. A monoclonal antibody (UB20) capable of recognizing a specific glycoprotein on conidial surface was used to localize the glycoprotein. The use of UB20 revealed that this glycoprotein was concentrated on the outer surface of the spore coat. UB20 IgG inhibited the attachment of conidia to hydrophobic surface in an antigen-specific manner, suggesting a role for the specific glycoprotein in the initial attachment of conidia (Hughes et al., 1999). *Colletotrichum lindemuthianum* produces two types of hyphae during infection of bean *(P. vulgaris)*. The biotrophic primary hyphae have a wide diameter and are entirely intracellular, whereas the necrotrophic secondary hyphae are narrow, either inter- or intracellular and are not surrounded by an extracellular matrix. Immunofluorescence labeling with a panel of MABs revealed that the glycoproteins present on conidia, germ tubes, appressoria, and primary hyphae were absent from the surface of secondary hyphae. However, chitin could be detected on the surface of the secondary hyphae, indicating that the fungal cell surface may be modified during necrotrophic growth, with none of the glycoproteins associated with earlier stages of infection process being produced (Perfect et al., 2001). The nature of the extracellular matrices surrounding the conidia, germ tubes, and appressoria of *Peronospora parasitica* (downy mildew disease), was studied using freeze-substituted transmission electron microscopy, lectin cytochemistry, and immunogold labeling. The cell surface carbohydrates present on the conidia were distinctly different from those present on germ tubes and appressoria. Germ tubes and appressoria produce two types of extracellular matrixes: (1) a fibrillar matrix containing β-(1,3)-glucans, confined to germling-substrate interface; and (2) another matrix containing protein spreading beyond the contact interface as a thin film. The tenacious adherence of both types of matrixes may contribute to germling attachment to host plant surface (Carzaniga et al., 2001).

Melanins derived from 1,8-dihydroxy-napthalene (DHN) are essential for pathogenicity and survival of fungal pathogens. Precise information on their localization within fungal cell walls is not available. Among 83 antibodies selected from phage display library, one antibody M_1 bound specifically to 1,8-DHN in competitive inhibition ELISA. Moreover, it did not show any cross-reaction with nine structurally related phenolic compounds. By employing immunogold labeling, using M_1 in the form of soluble scFv fragments, localization of melanins in the septa and outer (primary) walls of conidia of *Alternaria alternata* was observed. However, no melanin could be detected in the conidia of an albino mutant AKT 88-1. The M_1 antibody provides a new tool for detecting melanized pathogens in plants as well as in animals and for precisely mapping the distribution of the polymer within spores, appressoria, and hyphae of fungal pathogens (Carzaniga et al., 2002)

Germination of Spores

The germination of fungal spores such as conidia, ascospores, and basidiospores and asexual structures such as sclerotia is initiated when the environmental conditions are conducive. In addition, the exudates from leaves and roots of host plants that contain sugars, amino acids, mineral salts, phenols, and alkaloids may either stimulate or inhibit the germination of spores. Upon germination, spores produce a tubular structure known as germ tube which terminates as a thick-walled structure called an appressorium. Hutchison et al. (2000) developed a method of purifying appressoria of *C. lindemuthianum* (bean anthracnose disease). This method involves immuno-magnetic separation after incubation of infected bean leaf homogenates with an MAB that binds strongly to the appressoria. Preparation with a purity of >90 percent can be used for biochemical analysis of cell surface. The structure of the appressoria and their content play a vital role in the penetration of the host plant surface either directly or through natural openings. Fungal pathogens may produce several kinds of enzymes that are required for their metabolic and pathogenic activities.

Among them, hydrolases appear to be the most important, as they are involved in disease development.

Penetration of Host Plant

Fungal pathogens may enter into host tissue by mechanical pressure exerted through a cuticle consisting of layers of wax, cutin, pectin, and a network of cellulose fibrils impregnated with wall polymers that cover the epidermis. Plant cuticle is attached to the epidermal layer of cells via a pectinaceous layer. The cuticle is made of an insoluble biopolyester, known as cutin, which is embedded in a complex mixture of hydrophobic materials collectively known as wax. Cutin provides the principal physical obstacle for the direct penetration of the infection peg or hypha, which is formed from the appressorium, into the epidermis and the fungal cutinase seems to have a primary role in this process. Production of cutinase during infection by *N. haematococca* (*F. solani* f. sp. *pisi*) was observed through immunoelectron microscopic examination (Shaykh et al., 1977). Localization of cutinase in the germinating spores of *F. solani* f. sp. *pisi* and *C. gloeosporioides* was revealed by immunofluorescence technique (Podila et al., 1989). The activity of the cutinase could be inhibited by both polyclonal and monoclonal antibodies raised against the enzyme (Dickman et al., 1982; Kolattukudy 1985; Calvete, 1992). *Mycosphaerella* sp. can normally infect papaya fruits through wounds only. However, the transformant-expressing cutinase gene from *F. solani* f. sp. *pisi* could infect the intact papaya indicating a role for cutin for direct penetration by *Mycosphaerella* sp. after transformation. By using antibodies specific for the *Fusarium* cutinase the infection of the transformed *Mycosphaerella* sp. could be prevented (Dickman et al., 1989)

Evidence suggesting less or no significant role for cutinase in pathogenesis has also been reported in other pathosystems. Sweigard et al. (1992) showed that the transformants of *M. grisea* (rice blast disease), obtained by targeted disruption of cutinase gene, were as pathogenic as the parent type, indicating the noninvolvement of cutinase in pathogenesis. A residual cutinase activity observed in the transformants could have assisted in the penetration of host plants. Likewise, the cutinase A-deficient mutants of *B. cinerea*, lacking a functional *cutA* gene could penetrate and cause symptoms of infection on gerbera flowers and tomato fruits like the parent type (Van Kan et al., 1997). In addition, in the cucurbit-*F. solani* f. sp. *cucurbita* race 2 pathosystem loss of virulence or pathogenicity did not result following disruption of the *cutA* locus (Crowhurst et al., 1997). These results suggest that cutinase may or may not have a significant role in the penetration of the host by fungal pathogens depending on the host-pathogen combination.

In addition to cutinase, many other enzymes may act synergistically in the degradation of cuticle and cell wall; dissolution of the natural barriers of host plants cannot be entirely attributed to the activity of any one enzyme (Schäfer, 1994). *Phytophthora infestans* produces lipases in infected leaves. The lipases may cause apparent protoplast injury even before the disruption of cell walls as revealed by electron microscopy (Moreau and Rawa, 1984; Keon et al., 1987). The requirement of lipase activity during early stages of infection by *B. cinerea* on tomato leaves was inferred. The conidia of *B. cinerea* germinated on the leaves of tomato in the presence of antilipase antibodies. However, they could not penetrate the cuticle because of the inhibition of lipase activity by the antibodies (Comménil et al., 1998) (see Appendix section on detection of lipase activity). Likewise, the involvement of cellulases in the penetration of *Erysiphe graminis* f. sp. *hordei* (*Egh*) (powdery mildew of barley) was indicated by IF microscopy using two MABs specific to cellobiohydrolases from *T. reesei*. Antigen localization at the germ tube tips of *Egh* was observed. Cellobiohydrolase I was present at the primary germ tube tip, while cellobiohydrolase II was detected at the appressorial germ tube tip. Host penetration may be achieved by a combination of enzyme activity and mechanical pressure exerted by the appressoria of *Egh* (Pryce-Jones et al., 1999).

The differentiation of appressoria at the distal end of the germ tube is a critical step in fungal pathogenicity. A cyclic-AMP (cAMP) signaling mechanism appears to be involved in the differentiation of appressoria in *M. grisea*. A catalytic subunit of AMP-dependent protein kinase A encoded by *CPKA* gene seems to control appressorium formation by regulating cAMP signaling (Kronstad, 1997; Xu et al., 1997). A mitogen-activated protein kinase (MAP kinase) encoding gene, *PMK1,* was required for appressorium formation in *M. grisea* (Xu and Hamer, 1996). The presence of cellobiohydrolase I at the primary germ tube tip and cellobiohydrolase II at the appressorial germ tube tip of *E. graminis* was revealed by IF microscopy. The results suggested that host penetration by *E. graminis* was achieved by a combination of enzyme activity and mechanical force (Pryce-Jones et al., 1999).

Some fungal pathogens such as *Puccinia graminis tritici* (wheat rust) and *P. infestans* are able to penetrate both directly and through natural openings such as stomata, whereas others can enter through natural openings or through wounds only. Downy mildew pathogens such as *Plasmopara viticola* (grapevine downy mildew) enter plants exclusively through stomata. In such cases elaboration of any enzyme assisting the entry of the fungal pathogens may not be required. A solution rich in carbohydrates and amino acids is released when cortical cells of potato roots are disrupted during emergence of lateral roots which break through the cortex of parent roots. The zoospores of *P. infestans* are attracted by this fluid from roots and they encyst on the damaged cortical tissues and then penetrate into the exposed cells. The modification of cell wall components during penetration of epidermal cell walls by *U. fabae* (cowpea rust) was studied using antibodies raised against epitopes present in pectin, polygalacturonic acid, xyloglucan, and callose. The density of pectin and xyloglucan epitopes was found to be reduced, suggesting that the rust pathogen may degrade the plant cell wall at the penetration site by secreting the respective enzymes (Xu and Mendgen, 1997).

Colonization by Fungal Pathogens

Fungal pathogens seem to adopt different methods of colonization of host tissues after successful penetration into host plants. They produce different enzymes and toxins that alter structural features of host tissues and facilitate the absorption of nutrients required for their growth and reproduction. The importance of cell-wall degrading enzymes for successful colonization of host tissues has been well demonstrated in many pathosystems. In plant diseases characterized by a rapid and extensive degradation of cell wall, increase in permeability, and death of protoplasts, polygalacturonases (PGs) secreted by fungal pathogens have important roles in pathogenesis. The extracellular PG produced by *Fusarium moniliforme* can be detected by immunoblotting (de Lorenzo et al., 1987). The requirement and production of endopolygalacturonase (endo-PG) for both saprophytic growth of and infection by *C. lindemuthianum* has been indicated. Secretion of an endo-PG by *C. lindemuthianum* is governed by *CLPG1* and *CLPG2* genes. By using specific antibodies, the protein encoded by *CLPG1,* was detected *in planta* and was shown to be responsible for extensive degradation of host cell wall, indicating the importance of expression of *CLPG1* and consequent secretion of endo-PG by *C. lindemuthianum* (Centis et al., 1997). Likewise, an endo-PG encoded by the gene *Bcpg1* has an important role in the primary colonization of tomato tissues by *B. cinerea* (ten Have et al., 1998). *Fusarium oxysporum* f. sp. *lycopersici* also secretes an extracellular endo-PG encoded by the gene *pg1.* The expression of this gene in the infected roots of tomato has been detected and is essential for the pathogenicity of *F. oxysporum* (Pietro and Roncero, 1998). The crucial role of endo-PG produced by *Alternaria citri* (citrus black rot) distinguishes this pathogen from another morphologically similar pathogen *A. alternata* (brown spot). The endo-PG production is required for pathogenicity of *A. citri* which causes tissue maceration, whereas *A. alternata* induces necrotic spots by producing the host-selective toxin. The endo-PG with similar biochemical properties produced by *A. alternata* is not involved in pathogenesis. The endo-PG mutant of *A. citri* has reduced ability to colonize citrus peel segments. In contrast, the endo-PG mutant of *A. alternata* showed no change in its pathogenicity indicating that the cell wall-degrading enzymes may play different roles depending on the type of symptoms induced (Isshiki et al., 2001). *Botrytis cinerea* is known to secrete four exopolygalacturonases (exo-PGs). The polyclonal antiserum produced against one of the exo-PGs (60 kDa) recognizes two exo-PGs with molecular mass of 66 and 70 kDa. Using immunohistochemical analysis, the expression of exo-PGs was identified in cucumber leaves inoculated with spores of *B. cinerea.* The concentration of the exo-PGs steadily increased nine hours after inoculation, with lapse of time, indicating that exo-PGs may have an important role in pathogenesis at an early stage of infection as well as in tissue maceration of host plants (Rha et al., 2001).

Fusarium moniliforme causes rot of maize seedlings and organs such as roots, stalks, and ears. By using a highly specific monoclonal antibody, the endo-PG activation during infection was monitored. The endo-PG expressed by *F. moniliforme* in the very early stages of pathogenesis could be detected in its natural host plant (maize). Furthermore, differences in amounts of endo-PG produced by different and distinct *F. moniliforme* strains were recognized, indicating the existence of certain variability at the nucleotide and protein levels (Daroda et al., 2001).

Some cereal pathogens, such as *Rhizoctonia cerealis, Fusarium culmorum,* and *Pseudocercosporella herpotrichoides* produce enzymes that degrade components of hemicellulose (polysaccharides) (Cooper et al., 1988). These pathogens, when grown on wheat cell walls, elaborate large amounts of arabanase, xylanase, and laminarinase which degrade polysaccharides present in high cocentrations in wheat cell walls. Because wheat cell walls contain low concentrations of pectin, relatively small amounts of pectolytic enzymes are produced by these pathogens. Symptoms of rotting are associated with activities of arabinosidase, xylanase, and β-(1,3) glucanase as in the case of *Diplodia viticola* which causes bunch rot of grapes. Xylanase produced by *D. viticola* degrades xylan into a variety of xylan oligosaccharides (Strobel, 1963). The extracts of lesions induced by *R. solani* in bean hypocotyls contained enzymes to degrade xylan, galactan, galactomannan, polygalacturonic acid, and carboxy methyl cellulose (Albersheim et al., 1969). In grape berries, invertase (β-fructofuranosidase) activity was stimulated. A new invertase similar to that of *Botrytis* invertase (BIT) was detected. The PABs raised in chicken (anti-BIT-IgY) recognized the new invertase induced in infected berries in Western blotting. This anti-BIT-IgY was found to be highly specific in detecting the native BIT in the extracts of diseased berries (Ruiz and Ruffner, 2002).

Gaeumannomyces graminis var. *tritici* causes the destructive take-all disease in wheat. Production of cell wall-degrading enzymes such as cellulases, xylanases, and pectinases by this pathogen was indicated by enzyme-gold and immunogold-labeling techniques. Degradation of cell wall components, namely, cellulose, xylan, pectin, and lignin in the infected roots was visualized, indicating the activities of the cell wall-degrading enzymes on their respective substrates (Kang et al., 2000). Localization of β-(1,3)-glucanase and lignin over cell walls of wheat roots infected by *G. graminis* var. *tritici* was also observed (Huang et al., 2001). In addition, extracellular proteolytic activities in wheat leaves following infection by *Septoria tritici* showed marked changes. Both H_2O_2 production and chitinase activity increased when leaves were inoculated with viable conidia of *S. tritici*.

The extracellular serine proteinase extracted from the infected leaves inhibited spore germination, suggesting that the proteolytic activity of the leaf extracellular space may participate in the development of resistance in wheat plants (Segarra et al., 2002). In the case of *Phytophthora capsici* infecting pepper, the immunogold-labeling technique showed specific labeling of chitinase on the cell wall of the pathogen in both compatible (susceptible) and incompatible (resistant) interactions. The labeling was evenly distributed over the entire fungal wall in intimate contact with host cell wall. The chitinase protein appeared to be generally localized in the fungal cell wall (Lee et al., 2000) (Figure 8.1) (see Appendix section on detection of chitanase).

The presence of various storage proteins in the seeds of cereals has been observed (Kumar and Parameswaran, 1998; Krishnaveni et al., 1999; Duviau and Kobrehel, 2000). The cereal protein fractions—albumins, globulins, gliadins or prolamins, glutenins, or glutelins— present in wheat, barley, sorghum, rice, and maize exhibited varying degrees of sensitivity to extracellular fungal proteinases (EFPs) from *Aspergillus saitoi, A. sojae, A. oryzae, Rhizopus* sp., and *Mucor* sp. Wheat globulins showed high sensitivity to all EFPs. The proteinase of *A. saitoi* was the most effective in degrading the protein fraction. The proteinase of *A. oryzae* targeted more specifically albunins and globulin (>MW 50 kDa). The storage proteins of sorghum, both kafirns and glutelin, showed high resistance to EFPs, whereas rice storage proteins were degraded by all EFPs to the maximum extent (Duviau and Kobrehel, 2000).

In the rye-*Claviceps purpurea* pathosystem, by using immunogold-labeling technique, the presence of homogalacturonan (pectin) in the cell walls of rye ovaries was detected. *Claviceps purpurea* produced several enzymes such as PGs, pectin methyl esterases (PMEs), xylanases, and cellulases. The gold labels were increased at the host-pathogen interface indicating significant degradation of substrates by the action of the respective enzymes (Giesbert et al., 1998; Tenberge, 1999). In inoculated plants immunogold labeling at all infection stages was limited to the host-pathogen interface. The antigenic sites were present on the surface of external hyphae as well as in subcuticular and intercellular hyphae indicating that the enzyme β-(1,3)-glucanase was secreted by *C. purpurea* throughout the colonization process *in planta* (Tenberge et al., 1999).

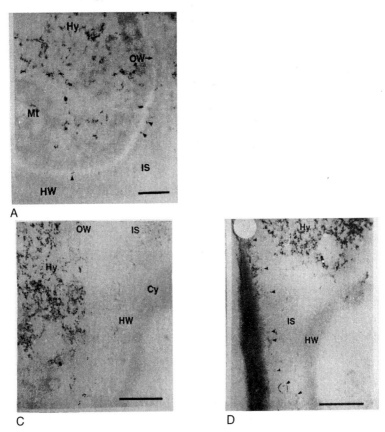

FIGURE 8.1. Electron micrographs of pepper stem inoculated with a virulent iso-late of *Phytophthora capsici* treated with tomato antichitinase antiserum labeled with goat antirabbit gold (10 nm) antibodies: (A) pathogen (Hy) growing inter-celluarly; gold particles deposited on fungal cell wall (OW), but not on host cell wall (arrowheads); (C) fungal cytoplasm (Cy) not labeled; a few gold particles labeled on cell wall (OW); (D) fungal hypha (Hy) growing in the intercellular space (IS); dense gold labeling chitinase in the intercellular space; but not on host cell wall (HW) (*Source:* Lee et al., 2000; Reprinted with permission by Elsevier.)

Enzyme-gold and immunolabeling techniques were employed to elucidate the extent of degradation of cell wall components of wheat spikes infected by *F. culmorum* (head blight disease) (Kang and Buchenauer, 1999). The inter-and intracellular development of *F. culmorum* in the ovary, lemma, and rachis of infected wheat spike resulted in significant changes in cell walls and middle lamella matrixes, in addition to marked modification of cell wall components. At three days after inoculation, some parts of host cell wall around the pathogen hyphae became loose and the deposition of gold particles over host cell wall around the hyphae was reduced and uneven (Figure 8.2A). The labeling density over the host cell walls in contact with the hyphae decreased considerably and some gold particles were observed over the electron-dense materials in the junction region between the host cells (Figure 8.2B). Labeling densities for cellulose, xylan, and pectin were markedly reduced in the cell walls of infected ovary, lemma, and rachis when compared with healthy tissues indicating the degradation of the cell wall components following infection. The host cell walls in contact with or close to the pathogen hyphae exhibited recognizable morphological changes and labeling density in these cells was appreciably reduced compared to those farther away from the pathogen. The degradation of pectin was comparatively greater than that of cellulose and xylan in the cell walls of the same infected tissues, suggesting that the pectinases may be secreted earlier or act more rapidly than cellulases and xylanase during early

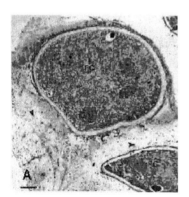
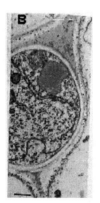

FIGURE 8.2. Cytochemical labeling for cellulose in the cell walls of tissues in wheat spikes infected by *Fusarium culmorum:* (A) lower intensity deposition of gold particles around the hyphae (FC) (arrowheads) in the outer epidermal cells of infected ovary (Bar = 1 μm); (B) reduced labeling density over host cell walls in contact with the hyphae (FC) (Bar = 0.5 μm); (C) the middle lamella between the cell and primary wall in contact with hypha showing either a few gold particles or no labels (arrowheads) (Bar = 0.5 μm) (*Source:* Kang and Buchenauer, 2000a; Reprinted with permission by Blackwell Wissenchafts-Verlag, Berlin.)

stages of pathogenesis (Figure 8.2C) (Kang and Buchenauer, 2000a) (see Appendix section on detection of cell wall components).

The extracellular hydrolytic enzymes essential for degrading preformed physical barriers present in host plants are produced by *Nectria haematococca* infecting peas. Pectate lyase (PL) genes *(pel)* of *N. haematococca* are inducible in the presence of pectin. When the *pel* genes *pelA* and *pelD* encoding the PLs were disrupted, virulence of the pathogen was dramatically reduced, indicating the synergistic action of both *pel* genes in pathogenesis (Rogers et al., 2000). In a study by Yakoby et al. (2000) PL was required for penetration and colonization of watermelon by *Colletotrichum magna. Colletotrichum magna,* a weak pathogen, when transformed with a *pel* clone from *C. gloeosporioides,* became more virulent and caused more severe damping-off symptoms on watermelon seedlings.

Production of mycotoxins by fungal pathogens has been observed in infected grains. The presence of fumonisin produced by *Gibberella fujikuroi* in maize grains has drawn worldwide attention because of the adverse effects of mycotoxicoses in animals and human beings. *Aspergillus flavus,* another dreadful pathogen of maize kernels, produces high levels of aflatoxin. The development of *A. flavus* is significantly curtailed by the presence of a 14 kDa protein, a trypsin inhibitor in the resistant cultivars (Chen and Chen, 1998). By using the immunogold-labeling technique, distinct differences in the toxing concentrations of deoxynivalenol (DON), a trichothecene derivative, secreted by *F. culmorum* in infected wheat spikelets, were detected. At early stages of infection the labeling densities for DON in resistant cultivars were significantly lower than those in susceptible ones, indicating greater accumulation of DON in susceptible cultivars (Kang and Buchenauer, 2000b).The detection of mycotoxins and their effects on humans and animals are discussed in Chapter 13.

Some host plant species may have preformed defense molecules and antifungal compounds that may restrict the development of pathogens. However, some pathogens are capable of secreting enzymes that may convert the antimicrobial compounds into nontoxic compounds. Such a situation has been observed in *Gloeocercospora sorghi* which infects sorghum. This pathogen produces the enzyme cyanide hydratase which converts cyanide released by sorghum tissues into nontoxic formamide (Wang et al., 1992). *Gaeumannomyces graminis* var. *tritici* was unable to infect oats because of its inability to enzymatically metabolize avenacin present in oats, while *G. graminis* var. *avenae* was able to do so. Disruption of avenacinase gene in *G. graminis* var. *avenae* led to loss of resistance to avenacin and consequent dramatic reduction in the pathogenicity of the transformants on oats (Schäfer, 1994). Moreover, the ability to tolerate and degrade the antifungal avenacin A1 seems to possibly affect the ability of fungi to colonize roots of oats. The fungi capable of tolerating avenacin could colonize oat roots, whereas the fungi sensitive to avenacin were eliminated from the rhizosphere of oats (Carter et al., 1999).

Production and involvement of enzymes or proteins in the development of diseases caused by obligate fungal pathogens have not been clearly demonstrated, possibly because of the nonavailability of suitable sensitive techniques. Many pathogens produce haustoria inside host cells for absorption of nutrients. Certain genes known as *in planta*-induced genes (PIGs) present in the haustoria of *U. fabae* are highly expressed. The *PIG2* gene encoding a protein with high homologies to fungal amino acid exporters was identified in the haustorial DNA of *U. fabae*. Localization of a putative amino acid exporter protein formed due to the expression of *PIG2* mRNA in the plasma membrane of the haustoria of *U. fabae* was visualized by IF technique indicating the involvement of haustoria in the uptake of nutrients from infected cells of host plants (Hahn and Mendgen, 1997; Hahn et al., 1997) (see Appendix section on detection of amino acid exporter protein). In response to inoculation of barley leaf epidermis with the powdery mildew pathogen *(E. graminis)*, the 14-3-3 proteins which form a family of highly conserved proteins with an important role in signaling networks, are induced. These proteins bind to and activate the plasma membrane H^+-ATPase. Possibly, the 14-3-3 proteins are involved in the epidermis-specific response to the pathogen via an activation of the plasma membrane H^+-ATPase (Finnie et al., 2002).

The presence of a stress-induced protein designated as FISP 17 induced by *F. solani* f. sp. *glycines* (Fsg) in stem exudates was detected by SDS-PAGE. This protein (17 kDa) had an N-terminal amino acid sequence similar to a starvation-associated message 22 protein (100 percent homology) and bean PR proteins (78 to 80 percent similarity). A polyclonal antibody specific against the synthetic peptide that was designed based on the N-terminal sequence, reacted with a

17kDa protein exudate from Fsg-infected plants. There was no reaction with exudates from healthy plants and also with culture filtrate of Fsg, indicating that FISP 17 was produced due to the interaction between the host and the pathogen (Li et al., 2000).

Fungal Elicitors

Fungal pathogens not only produce enzymes and toxins that facilitate their successful establishment in host plants, but also secrete elicitors that initiate defense responses in host plants. Elicitors can be defined as molecules present at the cell surface or secreted into the culture medium of plant pathogens which have the same ability to induce the host plant defense system as the pathogens they are derived from (Darvill and Albersheim, 1984). The concept of elicitors was later modified to include plant endogenous compounds, especially oligogalacturonides (OGA) derived from cell wall. Plants also contain their own elicitors of defense.

When *Phytophthora cryptogea,* a nonpathogen, was inoculated on tobacco stem, necrotic symptoms characteristic of hypersensitive response developed locally on the stem, as well on the leaves at a distance. The active molecule, a holoprotein, named cryptogein (CRY), induced defense-related responses, in addition to necrosis, such as production of ethylene, phytoalexin, and PR-1 proteins (Bonnet, 1988; Milat et al., 1991). The tobacco plants treated with CRY showed resistance to the pathogen *P. parasitica* var. *nicotianae* (black shank disease) (Ricci et al., 1989). Other *Phytophthora* spp. also have been reported to secrete similar proteins in cultural media. These proteins have been designated elicitins. They have high structural homology, but have significant differences in pI and biological effects on tobacco. These proteins can be detected and characterized by immunological techniques. The usefulness of immunoassays for a comparative study of elicitors has been demonstrated by Devergne et al., 1994) (see Appendix section on the detection of elicitins). The elicitins cryptogein from *P. cryptogea,* cinnamomin from *P. cinnamomi,* capsicein from *P. capsici,* and parasiticein from *P. parasitica* were detected by Western blot analysis (Figure 8.3). The presence of capsicein in tomato plantlets infected by *P. capsici* was detected on different days after inoculation by DAS-ELISA (Figure 8.4). Addition of cryptogein to a suspension of tobacco cells resulted in the aggrega-

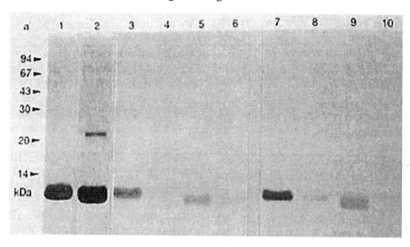

FIGURE 8.3. Detection of elicitins from *Phytophthora* spp. by Western blotting: (Lanes 3, 7) preparations of cryptogein; (Lanes 4, 8) cinnamomin; (Lanes 5, 9) capsicein; (Lanes 6, 10) parasiticein; (Lanes 1, 2) purified cryptogein; two different monoclonal antibodies (Lanes 1-6 and Lanes 7-10) (*Source:* Devergne et al., 1994; Reprinted with permission by Blackwell Scientific Publications, Oxford, United Kingdom.)

tion of cells in clusters. Immunocytochemical detection of pectin epitopes showed that the fibrillar material surrounding the elicited (treated) cells contained mostly low-methylated galacturonan sequences. The presence of a calcium pectate gel on the modified cell walls of treated cells was observed and this modification might be due to the reorganization of pectin in the middle lamellae (Kieffer et al., 2000).

Plant cell walls may contain complex OGAs, which may suppress the expression of hydroxy protein-rich glycoprotein (HRGP) genes in bean seedlings. This may facilitate local degradation of cell wall pectic polysaccharides by the endo-PG of *C. lindemuthianum* and penetration into the host plant (Boudart et al., 1995). The plant glucanohydrolase β-(1,3)-glucanase (βGlu) may release oligosaccharides from cell walls of fungal pathogens and these may act as elicitors of plant defense responses (Boller, 1995). Following infection of soybean by *Phytophthora megasperma* f. sp. *glycinea*, soybean βGlu releases β-glucans from the cell wall of the pathogen resulting in the accumulation of the phytoalexin (a defense compound)

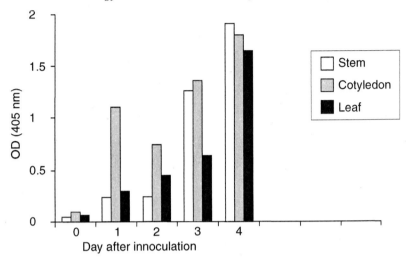

FIGURE 8.4. Detection of elicitins by DAS-ELISA in extracts of tomato plantlets inoculated with *Phytophthora capsici* (*Source:* Devergne et al., 1994; Reprinted with permission by Blackwell Scientific Publications, Oxford, United Kingdom.)

glyceollin. Umemoto et al. (1997) isolated a cDNA for a β-glucan elicitor binding protein (GEBP) from the plasma membrane of soybean root cells. Purified soybean βGlu was involved in the release of active β-glucan elicitors from fungal cells (Ham et al., 1995). On the other hand, *C. lindemuthianum* was found to secrete a protein capable of inhibiting an endo-βGlu of its host, French bean *(P. vulgaris)* (Albersheim and Valent 1974). Later, selective inhibition of plant βGlu by protein secreted by fungal pathogens was demonstrated. A β-(1,3)-glucanase inhibitor protein (GIP-1), which specifically inhibits the pathogenesis-related βGlu from soybean seedling, was isolated from the culture fluid of *Phytophthora sojae* f. sp. *glycines* (= *P. megasperma* f. sp. *glycinea*) (Ham et al., 1997).

Fungal pathogenesis is a result of a complex interplay between host βGlu and βGlu inhibitors (Leubner-Metzger and Meins Jr., 1999). This assumption arises out of the results of experiments on different pathosystems. A constitutive 34 kDa chitinase present in aleurone protoplasts of barley was not induced following elicitor treatment (Sheba et al., 1994). In cucumber, only one of the three acid class III chitinase genes *(Chi 2)* was induced by biotic and abiotic stresses

(Lawton et al., 1994). However, even when induced, all chitinases did not exhibit antifungal activities. A 35 kDa chitinase was strongly induced in rice suspension-cultured cells following treatment with the elicitor of *R. solani*. This inducible chitinase inhibited the mycelial growth and caused lysis of hyphae of *R. solani*. The purified chitinase from rice suspension-cultured cells was analyzed by Western blotting using a bean chitinase antiserum (Velazhahan et al., 2000) (Figure 8.5).

Both pathogens and nonpathogens can induce defense responses (Van Loon, 1999). However, nonpathgens induce them very early and more strongly. In rice, *Bipolaris sorokiniana,* a nonpathogen, induced the accumulation of transcripts of PR proteins more strongly than pathogen *Magnaporthe oryzae* (rice blast disease) (Manandhar et al., 1999). The activities of chitinase and β-(1,3)-glucanase were increased in rice plants inoculated with a nonpathogen *(Pestalotia palmarum)* to a greater level (incompatible interaction) than when inoculated with rice pathogen *(R. solani)* (compatible interaction). Appearance of two proteins (33 and 35 kDa) reacting with barley chitinase antibodies infected in rice plants was observed in Western blots (Deborah et al., 2001).

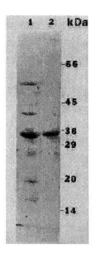

FIGURE 8.5. Western blot analysis of chitinase purified from elicitor-treated rice cells: (Lane 1) elicitor-treated rice cells; (Lane 2) purified chitinase (*Source:* Velazhahan et al., 2000; Reprinted with permission by American Chemical Society.)

Infection of maize seedlings by *F. moniliforme (G. fujikuroi)* led to the formation of appositions on the outer host cell wall surface, the occlusion of intercellular spaces, and formation of papillae. Immuno-localization studies revealed that PRms (pathogenesis-related maize seed) proteins accumulated at very high levels in the cell types involved in the formation of barriers for fungal penetration such as the aleurone layer of germinating seeds and the scutellar epithelial cells of isolated embryos. The presence of abnormal fungal cells in large numbers showing PRms-specific labeling and accumulation of PRms in clusters over the fungal cell wall were observed (Murillo et al., 1999). Also in this study, the activation of a calcium-dependent protein kinase *(CPK)* gene *(ZmCPK10)* transcriptionally in response to both fungal infection and treatment with fungal elicitors was observed. Activation of the *ZmCPK10* gene was extremely rapid and reached the maximum levels at 30 minutes after treatment. The activation was accompanied by enhanced levels of PRms mRNA. Furthermore, the expression of the *ZmCPK10* gene occurred only in those specific cell types in which the PRms gene was also expressed (Murillo et al., 2001). In response to treatment with the elicitor derived from *Fusarium graminearum (Gibberella zea)* two novel cell wall structural proteins were rapidly deposited in the cell wall matrix in spring wheat *(Triticum aestivum)* (El-Gendy et al., 2001).

Symptom Expression

Soft rot symptoms induced by fungal pathogens may be due to the maceration of the nonlignified tissues, dissolution of middle lamellae, and death of macerated cells. Infection of persimmon by *Gloeosporium kaki* leading to anthracnose disease has two phases: dry rot and soft rot. The dry rot symptom noted in the early stages was due to the action of pectolytic enzymes, whereas the soft rot symptom, which developed later, was induced by the tissue-macerating enzymes secreted by the pathogen (Tani, 1965, 1967). The cell wall-degrading enzymes produced by *R. solani* are directly involved in the formation of lesions on the bean hypocotyls (Bateman et al., 1969). In soft rots caused by *Verticillium albo-atrum*, PGs are initially produced, followed by the production of hemicellulases and cellulases (Cooper and Wood, 1975). The involvement of pectic lyases produced by *V. albo-atrum* in induction of wilt symptoms was revealed by the inability of the mutants to induce wilt symptoms. The mutants lacked the ability to produce pectic lyases (Durrands and Cooper, 1988). Generally, fungal pathogens do not induce cell death (necrosis) in the compatible host plant, but they produce cell wall-degrading enzymes that affect structural characteristics of the tissues or organs. However, some pathogens, for example, *Phytophthora cactorum* produces a novel protein factor named PcF that is capable of inducing leaf necrosis in strawberry. This protein had a molecular mass of 5622 Da (Orsomando et al., 2001).

Toxins produced by fungal pathogens have been shown to cause certain specific and nonspecific symptoms in susceptible plants. These toxins may be grouped as host-specific and nonspecific. Host-specific toxins (HST), also called selective toxins or pathotoxins, induce all or major symptoms of disease in susceptible host plants. The resistance or susceptibility of the host is correlated with the insensitivity or sensitivity to toxin. Almost all HST-producing pathogens are included in the genus, *Helminthosporium* (four species) and *Alternaria alternata* (seven pathotypes). Other fungal pathogens producing HST are *Periconia circinata, Phyllosticta maydis,* and *Corynespora cassiicola.*

Helminthosporium victoria infecting the oat cultivar Victoria, produces HV toxin (victorin) and a 100 kDa protein; victorin-binding protein was isolated from susceptible cultivar. However, the victorin-binding protein present in the resistant variety did not bind to victorin (Wolpert and Macko, 1989a,b). The peptide portion of the toxin molecule is nontoxic, but it seems to contribute to the toxin specifity (Macko et al., 1989). The HSTs produced by different pathotypes of *A. alternata* differ structurally and in the mode of action. They directly or indirectly cause dysfunctions of host cell plasma membrane leading to electrolyte leakage, cell collapse, and ultimately necrosis of the tissues. The toxin secreted by *A. alternata* f. sp. *lycopersicon* (AAL) (tomato stem canker) contributes to the pathogenicity, since the tomato cultivars susceptible to the pathogen are highly sensitive to the AAL toxin. The AAL toxin was demonstrated to function as an inhibitor of ceramide synthase (Wang et al., 1996).

Nonhost-specific toxins are not involved in pathogenesis and they may affect some fundamental physiological process of hosts as well as nonhost plants. The barley pathogen *Rhynchosporium secalis* produces a toxin consisting of phytotoxic peptides that can induce necrosis on leaves of barley and other cereals and some dicots (Wevelsiep et al., 1993). NIP1, a necrosis-inducing protein produced by this pathogen, appears to be the product of a fungal avirulence gene involved in a race-specific defense response in barley carrying the resistance gene *Rrs1* (Hahn et al., 1993). Fusicoccin produced by *Fusicoccum amygdali* (bud canker of almond and peach) induces wilting and drying of leaves. Biologically active fusicoccin can be detected by radioimmunoassay (Graniti, 1989).

The fungal pathogens, in addition to enzymes and toxins, may produce growth-regulating compounds that may be responsible for some characteristic symptoms. *Gibberella fujikuroi,* causing rice bakanae disease, produces gibberellin A and B which induce the abnormal elongation of internodes, chlorosis of leaves, and reduction in tillering (Yabuta and Hayashi, 1939). Some fungal pathogens produce high amounts of auxins and cytokinins which may be responsible for hyperplastic (excessive growth) development of organs/plants. *Taphrina* spp., causing leafcurl and witch's broom symptoms, has been shown to synthesize indole acetic acid (IAA) (Yamada et al., 1990). *Phytophthora nicotianae* induces water deficiency symptoms. Direct antigen-coating (DAC) ELISA was used to monitor the progress of the disease. Significant differences in the proline contents of infected and healthy plants were observed. Higher nutrient supply and light intensity increased the proline contents of the leaves in addition to acceleration of symptom expression. The proline content of leaves was suggested as a suitable marker for both biotic and abiotic stresses (Grote and Claussen, 2001).

SUMMARY

The identity of different fungal pathogens that cause a variety of crop diseases can be established with certainty by following various immunodiagnostic techniques. Furthermore, these techniques have been extensively employed to study different phases of disease development. Different kinds of enzymes secreted by fungal pathogens have been detected in culture media and *in planta*. These enzymes assist pathogens during penetration and colonization of tissues of susceptible plants. Furthermore, the constitutive expression and induction of defense-related proteins that hamper the development of fungal pathogens has been studied using immunological methods.

APPENDIX: IMMUNOLOGICAL PROTOCOLS FOR DETECTION OF FUNGAL PATHOGENS

Detection of Botrytis cinerea by ELISA (Bossi and Dewey, 1992)

Preparation of Immunogen

1. Cultivate the fungal pathogen *(B. cinerea)* on potato dextrose agar (PDA) in petri dishes or on slants at 21°C for 17 to 20 days; prepare the surface washings of solid cultures using 5 ml/petri dish or 2 ml/ slant of phosphate buffered saline (PBS) consisting of NaCl (0.8 percent), KCl (0.02 percent), Na_2HPO_4 (0.115 percent), KH_2PO_4 (0.025 percent), pH 7.2 and remove the wash fluid by suction.
2. Centrifuge the wash fluid at 13,000 g for three minutes to remove fungal cell debris and dilute the supernatant with PBS to have tenfold dilution.
3. Pass the cell-free wash fluid through a Centricon 30 kDa filter (Amicon No. 4208) to remove the high molecular weight carbohydrates and glycoproteins that may induce nonspecific antibodies; freeze-dry the filtrate and redissolve the contents in 1 ml of distilled water and use it as immunogen for the production of monoclonal antibodies.

Enzyme-Linked Immunosorbent Assay (ELISA)

1. Coat the wells (in triplicate) in the microtiter plates with PBS surface washing fluid (50 µl/well) overnight; wash the wells four times allowing two minutes for each washing and finally wash with distilled water briefly.
2. Air-dry the plates in a laminar flow hood and seal them in a polyethylene bag, sealed and stored at 4°C.
3. Incubate wells successively with hybridoma supernatants for 1 hour with a 1/200 dilution of a commercial goat antimouse polyvalent (IgG + IgM) peroxidase conjugate PBS with 0.05 percent Tween-20 (PBST) for one hour more.
4. Add the substrate solution containing tetramethyl benzidine (100 µg/ml) for thirty minutes.
5. Maintain the controls incubated with tissue culture medium containing 5 percent fetal bovine serum (FBS) in place of hybridoma supernatant.
6. Stop the reaction by adding 3 M H_2SO_4 (50 µl/well); record the color intensity using ELISA reader at 450 nm; absorbance readings three times greater than the control values are considered a positive reaction.

Detection of Spore Balls of Spongospora subterranea by ELISA (Harrison et al., 1993)

Preparation of Immunogen

1. Scrap the spore balls along with potato tuber debris from powdery scab lesions of washed tubers; air-dry and store for use as immunogen or fungal samples for ELISA tests.
2. Homogenize the scrappings (200 mg fresh weight) with 5 ml of sodium phosphate buffer (0.033 M), pH 7.2 in a glass tissue grinder; prepare aliquots of 0.7 ml (containing approximately 1 mg of protein) and store at $-40°C$.

Preparation of Antiserum

1. Inject rabbits intramuscularly with 0.7 ml of immunogen, with Freund's complete adjuvant; after 3 weeks interval inject the immunogen with Freund's incomplete adjuvant as done earlier.
2. Cross-absorb the blood serum with the extracts of pathogen-free healthy potato tubers to eliminate antibodies against tuber tissues; mix 1 ml each of antiserum and undiluted sap squeezed from healthy tubers; incubate the mixture successively for one hour at 37°C and for another one hour at 5°C.
3. Centrifuge at 8,000 g_n for 10 minutes; reject the precipitate; repeat incubation/centrifugation cycle five times until no more precipitate is formed after centrifugation.
4. Precipitate the γ-globulin fraction by adding equal volumes of neutralized saturated ammonium sulfate; centrifuge at 8,000 g_n for 10 minutes; resuspend the γ-globulin (precipitate) in 1 ml of phosphate buffer (0.1 M), pH 7.0 containing NaCl (8.5 g/l) (PBS) and store with NaN_3 (0.05 percent w/v) at 4°C.

Plate-Trapped ELISA (PTA-ELISA) Format

1. Prepare tuber samples by extracting in carbonate buffer containing 10 g/l Polyclar AT (BDH chemicals, Merck Ltd., United Kingdom) and $Na_2S_3O_5$ (0.1 M) and dilute appropriately with this extraction buffer.
2. Dispense aliquots of 0.1 ml to each well maintaining duplicates for each sample; incubate for 16 hours at 4°C and wash the wells with PBST thrice.
3. Add γ-glubulin (0.5 µg/ml) fraction; incubate for 3 hours at 37°C and wash the wells as done earlier.

4. Add goat antirabbit globulin conjugated to alkaline phosphatase and incubate for 3 hours at 37°C.
5. Add the substrate *p*-nitrophenyl phosphate and incubate for six to ten hours at room temperature.
6. Record optical absorbance at 405 nm using ELISA reader.

Determination of Biomass of Cladosporium fulvum in Tomato Leaves by ELISA (Karpovich-Tate et al., 1998)

Plate-Trapped Antigen (PTA)-ELISA

1. Extract the fungal antigens from frozen tomato leaf powder with PBS consisting of NaCl (0.8 percent), KCl (0.02 percent), Na_2HPO_4 (0.115 percent), and KH_2PO_4 (0.02 percent), pH 7.2 (1/5 w/v), and centrifuge at 12,000 g_n for 15 minutes.
2. Coat the wells (in triplicate) of the microtiter plates with PBS extracts of samples (100 μl/well); incubate overnight at 4°C; wash the wells with PBS four times and air-dry at 24°C in a laminar flow hood.
3. Block the nonspecific sites with 200 μl of 0.3 percent casein in PBS (w/v)/well.
4. Incubate after addition of specific monoclonal antibody for one hour and then with goat antimouse polyvalent peroxidase conjugate (Sigma A-0412) for one more hour.
5. Add tetramethyl benzidine substrate solution and incubate for 30 minutes.
6. Stop reaction by adding 100 μl of 3 M H_2SO_4.
7. Record the absorbance values at 405 nm using an ELISA reader.

Detection of Resting Spores of Plasmodiophora brassicae by ELISA (Orihara and Yamamoto, 1998)

Preparation of Immunogen

1. Homogenize club root-infected roots and hypocotyls (850 g) in distilled water for five minutes using a blender; filter the homogenate through gauze (eight layers thick); centrifuge the filtrate at 3,000 rpm for 20 minutes; resuspend the pellet in distilled water and repeat centrifugation cycle five times.
2. Prepare a sucrose gradient column (with 20 percent and 40 percent sucrose solutions) in a transparent centrifuge tube; overlayer the final suspension containing resting spores and plant cell debris on sucrose gradient column and centrifuge at 3,000 rpm for 20 minutes.
3. Collect the layer containing resting spores; wash with distilled water five times and store at −20°C.

Preparation of Antiserum

1. Inject the rabbit intramuscularly with 0.5 ml of immunogen (6×10^7 purified resting spores/ml of 0.85 percent NaCl solution); inject again intramuscularly with a mixture of 1 ml of immunogen and 1 ml of Freund's complete adjuvant after an interval of 2 weeks; administer additional dose of immunogen (0.5 ml containing 5.4×10^7 resting spores/ml) intravenously.
2. Collect the blood serum after a rest period of two weeks; purify the antibodies by ammonium sulfate precipitation and DEAE-cellulose column chromatography (see Chapter 3, Purification of Immunoglobulins and Antibody Fragments).

Indirect ELISA

1. Collect the club root-infected roots and hypocotyls and similar healthy tissues and store at $-20°C$.
2. Homogenize the samples separately in distilled water for five minutes; filter as done earlier (step 1, Preparation of Immunogen) and adjust the spore concentration to 1×10^6 spores/ml and dilute healthy samples to the same volume.
3. Suspend the sample extracts in coating buffer (carbonate) and dilute to required level and transfer 200 µl of each sample to two wells of microtiter plates and incubate overnight at 4°C.
4. Wash the wells thrice with PBS-containing polyvinyl pyrrolidone (2 percent) and BSA (2 percent) and incubate for one hour at room temperature.
5. Dispense to each well PBS containing 2 µg/ml of antiresting spore IgG; incubate at 37°C for four hours and wash the wells as done earlier.
6. Add PBS containing goat antirabbit IgG-alkaline phosphatase conjugate at a dilution of 1/2000; incubate for four hours at 37°C and wash the wells as done earlier.
7. Add 1 ml of diethanolamine (10 percent), pH 9.8 containing *p*-nitrophenyl phosphate (enzyme substrate) to each well; incubate for five minutes at 37°C in the dark and record the absorbance at 405 nm using ELISA reader.

Detection of **Colletotrichum falcatum** by Indirect-ELISA Technique (Viswanathan et al., 2000)

1. Grind the mycelium grown in Czapeck's Dox liquid medium at $28° \pm 2°C$ in ice-cold 0.1 M phosphate-buffered saline (PBS), pH 7.2 and centrifuge at 5,000 g_n for ten minutes at 4°C.

2. Assess the protein concentration of the supernatant using Brilliant Blue dye R-250 (Sigma) as per the procedure of Bradford (1976) and adjust the protein concentration to have 10 to 50mg/ml using carbonate buffer, pH 9.6.
3. Grind the plant tissues in carbonate buffer, pH 9.6; dilute to have concentration of 1:10 to 1:500 and use the extract for detection of the pathogen.
4. Dispense aliquots of 200 µl of the mycelial extract or plant tissue samples to each well of the microtiter plate; incubate overnight at 4°C and empty the wells.
5. Wash the wells thrice with PBS, pH 7.2 containing 0.05 percent Tween-20 (PBST); transfer 200 µl of antiserum (against *C. falcatum*) diluted suitably (1:1000 or 1:2000) in PBST, polyvinylpyrrolidone (2 percent) and ovalbumin in PBST to each well and incubate for three hours at 37°C.
6. Wash the wells as done earlier; place aliquots of 200 µl of goat anti-rabbit γ-globulin-alkaline phosphatase (ALP) conjugate (1:8000 in PBST) and incubate the plates for three hours at 37°C.
7. Wash the wells as done earlier; add aliquots of 200 µl of substrate solution containing paranitrophenyl phosphate (1 mg/ml) in diethanolamine (10 percent) buffer, pH 9.8.
8. Incubate for a predetermined time (about one hour) for development of color to the required level and stop the reaction by adding NaOH (3 M) at 50 µl/well.
9. Record the color intensity at 405 nm with an ELISA reader.

Detection of Resting Spores of Plasmodiophora brassica by Dot Immuno Binding Assay (DIBA) (Orihara and Yamamoto, 1998)

1. Prepare plant samples as per the procedure detailed in the section on detection of resting spores.
2. Spot 2 µl samples onto a 40 cm^2 nitrocellulose membrane sheet (Trans-Blot, BIO-RAD, USA); air-dry and block the nonspecific binding sites by immersing the membrane in a buffer solution consisting of Tris-HCl (20 mM), NaCl (500 mM), and Tween-20 (0.05 percent) (TTBS), pH 7.5, polyvinyl pyrrolidone (PVP) (2 percent), and BSA (2 percent).
3. Treat the membrane with 0.1 to 0.2 µg/ml of antiresting spore IgG in TTBS containing PVP (2 percent) and BSA (0.2 percent) (TTBSPB) for one hour at room temperature.
4. Treat the membrane with alkaline phosphatase conjugated goat anti-rabbit IgG in TTBSPB for one hour and then with buffer consisting of

Tris-HCl (0.1 M), NaCl (0.1 M), and $MgCl_2$, pH 9.5 containing nitro-blue tetrazolium substrate (0.33 mg/ml) and 5-bromo-4-chloro-3-indolyl phosphate *p*-toluidine salt prediluted with *N,N*-dimethylform-amide.

Direct Tissue-Blotted Immunobinding Assay (DTBIA)
Detection of **Fusarium** *spp. (Arie et al., 1995)*

1. Prepare cross sections of stems of infected plants (tomato/cucumber) (3 mm thick); place them on nitrocellulose membrane (0.45 μm) pore size, (Trans-Blot Transfer Medium. BioRad Laboratories, Richmond California) saturated with Tris-buffered saline (TBS), pH 7.0 for 10 to 30 minutes for direct tissue blotting.
2. Immerse the membrane in a blocking solution containing fetal calf serum (FCS) (10 percent v/v) and BSA (1 percent w/v) in TBS (FB-TBS) for one hour at room temperature.
3. Incubate with specific monoclonal antibody MoAb AP19-2 diluted in FB-TBS for one hour at room temperature and wash the membrane thrice in TBS containing Tween -20 (0.1 percent) (TBST).
4. Incubate the membrane with a mixture of biotinylated antimouse IgM-goat IgG (The Binding Site, Birmingham, United Kingdom), diluted 1:500 in FB-TBS and horseradish peroxidase-avidin D conjugates (diluted 500 times) (Vector Laboratories, Burlingame, California) for one hour at room temperature.
5. Wash the membrane with TBST for five minutes and repeat washing twice; immerse the membrane in substrate solution containing 4-chloro-1-naphthol and 0.02 percent hydrogen peroxide (v/v).
6. Observe for the development of blue color indicating positive reaction.

Immunofluorescence Assay (IFA) (Arie et al., 1995)

1. Cut cross sections (0.2 mm thick) of fresh stems, crowns, or roots of plants using a sharp razor blade and immerse the sections in a blocking solution consisting of BSA (10 percent) and gelatin (1 percent) in PBS (0.1 M), pH 7.0 for two hours at room temperature.
2. Soak the sections in the solution of specific monoclonal antibody MoAb 19-2 in PBST for two hours at room temperature.
3. Incubate the sections with fluorescein isothiocyanate (FITC) -labeled goat antimouse IgM; dilute 500 times with PBST.
4. Observe under a reflecting fluorescence microscope by β-excitation.

Detection of Botrytis cinerea by Protein A-Gold Labeling Technique (Svircev et al., 1986)

Preparation of Immunogen

1. Treat the mat of spores and mycelia with 0.5 percent formalin; centrifuge the suspension of fungal cells at 1,700 g_n for 10 min; resuspend the pellet in distilled water; repeat washing and centrifugation cycle three times and resuspend the mycelial mass in 2 ml of Freund's complete adjuvant to have a concentration of 10^6 cell/ml.
2. Inject the mycelial suspension intramuscularly and give the booster injection (second) after an interval of two weeks.
3. Collect the blood serum and store at −20°C.

Preparation of Protein A-Gold Label

1. Prepare the colloidal gold particles (15 nm diameter) by adding 4 ml of aqueous sodium citrate (1 percent) to 100 ml of a boiling solution of chloroauric acid (0.01 percent); cool the mixture for five minutes until a wine red color develops and store at 4°C in the dark.
2. Prepare the protein A-gold complex by adjusting the pH of colloidal gold suspension (10 ml) to pH 6.9 using potassium carbonate and add protein A (0.3 mg) (Sigma, St. Louis, Missouri) in 0.2 ml of distilled water.
3. Centrifuge at 48,000 g_n at 4°C to remove the excess unbound protein A.
4. Resuspend the dark red protein A-gold pellet in 10 ml of PBS (0.01 M), pH 7.4 and store at 4°C; the protein A-gold label may be stable for six to eight weeks.

Protein A-Gold Labeling

1. Float thin sections of plant tissue to be tested onto a saturated sodium periodate solution to remove osmium tetroxide used as a fixative for two to three minutes.
2. Wash the sections with distilled water three times and treat with ovalbumin (1 percent) for five minutes to block nonspecific binding sites.
3. Float the sections on drops of specific antiserum placed on coated grids and wash the sections thoroughly by passing the grid through a series of water drops.
4. Treat the sections with protein A-gold solution for 30 minutes; wash with drops of water as done earlier and stain with uranyl acetate (3 percent) for 20 minutes.
5. Observe under electron microscope.

Detection and Assay of Lipase Activity of Botrytis cinerea in Tomato Leaves by Immunoblotting (Comménil et al., 1998)

Production of Antiserum

1. Inject New Zealand White rabbits subcutaneously with an emulsion of purified protein (lipase) (160 µg) mixed with Freund's complete adjuvant; give two more injections through the same route using emulsions of antigen along with Freund's incomplete adjuvant at an interval of 21 days.
2. Collect the blood after a rest period of 14 days after final injection.
3. Purify the polyclonal antibodies by Protein-A Sepharose affinity chromatography; elute the antibodies using 0.5 M phosphate buffer, pH 3.0; neutralize the excess acidity of the proteins with 1 N NaOH (0.2 ml), pH 8.0 and dialyze against deionized water.
4. Add the antibodies to 2 percent ampholytes 3-10 (Bio-Rad); place in a Rotofor Cell of the isoelectric focusing apparatus (Bio Rad) and run at 12 watts for four hours.
5. Collect the solution from each compartment; dialyze against deionized water and use the antibodies for the assays.

Immunoblotting

1. Separate the proteins in the samples by performing sodium dodecyl-sulfate-polyacrylamide gel electroporesis (SDS-PAGE) protocol.
2. Transfer the proteins in the gel onto a nitrocellulose membrane by running at 100 V for one hour in a Bio-Rad Trans Blot electrophoresis transfer cell, using the electrotransfer buffer containing Tris (25 mM), glycine (120 mM) in methanol (20 percent).
3. Block the nonspecific binding sites in the membrane by incubating in Tris-HCl (25 mM), pH 7.4, NaCl (0.5 M) in polyvinylpyrrolidone (PVP) (2 percent) overnight at room temperature.
4. Treat the membrane with primary antibodies (specific to *B. cinerea* lipase) in a solution consisting of Tris-HCl (25 mM), pH 7.4, NaCl (0.5 M), and Triton X-100 (0.1 percent) (TBST) containing PVP (1 percent) for two hours.
5. Wash for ten minutes three times with TBST and incubate with solution containing the secondary antibody goat antirabbit IgG diluted to 1:2000.
6. Wash three times as done earlier with TBST.
7. Detect lipase using Bio-Rad alkaline phosphatase conjugated antibody kit.

Detection of Chitinase of **Phytophthora capsici** by Immunogold Labeling Technique (Lee et al., 2000)

Preparation of Ultrathin Sections of Pepper Stem Tissue

1. Freeze the infected and healthy stems under high pressure using a HPM 010 device (BAL-TEC, GmbH, Balzer, Liechtenstein).
2. Cut ultrathin sections using a Diatome diamond knife and mount the section on pioloform and carbon-coated nickel grids (100 mesh).

Immunogold Labeling

1. Treat the ultrathin sections placed on grids with lysine, gelatin, and bovine serum albumin (BSA) to saturate nonspecified binding sites.
2. Wash the sections with PBS and treat the sections with polyclonal antibodies raised against tomato chitinase.
3. Wash the sections with PBS and stain them with goat antirabbit IgG conjugated with colloidal gold (10 nm) for three hours at room temperature.
4. Stain the sections with uranyl acetate (8 percent) for ten minutes and then with Reynold's lead citrate for ten minutes.
5. Observe under the electron microscope.

Detection of Cell Wall Components by Enzyme-Gold and Immunogold Techniques (Kang and Buchenauer, 2000a)

Preparation of Enzyme-Gold Probes

1. For preparing cellulose-gold probe, add colloidal gold (15 nm diameter) solution (pH 5.4) to cellulase (1 mg) of *Trichoderma reesei* (Worthington Biochemical Corporation, Freehold, New Jersey) and incubate.
2. Centrifuge the mixture at 17,000 g_n for 30 minutes; resuspend the pellet in 10 ml of citrate buffer (10 ml), pH 5.6 and centrifuge again.
3. Resuspend the residue in 1 ml of citrate buffer containing 0.02 percent (w/v) of polyethylene glycol (20,000MW).
4. For preparing xylanase-gold probe, conjugate xylanase (800 µg) purified from *T. viride* (Sigma Chemical Corporation, St. Louis, Missouri) with gold solution (10 ml), pH 5.5.
5. Rinse the conjugate in acetate buffer (50 mM), pH 5.0 and resuspend the xylanase-gold probe in acetate buffer (1.0 ml).

Cytochemical Labeling

1. Quench the ultrathin sections of healthy and infected spikelet tissues with citrate buffer (50 mM, pH 5.4) containing 1 percent bovine serum albumin (BSA) and gelatin (0.5 percent) for 20 minutes and incubate with cellulase-gold solution (1:10 dilution) for 20 minutes.
2. For labeling the xylan, treat the ultrathin sections with acetate buffer (50 mM, pH 5.0) containing BSA (1.0 percent) and gelatin (0.5 percent) for 20 minutes and incubate with xylanase-gold solution (1:10 dilution) for 20 minutes.
3. For detection of pectin by immunogold technique, block the ultrathin sections with 1.0 percent of BSA in Tris-buffered saline consisting of Tris HCl (10 mM) and NaCl (150 mM) for 20 minutes and incubate with hybridoma culture supernatant of the monoclonal antibody JIM7 diluted 1:1 with TBS for 14 hours at 4°C.
4. Rinse the sections in TBS; incubate with goat antirabbit IgG conjugated with colloidal gold particles (15 nm) diluted in TBS to 1:30.
5. Stain the sections with uranyl acetate and lead citrate and examine under electron microscope.

Assessment of Labeling Density

1. Determine the number of gold particles per square micrometer over specified cell wall areas on 18 to 28 microphotographs.
2. Calculate the difference in the number of gold particles between healthy and infected host tissues and analyze statistically by the paired t-test.

Detection of Amino Acid Exporter Protein of Uromyces fabae by Immunofluorescence Technique (Hahn et al., 1997)

Antibody Generation

1. Synthesize peptides from the desired coding region; couple the peptides as a mixture via their N-terminal cysteines to keyhole limpet hemocyanine and inject the conjugate into the rabbits.
2. Collect the blood serum and purify the antibodies using Sepharose 4B column.

Immunofluorescence Microscopy

1. Inoculate leaves of broad bean plants *(Vicia faba)* by spraying rust spore suspension; eight days after inoculation, vacuum-infiltrate the infected leaf pieces with 8 percent methanol in water under high pres-

sure frozen with an HPM 010 high pressure instrument (Balzers, Liechtenstein).

2. Freeze-substitute the specimens in acetone at −90°C and embed in a mixture of butyl-methacrylate (75 percent) and methyl methacrylate (25 percent), benzoinethylether and dithiothreitol (10 mM) in mixtures with acetone (25, 50, 75, and 100 percent) for one day each at 4°C for polymerization under UV light.

3. Place sections (1 to 2 μm thick) of specimens on microscope slides coated with Biobond (British Biocell, Cardiff, UK) and allow to dry at 30°C.

4. Treat the sections three times for 15 minutes each with blocking buffer consisting of BSA (1 percent), autoclaved yeast cell walls (1 percent) in TBS containing Tris-HCl (10 mM), NaCl (150 mM), pH 7.4.

5. Incubate sections with primary antibodies (preimmune serum or affinity purified specific antibodies) diluted with TBS (1:50) and wash three times for 15 minutes each with TBS.

6. Incubate the sections with the secondary antibody-fluorscein-conjugated goat antirabbit IgG for one hour at 20°C and rinse the sections three times for 3 minutes each with TBS and flush with distilled water.

7. Mount the sections in Citifluor (Citifluor Inc., London, United Kingdom) and observe under the Zeiss Axioscop microscope equipped for epifluorescence (filters BP490, FT510, LP 565).

Detection of Elicitins by Immunodiagnostic Techniques (Devergne et al., 1994)

Preparation of Antigen

1. Separate the mycelium of the pathogen cultured in appropriate medium for 10 to 21 days by Buchner filtration and sterilize the filtrate by passing through a Millipore membrane (0.45 μm).

2. Analyze the crude filtrate by ELISA or Western blots directly or after concentration (tenfold) by vacuum.

3. Purify elicitin preparations by ion-exchange chromatography on Zeta-Prep (CunoInc., Mereden, USA); use a sulpho-propyl matrix for cryptogein and cinnamomin and elute the fractions using 0.5 M NaCl; use a quartnary amino ethyl matrix in the case of capsicein and parasiticein and elute the fractions with 0.4 M sodium acetate at pH 4.0.

4. Dialyze the elicitin fractions against distilled water and store after lyophilization.

Production of Antisera

1. Immunize the rabbits with elicitin (E) preparation (2 mg) in water emulsified with complete Freund's adjuvant (v/v) by subcutaneous injections on the back (ten sites).
2. After two or three weeks, provide a second injection with elicitin emulsified with incomplete Freund's adjuvant, and give a booster injection at monthly intervals with the same immunogen.
3. Assay the antiserum titre, at seven to ten days after each booster injection, by indirect antigen-coated plate (ACP)-ELISA, using elicitin preparation (60 ng/ml) as coating.
4. Purify the immunoglobulins from the antiserum by the revanol precipitation method (Chapter 3).

Enzyme-Linked Immunoassay (ELISA) Formats

1. Use 96 flat-bottomed well polyvinyl plates (Falcon 3912, Microtest III, Becon Dickinson, Osnard, USA) because of their distinct capacity to adsorb elicitins in (ACP)-ELISA format and dispense reagents (100 μl/well) in duplicate wells for each sample.
2. For (ACP)-ELISA (1) the successive steps are as follows: add antigen (elicitin) followed by monoclonal antibody and antimouse Ig-alkaline phosphatase (ALP) conjugate (GAM-E).
3. For DAS-ELISA (2) the steps are as follows: add polyclonal antibody (Pab) followed by blocking with BSA, addition of elicitin, then monoclonal antibody (mAB) and GAM-E.
4. For DAS-ELISA (3) the successive steps are as follows: add antibody against crystallizable fragment (Fc) of Ig, followed by addition of BSA, mAB, elicitin, Pab, and antirabbit antibody raised in goat-ALP conjugate.
5. For coating the wells, use the reagents dissolved in carbonate buffer (0.05 M), pH 9.6 and incubate the plates overnight at 4°C.
6. Dilute the goat antimouse-Fc-specific antibody (1/750) in the carbonate buffer; add BSA (1.0 percent) in 0.05 M Tris buffer saline; incubate for one hour prior to incubation with the mAbs, in the case of DAS-ELISA (3) procedure; for diluting antigens and antibodies, use Tris buffer saline containing Tween (0.05 percent) and BSA (1 percent) in subsequent steps and incubate at 37°C for two hours.
7. Add *p*-nitrophenyl phosphate (1 mg/ml) in diethanolamine buffer (0.01 M), pH 9.8 as substrate and record the absorbances at 405 nm using Twinreader photometer (Flow Labs, SA, Puteaux, France).

Preparation of Plant Extracts

1. Use plants raised in growth chambers (24°C, 16 hour day length); wash their roots with tap water and cut off their apixes.
2. Place the plants in slanting petri dishes with their roots in contact with wet filter paper and inoculate by placing agar plug or mycelium of the pathogen on the cut root tips.
3. At different periods after inoculation (one, two, three, and four days) take stem or leaf samples (about 1 g each) from the inoculated and healthy plants; freeze the samples immediately in liquid nitrogen and store at −20°C.
4. When required, grind the samples with mortar and pestle in Tris buffer saline(1 ml); centrifuge at 3,500 g_n for 15 minutes and dilute the supernatant with the same buffer to the required level just before transfer to the wells in microtiter plates.

Chapter 9

Plant-Bacterial Pathogen Interactions

Bacterial pathogens cause various kinds of symptoms in infected plants. Some of the visual symptoms may be characteristic of bacterial infections. Formation of water-soaked symptom in the early stages and production of cloudy ooze from infected tissues are associated with many bacterial diseases. However, these symptoms are not sufficient to identify the pathogen and to diagnose disease. Conventional methods of isolating bacterial pathogens in suitable culture media and performing different biochemical tests are required to diagnose disease. However, these methods require time and sometimes it may be difficult to establish the precise identity of the pathogen in question (Narayanasamy, 2001). In some plants, bacterial infection may remain latent without any visible symptoms, as in tomato bacterial canker and potato ring rot diseases (De Boer et al., 1996). When a large number of samples has to be handled rapidly, such as in plant quarantines and certification programs, the use of molecular diagnostic techniques is preferred. Immunodiagnostic methods are sensitive, rapid, and reliable for early detection of bacterial pathogens and also for the differentiation of pathovars within a morphologic species. Phytoplasmas, which cause several crop diseases, have structural similarity with bacteria. The serodiagnosis of phytoplasmas is discussed in this chapter.

IMMUNODIAGNOSTIC TECHNIQUES FOR BACTERIAL PATHOGENS

Most plant pathogenic bacteria are single-celled entities. They do not produce different kinds of spore forms as do the fungal pathogens. They are much smaller in size and have well-defined cell walls on which a variety of specific antigenic determinants (epitopes) are present. These antigenic determinants are readily accessible to probes by specific antibodies, resulting in the recognition of the variations in the composition of bacterial cell walls. Bacterial pathogens are more readily characterized by immunological assays compared to fungal pathogens. Both polyclonal and monoclonal antisera have been raised against several bacterial pathogens and have been reported to be effective for the detection of bacterial pathogens in seeds, clonal materials, plants, and soils.

The sensitivity and specificity of polyclonal antisera can be increased by either diluting the cross-reacting antibodies or by cross-absorbing such antibodies present in the antiserum. Marked increase in the specificity of serological reactions may be achieved by use of monoclonal antibodies (MABs), leading to differentiation of strains/ pathovars of a pathogen species. MABs raised against two German isolates of *Xanthomonas campestris* pv. *campestris* (*Xcc*) were employed to subdivide *Xcc* isolates into three serogroups S1, S2, and S3 (Rabenstein et al., 1999). Serological groups of *Pseudomonas syringae* pv. *coronafaciens* from oats were recognized based on the antigenic properties of the bacterial strains (Pasichnyk, 2000).

Immunodiagnostic assays are very effective in detecting bacterial pathogens in infected seeds, tubers, and other plant materials especially when symptoms of infection are not clearly visible. A highly sensitive, rapid, and easy method of detection of latent infection of *Ralstonia solanacearum* in potato tubers was developed by Priou et al. (1999). This technique may be adopted for seed and tuber quality control. Latent infection of groundnut (peanut) by *R. solanacearum* (wilt disease) was detected by ELISA (Shan et al., 1997). Two types of symptoms—black rot and blight—induced by *X. campestris* pv. *campestris* in cabbage seedlings are strain dependent. In seedbeds, most of the ELISA-positive seedlings were symptomless, indicating the effectiveness of the immunoassay employed for the detection of latent infection (Shigaki et al., 2000).

Antisera that recognize the bacterial cell wall antigens or specific components or metabolites have been used for the detection and characterization of bacterial pathogens. By employing the double diffusion test, the antisera raised against glutaraldehyde-fixed flagella of two strains, UQM 551 of *Pseudomonas syringae* pv. *pisi* and strain L of *P. syringae* pv. *syringae,* were shown to have high levels of specificity in the detection of respective strains. In addition, the antisera specific to heat-killed cells of *P. syringae* could be used to distinguish *P. syringae* from all other bacterial species and genera tested including strains of *P. fluorescens, E. coli, Agrobacterium,* and *Rhizobium* (Mazarei et al., 1992) (Figure 9.1). The flagellar extracts of *X. campestris* pv. *campestris* were used to raise the monoclonal and polyclonal antibodies. Polyclonal antibodies were employed to probe the immunoblots to identify bands of flagellar extracts separated by SDS-PAGE (Franken et al., 1992). The antibodies raised against flagella from strains representing most pathovars of *P. syringae* were used to differentiate flagella of serotypes H1 or H2 (Malandrin and Samson, 1999). MABs raised against the European potato strains of *Erwinia chrysanthemi* reacted with a fimbrial antigen present in all except two strains of *E. chrysanthemi* isolated from potato. This demonstrates their usefulness for the detection of pathogen not only in potato, but also in other host plant species (Singh et al., 2000) (see Appendix section on the ELISA detection of *E. chyrsanthemi*).

Gram-negative pathogenic bacteria produce lipopolysaccharides that form one of the structural components of the outer membrane of

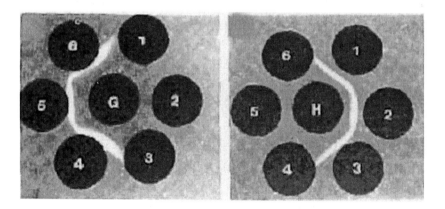

FIGURE 9.1. Reactions of antisera raised against glutaraldehyde-fixed flagella of *Pseudomonas syringae* pv. *syringae* (strain UQM 551) and *P. syringae* pv. *syringae* (strain L) as determined by double diffusion test: (G) antiserum to strain L; (H) antiserum to strain UQM 551; (1) whole untreated bacterial cells of strain UQM 551; (2) heat-killed bacterial cells of strain UQM 551; (3) sonicated cells of strain UQM 551; (4) whole untreated cells of strain L; (5) heat-killed cells of strain L; (6) sonicated cells of strain L (*Source:* Mazarei et al., 1992; Reprinted with permission by Blackwell Scientific Publications, Oxford, United Kingdom).

bacterial cells. Antisera specific for these lipopolysaccharides have been used for indexing plant materials for the presence of bacterial pathogens such as *Agrobacterium tumefaciens, Erwinia carotovora* ssp. *atroseptica, Xanthomonas axonopodis* pv. *citri, Ralstonia solanacearum,* and *P. syringae* (De Boer et al., 1996; Ovod et al 1999; McGarvey et al., 1999; Gallo et al., 2000).

Many strains of *Pseudomonas* produce phytotoxic lipodepsipeptides such as syringomycin, syringopeptins, and related compounds. The antiserum raised against a macromolecular derivative of syringopeptins (KLH-SP$_{25A+B}$) detected free syringopeptins (SP$_{22}$ and SP$_{25}$) and also syringopeptins in aqueous extracts of cotyledon of zucchini infected by *P. syringae* pv. *lachrymans* (Fogliano et al., 1999; Gallo et al., 2000). Other pathovars of *P. syringae* such as *atropurpurea* (Nishiyama et al., 1977) and *glycinea* (Mitchell and Young, 1978) produce coronatine (COR). This toxin is composed of two structural components—coronatic acid (CFA) and coronamic acid (CMA) which are linked by an amide bond. By using monoclonal antibodies prepared against COR, the toxin could be detected and quantitatively estimated (Jones et al., 2001).

The presence of bacterial pathogens in vector insects may be detected and their seasonal dynamics can be estimated with serological techniques. The proportion of the beetles *Acalymma vittata* harboring *Erwinia tracheiphila* (bacterial wilt of cucurbits) was estimated. Furthermore, strong serological evidence shows that *A. vittata* served as the primary overwintering reservoir of *E. tracheiphila,* which may infect the cucurbits (Fleischer et al., 1999). After 24 hours of feeding on infected plants, a high percentage of individual beetles (60 to 70 percent) tested positive for *E. tracheiphila* antigen by DAS-ELISA. The beetles retained the bacterial pathogen for about 30 days (García-Salazar et al., 2000).

Selective separation of bacterial pathogens present along with nontarget bacterial species in seeds and plant materials can be achieved by following immunomagnetic separation (IMS). This procedure is coupled with the polymerase chain reaction (PCR) to enhance the sensitivity of pathogen detection. PCR assay could detect *E. carotovora* ssp. *atroseptica* only if the concentration of bacterial cells in potato peel extract was at least 10^5 cells/ml. However, when IMS procedure was performed prior to PCR, the detection limit was 100 bacterial/ml, indicating the effectiveness of the IMS procedure in increasing the sensitivity of PCR (van der Wolf et al., 1996). The IMS procedure in combination with PCR was demonstrated to be useful in the identification of two leafhoppers *Graphocephala coccinea* and *G. versata* as the vectors of the bacterial leaf scorch disease of American elms (Pooler et al., 1997). The bacterial pathogen *Acidovorax avenae* ssp. *citrulli* causes the devastating bacterial fruit blotch (BFB) disease of watermelon. Adoption of IMS-PCR resulted in a 100-fold increase in the detection sensitivity over direct PCR. Because infected seeds form the primary sources of infection for *A. avenae* ssp. *citrulli,* use of IMS-PCR may be very effective in eliminating infected seedlots (Walcott and Gitaitis, 2000) (see Appendix section on immunomagnetic separation). For specific detection and selective isolation of *P. syringae* pv. *phaseolicola* (causing halo blight disease of beans), IMS and agglutination tests were performed utilizing PABs against whole bacterial cells and the exopolysaccharide fraction (Güven and Mutlu, 2000).

Among the various immunodiagnostic techniques employed, enzyme-linked immunosorbent assay (ELISA), immunofluorescence (IF) assay, immunoblotting assay, and the immunogold-labeling technique have been more frequently used, whereas virobacterial agglutination (VBA) tests and immunosorbent electron microscopy (ISEM) have been useful in some cases.

Enzyme-Linked Immunosorbent Assay (ELISA)

Different formats of ELISA have been widely used for the detection and differentiation of several plant pathogenic bacterial species, subspecies, and pathovars: *C. michiganensis* ssp. *sepedonicus* (Slack et al., 1996; De Boer and Hall, 2000), *E. carotovora* ssp. *atroseptica* (Perombelon and Hynan, 1995; Perombelon et al., 1998), *X. oryzae* pv. *oryzicola* (Zeng et al., 1996), *Xylella fastidiosa* (Berisha et al., 1998; Purcell et al., 1999), *Ralstonia solanacearum* (Elphinstone and Stanford, 1998), *X. campestris* pv. *campestris* (Rabenstein et al., 1999), *X. campestris* pv. *cyamopsidis* (Vijayanand et al., 1999), *X. axonopodis* pv. *dieffenbachiae* (Hseu and Lin, 2000), *C. michiganensis* ssp. *insidiosus* (Kokošková et al., 2000), *E. chrysanthemi* (Singh et al., 2000), and *X. axonopodis* pv. *citri* (Jim et al., 2001). Information on the detection of other bacterial pathogens is available in the literature (Narayanasamy, 2001).

The DAS-ELISA procedure was followed for the detection of *X. axonopodis* pv. *dieffenbachiae* using polyclonal antisera, the detection limit being 10^4 colony-forming units (CFUs)/ml of the pathogen (Hseu and Lin, 2000). *Clavibacter michiganensis* ssp. *insidiosis* (bacterial wilt of lucerne) could be detected and identified using PABs in DAS-ELISA and PTA-ELISA formats (Kokošková et al., 2000). Monoclonal antibodies (MABs) were used to differentiate serogroups of *X. campestris* pv. *campestris* (*Xcc*) and all MABs reacted with epitopes on LPS of *Xcc* in PTA-ELISA method (Rabenstein et al., 1999). A triply cloned strain of *X. fastidiosa* inducing citrus variegated chlorosis (CVC) in sweet orange was detected by DAS-ELISA (Li et al., 1999). The triple antibody ELISA test utilizing a specific MAB (6A6) was employed to detect *E. chrysanthemi* in potato stems and tubers. This test had a detection limit of 10^7 cfu/ml (Singh et al., 2000) (see Appendix section on ELISA detection of *E. chrysanthemi*). The ELISA procedure was advantageously adopted for the detection of *X. axonopodis* pv. *citri* in citrus fruits meant for export. ELISA was found to be more sensitive by detecting the bacteria in 100 percent of infected fruits. The bacteriophage test (BPT), in addition to being more laborious and time-consuming, detected the pathogen in 44 percent only (Jim et al., 2001).

Pathogenic bacterial species of *P. syringae* are known to produce bioactive toxic syringopeptins and the pathovars *atroseptica* and *glycinea* produce the phytotoxin, coronatine (COR). By using the polyclonal antiserum raised against purified syringopeptins, the presence of free syringopeptins (SP_{22} and SP_{25}) could be detected by indirect ELISA, the estimated detection limit being 0.05 µg/well. Furthermore, abundant quantities of the phytotoxin (approximately 0.1 mg of syringopeptins/g of fresh plant tissue) were detected in the cotyledons of zucchini plants infected by *P. syringae* pv. *lachrymans* (Fogliano et al., 1999) (see Appendix section on detection of syringopeptins). Competitive ELISA was shown to be effective in detecting COR. By using specific MABs the two structural compounds of COR could be recognized. MAB 11 B8 primarily recognized coronamyl amide features of COR, while 11G8 recognized the coronafacoyl amide of COR (Jones et al., 2001).

The sensitivity of the ELISA technique may be appreciably enhanced by inclusion of an enrichment step prior to the assay. Bacteria are concentrated by isolating them from infected seeds or plant tissue and incubating in a semi-selective enrichment broth (SSEB). *Xanthomonas campestris* pv. *undulosa* which is seed-borne in wheat, could be detected in seed samples with populations less than 5×10^2 CFU/ml when the enrichment step was followed prior to ELISA. In assays without the enrichment step, a population of 5×10^3 CFU was required for detection by ELISA (Frommel and Pazos, 1994). For the detection of *E. carotovora* ssp. *atroseptica,* enrichment of the bacterial pathogen in potato peel extracts was performed directly in microplate wells prior to (DAS)-ELISA (Perombelon et al., 1998). Likewise, *R. solanacearum* in latently infected potato tubers was more efficiently detected by a postenrichment ELISA on nitrocellulose membrane (NCM-ELISA). The tuber extract was plated on modified Kelman's medium and incubated for 48 hours in modified SMSA broth at 30°C. As few as ten bacterial cells/ml could be detected, demonstrating the sensitivity, reliability, and low-cost nature of the NCM-ELISA method which was equally effective as nucleic acid spot hybridization (NASH) technique (Priou et al., 1999).

Immunofluorescence Technique

The immunofluorescence (IF) technique has been used to detect and differentiate bacterial pathogens and their serovars using PABs and MABs. *Pseudomonas andropogonis* infecting carnations and other ornamental plants could be detected using monoclonal antibodies in the immunofluorescence microscopy technique (Li et al., 1993) (Figure 9.3A and B). The presence of *X. campestris* pv. *undulosa* in wheat seed lots was detected by indirect IF test (Bragard and Verhoyen (1993). The occurrence of three serovars of *X. albilineans* and their distribution in different countries were studied by Rott et al., (1994). IFAS and IFC procedures were adopted for the detection of *R. solanacearum* in potato tubers (Elphinstone and Stanford, 1998; van der Wolf et al., 1998). The use of MABs prepared from phage display library against the lipopolysaccharides of *R. solanacearum* was reported to be more efficient in pathogen detection (Griep et al., 1998). The flagellins present in the flagella of *P. syringae* pathovars were characterized by immunofluorescence staining, SDS-PAGE, and immunoblotting. Two serotypes were differentiated based on the nature of flagellin and the characteristics of flagellins were proposed as an

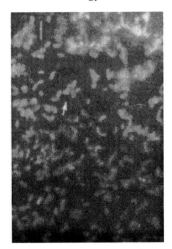 

FIGURE 9.3. (A) Detection of *Pseudomonas andropogonis* by indirect immuno-fluorescence staining using monoclonal antibodies; (B) detection of *P. andropogonis* in carnation leaf extracts (*Source:* Li et al., 1993; Reprinted with permission by Blackwell-Verlag.)

identification tool (Malandrin and Samson, 1999). The MABs against *E. chrysanthemi* reacted specifically, whereas the PABs cross-reacted with most of the *E. carotovora* strains tested (Singh et al., 2000) (see Appendix section on electron micorscopy detection of *E. chrysanthemi*).

An indirect immunofluorescence colony staining technique has been developed for the detection of important seed-borne pathogens of tomato such as *X. campestris* pv. *vesicatoria, P. syringae* pv. *tomato,* and *C. michiganensis* ssp. *michiganensis.* In this technique, a specific antiserum is used for initial binding of target bacteria and the positive colonies are visualized by applying commercially available secondary antibodies conjugated with fluorescein isothiocyanate (FITC) and observing under a fluorescence microscope. This procedure is economical, easy to perform, and results can be obtained in four to five days compared to 30 to 45 days required for conventional methods. If monoclonal or recombinant antibodies are used, this technique is highly sensitive and has the potential for detection and localization studies (Veena and Vuurde, 2002).

Dot Immunobinding Assay (DIBA)

Detection of bacterial pathogens in asymptomatic plants is important, since infected plants/plant materials can be eliminated resulting in the restriction of disease spread (Hsu et al., 1995). This situation was observed in the case of the citrus canker pathogen *X. axonopodis* pv. *citri*. The pathogen could be detected by DIBA in 38.4 percent of symptomless citrus plants (Wang et al., 1997). DIBA established the relationship between the bacterium causing coffee leaf scorch disease and the strain of *X. fastidiosa* inducing citrus variegated chlorosis disease. It demonstrated the cross-reaction with antiserum against *X. fastidiosa* (de Lima et al., 1998). The accuracy and reliability of tissue blot immunoassay for the detection of *Clavibacter xyli* ssp. *xyli* causing sugarcane ratoon stunting disease was also clearly demonstrated (Hoy et al., 1999).

Immunogold Labeling

The immunogold labeling technique has been useful for visualization of the bacterial pathogens per se or in situ in infected plant tissues. Ariovich and Garnett (1989) found the immunogold staining technique to be highly specific and rapid for the detection of the bacteria associated with citrus greening disease. Immunogold labeling of whole cell lysates of *P. andropogonis* was found to be useful in characterizing the epitopes located within the bacterial cell (Li et al., 1993) (Figure 9.4). Liu et al. (1999) employed the immunogold-labeling method for the detection of the causal agent of citrus disease called "huanglongbing" in local Chinese villages. Some pectolytic *Erwinia* spp. have nonflagellar appendages known as "fimbriae" which may be involved in adhesion of bacterial cells. By using fimbrial-specific MABs, the presence of fimbriae in bundles in *E. chrysanthemi* was observed by following the immunogold-labeling technique. A specific MAB targeted a fimbrial epitope. Using MABs specific to fimbrial antigens, all the European potato isolates of *E. chrysanthemi* could be detected (Singh et al., 2000) (see Appendix section on electron microscopy detection of *E. chrysanthemi*).

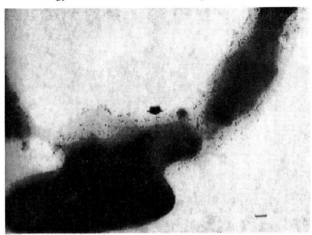

FIGURE 9.4. Immunogold labeling of whole cell lysates of *Pseudomonas andro-pogonis* with monoclonal antibodies; arrow indicating gold particles bound to bacterial lysate (*Source:* Li et al., 1993; Reprinted with permission by Blackwell-Verlag).

Agglutination Tests

Slide agglutination tests, probably the simplest among serological tests, were used in the early stages of development of immunodiagnosis. Because of the low level of sensitivity and the requirement of relatively high titer antiserum, an improvement was made by sensitizing the bacterial cells of *Staphylococcus aureus,* which have high concentrations of protein A on their surface. The PABs specific to the bacterial species to be detected are conjugated with the cells of *S. aureus.* The rapid slide agglutination test developed by Lyons and Taylor (1990) was used to detect *P. syringae* pv. *phaseolicola* in bean and *P. syringae* pv. *pisi* in pea. The PABs produced against *E. amylovora* were used to sensitize the cells of *S. aureus.* Detection of *E. amylovora* in plant sap and bacterial slime of hawthorns (*Grataegus* sp.) infected by *E. amylovora* was possible with this improved method and also more reliable and efficient (Mráz et al., 1999). *Xanthomonas campestris* pv. *campestris (Xcc)* was effectively detected in washings of seeds of *Brassica* by using the improved slide agglutination test (Mguni et al., 1999).

IMMUNODIAGNOSIS OF PHYTOPLASMAL DISEASES

Attempts to isolate phytoplasmas, except three helical *Spiroplasma* spp. in pure culture, have been unsuccessful. Characterization based on symptoms, host range, and modes of transmission has not resulted in reliable identification of causative phytoplasma. Hence the need for development of sensitive and reliable diagnostic techniques. Immunodiagnostic techniques have provided adequate levels of reliability and sensitivity for detection of phytoplasmas and for establishing the relationships between phytoplasmas. Both polyclonal and monoclonal antisera have been developed and employed in different serological methods. Among the several immunodiagnostic techniques, ELISA, IF, and ISEM have been commonly used.

Enzyme-Linked Immunosorbent Assay (ELISA)

Enzyme-linked immunosorbent assay (ELISA) has been widely used to detect and differentiate phytoplasmas causing many serious plant diseases: aster yellows (Lin and Chen, 1985; Errampalli and Fletcher, 1993), corn stunt (Davis and Fletcher, 1983; Henriquez et al., 1999), maize bushy stunt (Chen and Jiang, 1988), sesamum phyllody (Srinivasulu and Narayanasamy, 1995), sugarcane grassy shoot (Viswanathan and Alexander, 1995), sugarcane yellows (Aljanabi et al., 2001), tomato big bud (Hsu et al., 1990), X disease of stone fruits (Guo et al., 1998), citrus stubborn disease (Najar et al., 1998), brinjal (eggplant) little leaf disease (Das and Mitra, 1999), and apple proliferation (Loi et al., 2002).

Use of MABs was found to be more sensitive for the detection and differentiation of phytoplasmas. MABs specific to aster yellows (AY) phytoplasma, when tested by indirect ELISA, reacted specifically with AY-infected plants and differentiated AY phytoplasma from other phytoplasmas (Lin and Chen, 1986). Likewise, the MABs raised against *Spiroplasma citri* could differentiate its strains and were able to distinguish *S. citri* from corn stunt spiroplasma (Lin and Chen, 1985).

Errampalli and Fletcher (1993) prepared monospecific polyclonal antibodies against a protein associated with AY phytoplasma. Monospecific PABs were more efficient in detecting AY phytoplasma. The presence of a surface protein p89 in *S. citri* was detected by electrophoresis and Western blotting. By using ELISA, *S. citri* was detected in the vector leafhopper *Circulifer tenellus*. The surface protein p89 may have a role in the adherence of *S. citri* to the vector cells (Yu et al., 2000). An immunodominant membrane protein (IMP) associated with apple proliferation (AP) phytoplasma was detected using specific antibodies. The same protein (P-318B) was also detected by an antiserum raised against antigen preparations from AP-infected plants (Berg et al., 1999). By employing DAS-ELISA, PCR, and RT-PCR techniques (Aljanabi et al., 2001), sugarcane yellow leaf syndrome was demonstrated to be due to combined infection by phytoplasma sugarcane yellows (ScY) and sugarcane yellow leaf virus (ScYLV). AP phytoplasma could be detected in infected apple roots, stems, and leaves by employing two monoclonal antibodies in ELISA and IF tests (Loi et al., 2002).

Relationships between phytoplasmas may be established by ELISA technique. The antiserum specific to white leaf disease (WLD) phytoplasma infecting sugarcane cross-reacted with the antigen preparation from grassy shoot disease (GSD)-infected sugarcane suggesting a relationship between the phytoplasmas occurring in India and Taiwan (Viswanathan, 1997). The MAB specific to X-disease phytoplasma from chokecherry *(Prunus virginiana)* reacted with five phytoplasmas included in the X-disease phytoplasma cluster (Guo et al., 1998).

PHASES OF DEVELOPMENT
OF BACTERIAL DISEASES

Penetration Through Natural Openings

Bacterial pathogens do not undergo the process of germination as the fungal pathogens do. Bacterial pathogens are essentially single-celled organisms and do not have any other morphologic structures that may aid them during pathogenesis. They possess cell surface-anchored structures such as pili, flagella, lipopolysaccharides, expolysaccharide slime layers, and outer membrane proteins that may help in the entry into and survival within host plants. Furthermore, they produce several extracellular enzymes and a variety of compounds that may be essential for their virulence.

Most bacterial pathogens depend on a simple passive process of moving through a film of water to gain entry into host plants. No other special mechanism seems to operate. Natural openings such as stomata, hydathodes, lenticels, and wounds or injuries occurring during agricultural operations, harvesting, transport, and storage form the entry points for bacterial pathogens.

Colonization of Host Tissues

Bacterial pathogens secrete several enzymes and macromolecules required for nutrient uptake and development of diseases. Pathogens such as *E. chrysanthemi,* causing soft rot diseases, produce proteases which are transported from the bacterial cytoplasm directly into the extracellular environment. This type of one-step secretion (type I pathway) is governed by the secretory apparatus consisting of three accessory proteins (Prt D, E, and F) (Pugsley, 1993). Through the two-step *sec* dependent (type II) pathway, also known as general secretory pathway (GSP), pectinases and cellulases are secreted by the bacterial pathogens. The enzymes, pectates, lyases, polygalacturonases, and cellulases are first exported to the bacterial cell periplasm, but they are unable to move through the outer membrane. The perisplasmic forms of these enzymes move through the outer membrane and then into the extracellular environment, thus requiring a second step. The pectolytic enzymes produced by the bacterial pathogens cause a breakdown of pectic substances such as calcium pectate, which forms cementing material and keeps the cells intact.

The pectate lyase proteins are involved in the degradation of plant cells and proteins PelA, PelB, PelC, and PelE have been purified and the modes of depolymerization of polygalacturonase have been studied by Preston et al. (1992). Purified pectate lyase A (PelA) is composed of two protein molecules of 38 kDa (Doan et al., 2000). The cell-wall degrading enzymes produced by *E. carotovora* ssp. *carotovora (Ecc)* are considered as the principal virulence determinants and their synthesis is coordinately regulated by a complex network. The interaction between the EXP S and EXP A proteins encoded by *expS* and *expA* genes may control the expression of virulence gene of this pathogen (Eriksson et al., 1998). In the case of *E. chrysanthemi,* production of the cell-wall degrading enzymes in large quantities is required for inducing disease symptoms. The H-NS protein associated with nucleoid of *E. chrysanthemi* appears to have a role in pathogenesis. The *hns* mutant showed dramatic reduction in Pel production as well as reduced virulence, suggesting that H-NS protein has a critical role in the expression of virulence and pathogenicity of *E. chrysanthemi* (Nasser et al., 2001).

Additional factors such as flagellins (flagellar proteins) may also be required for colonization of host tissues. The flagellins from *P. avenae* strains N1141 (incompatible with rice) and H8301 (compatible with rice) were purified and a strain-specific antiserum was raised against N1141 cells. By using the antiserum, the flagellin protein of N1141 strain was detected. When the cultured rice cells were treated with the N1141 flagellin protein, hypersensitive cell death was observed within six hours of treatment. In contrast, the flagellin protein from the compatible strain H 8301 did not induce such response in rice cells. Furthermore, the hypersensitive cell death could be prevented by pretreatment of rice cells with anti-N1141 flagellin antibody indicating the role of flagellin protein in the susceptible and resistant response of the host plants (Che et al., 2000).

The pathogenicity and elicitation of resistance by bacterial pathogens may be controlled by *hrp* (for **h**ypersensitive **r**esponse [HR] and **p**athogenicity) genes, the functions of which are characteristic features of the type III apparatus. The *hrp* genes have three important biochemical functions: gene regulation, protein secretion, and induction of HR. Mutation of type III protein secretion has been shown to abolish virulence of bacterial pathogens. As bacterial pathogens establish contact with host cells, type III secretion is activated which leads to the secretion of virulence proteins followed by leakage of plant nutrients in the extracellular space (apoplast) (Lindgren, 1997; He, 1998).

Erwinia amylovora, causing fire blight of apple, depends on a functional Hrp type III secretion system for its pathogenicity. *Erwinia amylovora* produces harpin, a protein with a pivotal role in its pathogenicity, and exports it in vitro via the type III secretion system. An antiserum raised against harpin was employed to detect its extracellular localization. Harpin was associated with the bacterial cells, but not with host plant cells (Perino et al., 1999). The Hrp system in *P. syringae* pv. *tomato (Pst)* secretes Hrp A, HrpZ, HrpW, and Avr Pto proteins and assembles a surface appendage called the Hrp pilus in *hrp* gene-inducing minimal medium. An antibody specific to HrpA efficiently labeled Hrp pili, whereas the antibodies against HrpW and HrpZ did not label the Hrp pili, indicating that HrpA is the major structural protein of Hrp pilus and that Hrp pili are assembled both in vitro and *in planta* (Hu et al., 2001). The formation of pili from *Pst* was observed by electron microscopy and immunocytochemistry. When labeled antibodies to HrpA were used, pili developed rapidly in a nonpolar manner, shortly after the detection of the *hrpA* transcript. Structures at the base of the pili were clearly differentiated from the basal bodies of flagella. The HrpZ protein was also detected on the pili by immunogold-labeling technique, contrary to the finding of Hu et al. (2001). The functional pili crossed the plant cell wall to generate tracks of immunogold labeling for HrpA and HrpZ (Brown et al., 2001).

Both pathogenicity factors and avirulence proteins such as AvrBs3 (inducing resistance in incompatible interaction) are secreted via the type III secretion system of *X. campestris* pv. *vesicatoria.* The Avr Bs3 protein contains a nuclear localization signal (NLS) and it acts inside plant cells. The Avr Bs3 protein secreted via the type III system may be recognized in the cell nucleus of pepper plants carrying corresponding *Bs3* resistance genes leading to initiation of HR response (Bonas et al., 2000).

The type IV secretory pathway operates in *A. tumefaciens* which causes crown gall diseases in several dicotyledonous plant species. Crown gall tumors are formed due to the transfer of a nucleoprotein complex into susceptible plant cells that is mediated by the virulence (*vir*) region-encoded transport system. The *vir* region, containing eight operons (*vir A* to *vir H*) is present in the Ti plasmid of *A. tumefaciens.* One half of the Ti plasmid contains *vir* genes (*vir A* to *vir H*) that control virulence and tumor formation and the other half has genes for replication, opine catabolism, and conjugation. The transport of the T-DNA of Ti plasmid is mediated by the Vir B membrane proteins required to form a pore in the bacterial membranes. By employing the IF and ISEM techniques, localization of the VirB8, VirB9, and VirB10 proteins primarily to the inner membrane, outer membrane, and periplasm, respectively, could be observed. These proteins may form the components of the pore through which the T-DNA may be transported (Kumar et al., 2000).

Agrobacterium tumefaciens, A. rhizogenes, and *Acinetobacter radiobacter* produce straight microfibrils, not only when in contact with wheat seedling roots, but also when in contact with one another. The majority of microfibrils are susceptible to attack by cellulase. Using colloidal gold-conjugated antibodies against 0-specific lipopolysaccharide, Vir proteins, and cellulase, the cell surface structure of *A. tumefaciens* was examined. The antibodies against Vir B2 proteins interacted with a tuft of thin microfibrils located on one pole of the bacterial cells whose *vir* genes were induced by acetosyringone (Chumakov et al., 2001). The Vir proteins Vir E2 and VirF are directly secreted into plants cells which are transformed into tumor cells. The Vir F protein seems to be the first prokaryotic protein with an F box by which it can interact with plant homologues of yeast Skp1 protein (Schrammeijer et al., 2001).

Symptom Expression

Many bacterial pathogens induce characteristic water-soaked symptoms in the infected leaves. The extracellular polysaccharides (EPS) produced by the pathogens cause such symptoms. Tissue degradation associated with soft rot disease is due to the action of cell wall-degrading enzymes. In many pathosystems, production of EPS and cell wall-degrading enzymes is correlated with the development of disease symptoms. The bacterial cells produce slimy polysaccharides that form a capsule protecting them against recognition by the host defense system. In the case of vascular wilts, the breakdown products of host origin and bacterial matabolites may clog the vessels resulting in partial or total obstruction of translocation of water and nutrients from the roots to the aerial plant parts.

Formation of chlorosis, necrosis, and leaf spots of various sizes and shapes is seen in plants infected by bacterial pathogens. The toxic metabolites produced by the pathogens are responsible for such symptoms. *Pseudomonas syringae* pv. *tabaci* causing tobacco wild fire disease produces a nonspecific toxin known as tabtoxin that consists of tabtoxinine β-lactam linked to either theronine or serine. The toxicity is due to the breakdown product tabtoxinine β-lactam and not due to tabtoxin itself. Chlorosis of leaf caused by tabtoxin could be reversed by L-glutamine, since tabtoxin inhibits glutamine synthetase (Sinden and Durbin, 1968). Tabtoxinine β-lactam inhibits the conversion of glutamic acid to glutamine by irreversibly inactivating glutamine synthetase (Thomas et al., 1983). The phaseolotoxin produced by *P. syringae* pv. *phaseolicola* also induces a green-yellow chlorotic zone at the site of infection in beans as ornithine accumulates. Phaseolotoxin inhibits ornithine transcarbamylase which synthesizes citrulline from ornithine, resulting in the accumulation of ornithine (Patil et al., 1970).

The phytotoxin coronatine (COR) produced by *P. syringae* pv. *atropurpurea* induces chlorosis and browning in Italian rye grass (Nishiyama et al., 1977). The COR produced by *P. syringae* pv. *tomato* DC 3000 causes necrotic lesions on tomato leaves (Bender, 2000). The COR-specific MAB 11B8 was employed to detect COR in various host plants infected by *P. syringae*, using a modified indirect competitive ELISA, the detection limit being 50 pg/well. Immunofluorescence (IF) microscopy revealed the association of the COR-specific MAB 8H3G2 with chloroplasts and proteinaceous structures in the vacuole of COR-treated tomato tissue. When COR-treated tissue was incubated with antiserum to proteinase inhibitor I followed by a secondary antibody conjugated to gold, the vascular proteins were densely labeled, indicating the localization of COR in specific sites in treated tomato tissues. Relatively high concentrations of COR (0.5 to 3.5 µg COR/g of fresh tissue) were detected in plants infected by *P. syringae* pv. *tomato* DC 3000 and *P. syringae* pv. *glycines* (Zhao et al., 2001).

Agrobacterium tumefaciens produces growth-regulating compounds such as indoleacetic acid (IAA) and cytokinins. The genes *tms-1* and *tms-2* of the pathogen encode the enzymes tryptophan monooxygenase and indoleacetamide hydrolase, respectively. These enzymes are involved in the conversion of tryptophan to IAA (Yamada et al., 1985). The tumors are formed due to the stimulation of host cells to undergo enlargement (hypertrophy) and excessive cell division (hyperplasia).

SUMMARY

Bacterial pathogens produce certain visual symptoms in infected plants. To identify these pathogens, pathogenicity tests and biochemical tests may be performed. However, these tests require long periods and lack the level of sensitivity and specificity needed to differentiate related strains and pathovars. Immunodiagnostic assays using polyclonal or monoclonal antibodies have provided required reliability, sensitivity, and specificity for the detection, identification, quantification, and differentiation of bacterial pathogens. Immunoassays, particularly ELISA, have been particularly useful in diagnosing latent infections in planting materials such as tubers and seeds. Bacterial pathogenesis and progress of disease development may be studied by using different serological techniques. The presence and production of various enzymes and toxins by bacterial pathogens *in planta* have been detected and visualized by ISEM and immunogold-labeling techniques. These immunological techniques have also provided critical information for understanding the role of enzymes and toxins in inducing different symptoms in susceptible host plants.

APPENDIX: DIAGNOSTIC TECHNIQUES
FOR THE DETECTION OF BACTERIAL PATHOGENS

Detection of Erwinia chrysanthemi *Using Fimbrial-Specific Antibody-Based ELISA (Singh et al., 2000)*

Production of Antibodies

1. Inject a New Zealand white rabbit intramuscularly with glutaraldehyde-fixed cells of *E. chrysanthemi* (EC) as immunogen, five times at two-week intervals and collect the blood by cardiac puncture and separate serum fraction.
2. For the production of monoclonal antibodies, immunize the mice by intraperitoneal injection of glutaraldehyde-fixed EC cells.
3. Fuse the splenocytes from immunized mice with FOX-NY cells using polyethylene glycol (50 percent) and dimethyl sulphoxide as fusogens.
4. Use Dulbeccos modified Eagles Medium supplemented with hypoxanthine, aminopterin, and thymidin to select suitable hybridomas for screening for specificity of reaction.

Screening Hybridomas for Specific MABs

1. Wash 24-hour-old EC cells from pure cultures; suspend in 0.01 M phosphate-buffered saline (PBS), pH 7.2, and adjust the concentration of cells to an OD of 1.0 at 620 nm.
2. Transfer 100 µl of bacterial cells (diluted to 1:1 in 0.1 M carbonate buffer, pH 9.6) to each well of the microtiter plate and incubate overnight at 4°C.

3. Block the nonspecific sites with skim milk (5 percent) for 20 minutes, wash, and add hybridoma fluid containing MABs.
4. Perform all incubation and washing steps as in standard ELISA protocol (one hour at 37°C)
5. Detect the reactive MABs using goat antimouse antibodies conjugated with alkaline phosphatase and *p*-nitrophenol phosphate at 0.5 mg/ml in 1 M diethanolamine buffer, pH 9.8, and incubate for 45 to 60 minutes.
6. Determine the color intensity at 405 nm using an ELISA reader.

Detection of EC in Plant Tissues

1. Prepare the plant extracts (1 percent) using sample buffer consisting of KH_2PO_4 (2.0 g), Na_2HPO_4 (11.5 g), disodium EDTA (0.14 g), thimerosal (0.02 g), and lysozyme (0.2 g) in water (1 l).
2. Coat the wells of the microtiter plates with polyclonal antiserum (1:5000 dilution of raw serum in 0.1 M carbonate buffer, pH 9.6).
3. Transfer the plant extracts to each well to capture the antigen by PABs present in the wells.
4. Follow the steps as per standard ELISA protocol.

Immunomagnetic Separation (IMS) for Acidovorax avenae ssp. citrulli (Walcott and Gitaitis, 2000)

Production of PABs

1. Treat the bacterial cells at 100°C for two hours; pellet the lysed cells by centrifugation at 20,000 g_n for 15 minutes at 4°C; collect the supernatant and adjust the pH 7.0.
2. Add ammonium sulfate until saturation to induce precipitation of bacterial cell surface antigen and incubate the suspension at 4°C overnight.
3. Pellet the precipitate by centrifuging at 20,000 g_n for 15 minutes at 4°C; resuspend the pellet in sterile distilled water and dialyze using 50,000 MW cut-off membrane, against sterile distilled water to remove ammonium sulfate and adjust the final antigen concentration to 300 µg of protein/ml.
4. Immunize the rabbits and collect the blood serum.
5. Purify the IgG fraction of anti-*Acidovorax avenae* ssp. *citrulli* (AAC) serum using an Avid Chrom Protein Antibody Purification Kit (Sigma-Aldrich, St. Louis) and adjust the final protein concentration to 2 mg/ml.

Indirect ELISA

1. Incubate AAC bacteria (5×10^6 CFU) with 10 µl of coupling buffer consisting of $NaHCO_3$ (2.93 g), $NaCO_3$ (1.59 g), and distilled water (800 ml) in individual wells in microtiter plates for 30 minutes.
2. Add blocking buffer (40 µl/well) consisting of 0.01 M PBS with 0.1 percent bovine serum albumin (PBS-BSA) to each well; agitate the plates gently and incubate at room temperature for 30 minutes.
3. Rinse the wells three times with wash buffer (PBS containing 0.2 percent Tween-20, PBST); add diluted anti-AAC (1:1000) at 20 µl / well; mix the reagents gently for 30 seconds and incubate for 30 minutes at room temperature.
4. Wash the wells three times as done earlier; add secondary antiserum-goat antirabbit IgG conjugated with alkaline phosphatase (20 µl/well) and incubate for 30 minutes at room temperature.
5. Rinse the wells three times with wash buffer as done earlier; remove the excess fluid from the wells; add the substrate *p*-nitrophenyl phosphate (20 µl/well) and incubate at room temperature until yellow color appears indicating positive reaction in positive control.
6. Stop the reaction by adding 1 N HCl (20 µl/well) and record the color intensity using an ELISA reader (Bio-Tek Instruments Inc., Winooski, Vermont).

Coating of Immunomagnetic Beads (IMBs)

1. Use super-paramagnetic beads, precoated with sheep antirabbit antibodies (Dynabead M280 sheep antirabbit Dynal, Oslo, Norway) and coat these beads with purified IgG fractions of anti-AAC (according to manufacturer's instructions).
2. Agitate the IMBs for two minutes; remove 500 µl of the suspension containing approximately 2×10^8 beads and place them into a 4 ml vial and wash the IMBs four times with PBS while the beads are held by a magnetic particle concentrator (MPC, Dynal, Oslo, Norway).
3. Suspend the IMBs in 3 ml of PBS; add purified IgG protein (80 µg) and incubate on a tilt-shaker for 24 hours at 4°C.
4. Rinse IMBs four times with 4 ml of PBS-BSA for each rinsing and resuspend the IMBs in 3 ml of PBS-BSA to have a final concentration of approximately 10^7 beads/ml.
5. To check proper coating of IMBs, incubate 100 µl of IMBs as processed previously, with 100 µl of a 1:1000 dilution of goat antirabbit antibodies conjugated with alkaline phosphatase (Sigma-Aldrich) for one hour and wash four times with PBS-BSA.

6. Add 100 μl of para-nitrophenyl phosphate (PNPP) as substrate; incubate for 20 minutes and remove the IMBs with magnetic particle concentrator (MPC).
7. Determine the optical density of the buffer at 405 nm using an ELISA plate reader; values twice that of a negative control (containing uncoated beads) indicate proper coating of IMBs.

Determination of Threshold of Bacterial Cell Recovery

1. Prepare serial dilutions of AAC ranging from 0 to 10^3 CFU/ml using 2 ml aliquots of PBS or PBS containing seed debris for each dilution; for preparing seed debris washing, shake 25 g of watermelon seeds in sterile PBS at 150 rpm on a rotary shaker for two hours; remove the seeds and sterilize the supernatant by autoclaving at 121°C and 15 psi of pressure for 15 minutes.
2. To prepare AAC suspensions, add known levels of the bacterial cells to glass vials containing 2 ml of PBS or sterile PBS with seed debris.
3. Estimate the number of CFU per treatment by spread-plating 100 μl of each suspension onto King's medium B (KMB).
4. Determine the recovery efficiency by incubating 2ml of each sample with 75μl (7.5×10^5 beads) of anti-AAC-coated IMBs and gently mix them for 15 minutes at room temperature.
5. Rinse the beads three times with PBS-BSA; resuspend them in 50 μl of PBS-BSA and spread-plate onto KMB.
6. Incubate the plates at 28°C for 48 hours and count the colonies of AAC.
7. Repeat the assessment four times and use two-way analysis of variance (ANOVA) for comparison.

Detection of Syringopeptins Produced by Pseudomonas syringae pv. lachrymans by ELISA (Fogliano et al., 1999)

Preparation of Immunogen

1. Mix purified syringopeptins (SP_{25A} and SP_{25B} 2:1 w/w) and conjugate with a carrier protein BSA or keyhole limpet hemocyanin (KLH).
2. Prepare the cross-linking agent diazobenzidine tetrachloride (DAB), by allowing reaction between $NaNO_2$ (3.5 mg) and benzidine (3 mg) dissolved in 0.2 M HCl (1.0 ml) for two hours at 4°C.
3. Add the cross-linking agent (0.1 ml) to 0.1 M borate buffer, pH 9.0 (2.0 ml) containing the carrier protein (KLH [6 mg] or BSA [3 mg]

and SP_{25A+B} (3 mg); place the mixture in an ice bath for the reaction to continue for 24 hours and stop the reaction by adding 0.2 N NaOH to bring the pH to 9.0.

4. Dialyze the reaction mixture (now yellow in color) against 1,000 volumes of 20 mM phosphate buffer, pH 7.2, for two days and lyophilize the contents.

5. Check the coupling between SP_{25A+B} and the carrier protein using SDS-PAGE technique.

6. Add the substrate (0.1 ml/well) tetramethyl benzidine in 0.1 M citrate buffer (0.1 mg/ml) and 30 percent H_2O_2 (5 ml) and incubate for 15 minutes at room temperature.

7. Stop the reaction with 1 N H_2SO_4 (0.1 ml) and determine the OD at 450 nm.

8. Test each lipodepsipeptide (LDP) in triplicate in three independent sets of assessments.

For detection of LDPs in diseased plant tissue, aliquots of lyophilized aqueous extract (1 mg/ml) from infected tissue (zucchini cotyledons) are used for coating the wells in the competitive ELISA procedure.

Production of PABs

1. Immunize New Zealand white rabbits by injecting KLH-SP_{25A+B} emulsified with Freund's complete adjuvant (1:4, v/v) and provide two booster injections at four weeks after the first injection.

2. After a 20-day rest period, collect the blood by ear puncture; separate blood serum and store at $-20°C$.

Indirect ELISA

1. Add pure lipodepsipeptides (LDPs) from bacterial culture (0.1 ml) to each well of the microtiter plates; incubate overnight at 4°C and rinse the wells three times with 0.1 M PBS, pH 7.2.

2. Block the nonspecific sites with horse serum (10 percent in PBS) for one hour and rinse the wells three times with PBS as done earlier.

3. Dispense the antiserum containing PABs (0.1 ml/well) diluted in PBS (1:50 to 1:5000); incubate for one hour at 37°C and wash the wells with PBS three times as done earlier.

4. Add peroxidase-conjugated antirabbit IgG secondary antibody (diluted in PBS to 1:4000, v/v); incubate for one hour at room temperature and wash three times with PBS as done earlier.

Detection of **Erwinia chrysanthemi** *by Immunogold Labeling and Electron Microscopy (Singh et al., 2000)*

1. Suspend bacterial cells (24-hold) in PBS and adjust the concentration to 1.0 at OD_{620}; add equal volume of hybridoma culture fluid; mix well and incubate overnight at 4°C.
2. Wash the cells three times in PBS; suspend the cells in PBS (200 µl); add gold beads (15 µl) conjugated with goat antimouse IgG and IgM (Bio/Can Scientific, Mississagua, Ontario, Canada) and incubate for one hour at room temperature with agitation.
3. Wash the cells three times with PBS and resuspend them in PBS (200 µl).
4. Transfer the bacterial cell suspension to coated grids, air dry and stain with 1 percent phosphotuingstic acid (PTA).
5. Examine under electron microscope.

Chaper 10

Plant-Viral Pathogen Interactions

Plant viruses possess certain unique characteristics that distinguish them from other microbial pathogens such as fungi, bacteria, and phytoplasmas. Plant viruses are extremely small in size and hence they are capable of passing through the filters that retain fungi and bacteria. Viruses are essentially intracellular parasites and are not found outside susceptible plant cells. They do not have any metabolic activity and are not known to produce any extracellular enzymes or toxins as do fungal and bacterial pathogens. Plant viruses have simple structures made of a protein coat that encloses the viral genome-ribonucleic acid (RNA) or deoxyribonucleic acid (DNA). As such, they resemble plant cellular constituents rather than cells themselves. The pathogen mass accumulating in susceptible host plants may account for a very small proportion of plant biomass.

Because plant viruses show very little variation in morphological characteristics, the need for development of methods sensitive enough for their detection, identification, quantification, and differentiation has increased. Most of the immunodiagnostic techniques originally developed for detection of plant viruses (see Chapter 4) were later modified for the detection of fungal, bacterial, and phytoplasmal pathogens. Various serological techniques were used for the identification of 270 viruses infecting a wide range of plant species (Van Regenmortel, 1982). The list of viruses identified using serological techniques has steadily increased. In addition, quantitative assays of plant viruses, and establishment of the degree of relationship between viruses and diagnosis of virus infection in seeds, vegetatively propagated materials (such as cuttings, tubers, bulbs, and budwoods) have been effectively achieved by employing serological techniques (Slack and Shepherd, 1975; Simmonds and Cumming, 1979; Dolores-Talens et al., 1989). Identification of different isolates of barley yellow dwarf virus (BYDV) in field samples of maize was possible by

ELISA and maize was found to be a reservoir of BYDV in Spain (Comas et al., 1993). The usefulness of these techniques for quarantine, certification programs, and epidemiological studies has been clearly demonstrated by several studies (Chu et al., 1989; Van Regenmortel and Dubs, 1993; Fox, 1998; Narayanasamy, 2001).

IMMUNODIAGNOSTIC TECHNIQUES FOR PLANT VIRUSES

Plant viruses are both immunogenic and antigenic because of the presence of the coat protein (CP). The immunoidentification techniques based on agglutination and precipitation (see Chapter 4) were found to be very useful under field conditions and even in remote areas researchers could perform these simple tests because the antisera are portable. The antigen reacts with the antibody resulting in the precipitation of the antigen-antibody complex in the solution. In agglutination tests, the particulate antigen is aggregated to form clumps because of the reaction between antigen and antibody. To enhance the sensitivity of the test, carriers such as bentonite, latex particles, or bacterial cells may be sensitized using appropriate antigen/antibody (Khan and Slack, 1978). The virobacterial agglutination (VBA) test (Walkey et al., 1992) and high-density latex flocculation (HDLF) test (Kawano and Takahashi, 1997) have been shown to be more sensitive for detection of viruses. Complement fixation procedure, more frequently used in human medicine, has been employed in a restricted manner. The advantage of complement fixation assays over precipitation tests is the greater accuracy in measuring small differences in antigen and antibody concentration, as demonstrated in the case of southern bean mosaic virus (SBMV). The extent of cross-reactivity between two strains of SBMV was found to be strictly proportional to antiserum titer (Tremaine and Wright, 1967).

Immunodiffusion tests are based on the diffusion of antigen and antibody through agar medium and formation of precipitin lines at positions where both reactants meet at optimal concentrations. The double diffusion test performed in petri dishes provides a distinct advantage of comparison of many samples under a set of identical conditions. Detection of a virus in different samples and establishing the relationship of a new (or unknown) virus with already-identified viruses are the important uses of the diffusion tests, especially for spherical viruses. The rate of diffusion of rod-shaped viruses in agar medium, however, is hampered. Hence the use of virions degraded by pyrrolidone and pyridine becomes necessary (Shepard, 1970). The mixtures of antigens may be separated by gel electrophoresis and the individual antigens may be tested to avoid cross-reactions.

Various labeled-antibody techniques have been developed by attaching labels or markers such as enzymes, fluorescent dyes, or radioactive materials to either antigens or antibodies. Currently, these methods are being used since their sensitivity, reliability, and rapidity have been improved with the use of automation to handle large number of samples. Among the labeled-antibody techniques, ELISA and its different formats have been extensively employed for the detection, identification, and differentiation of plant viruses (Clark and Bar-Joseph, 1984). Other methods used are dot immunobinding assay (DIBA), tissue blot immunoassay (TBIA), enzyme-linked fluorescent assay (ELFA), fluorescent antibody techniques, cytofluorimetric methods, and in situ immunoassay (ISIA). The immuno capillary zone electrophoresis (I-CZE) technique provides the specificity of a serological assay combined with sensitivity, rapidity, and automation (Eun and Wong, 1999). This technique is useful for mass-indexing and certification programs (see Chapter 4, Immunocapillary Zone Electrophoresis Technique). A combination of immunoassay and polymerase chain reaction (PCR) has been shown to enhance sensitivity and specificity compared to either of these techniques (see Chapter 4).

The availability of an antiserum containing either polyclonal (PAB) or monoclonal (MAB) antibodies is the first requirement for performing suitable immunodiagnostic techniques. The protocols for the preparation and purification of antibodies have been described in Chapter 3. PABs and MABs specific against many plant viruses have been prepared. The sensitivity and specificity of the immunodiagnostic tests can be enhanced by the use of MABs. Detection of as many as 75 viruses belonging to 23 genera has been achieved using MABs (Narayanasamy, 2001).

The efficacy of purified antibodies raised in chicken egg yolk (IgY) was compared with PABs and MABs from mammalians for the detection of potato viruses X and Y in two ELISA formats. Monospecific IgY was also produced by cross-absorption with a specific virus. The IgY was shown to be equally sensitive as IgG from mammalians in detecting PVX and PVY in infected tobacco leaves. The ease of producing and purifying larger quantities of IgY (by 40-fold) compared to IgG from rabbit antiserum is a clear advantage (Weilbach and Sander, 2000).

Grapevine leaf roll is a complex disease with which eight closteroviruses (GLRaV-1 to GLRaV-8) are associated (Monis, 2000). These viruses could be detected only late in the growing season when virus titers reach their peak, even if MABs are employed. A novel 37 kDa protein was associated with infection by GLRa V-8. The MAB raised against this protein (p37) reacted specifically with p37 associated with GLRaV-8, but not with p38 associated with GLRa V-1 indicating that specific MABs may be employed to detect and identify the components of virus complexes (Monis, 2000). If MABs are used, DIBA seems to give more satisfactory results than tests with PABs, but the sensitivity of ELISA is greater, irrespective of the nature of antibodies (PABs or MABs) (Sherwood et al., 1987). This suggests the need for selecting a suitable detection method depending on the availability of PABs/MABs.

Purification of plant viruses is a long and cumbersome process and it is often difficult to eliminate proteins of host plant origin that may interfere with serological reactions. In the absence of purified viruses or viral proteins, a novel procedure is followed for the production of antisera. The CP gene or nonstructural proteins such as the movement protein (MP) gene is cloned and expressed in the bacterium *E. coli*. The polypeptides produced in the bacterial cells are used as immunogens to generate antibodies in rabbits (Vaira et al., 1996; Helguera et al., 1997; Petrzik et al., 2001). Other proteins encoded by viral genomes have also been used as immunogens (Li et al., 1996; Arbatova et al., 1998). Antibodies raised in this way were found to be more specific and versatile in their reactions with target proteins or viruses as in tomato spotted wilt virus (Vaira et al., 1996), garlic virus A (Helguera et al., 1997), beet necrotic yellow vein furovirus (Li et al., 1996), potato virus Y (Arbatova et al., 1998), grapevine leaf roll-associated closterovirus -3 (Ling et al., 2000), grapevine virus B (Saldarelli and Minafra, 2000), and *Prunus* necrotic ringspot virus (Petrzik et al., 2001). Both the purified recombinant nuclear inclusion b (NIb) protein of potato virus Y and CP expressed in *E. coli* were used to prepare specific PABs that were employed to detect viral proteins in infected potato plants (Liu et al., 1999). The nonspecific reactions resulting from using complex viruses purified from infected host plant materials can be avoided by this procedure (see Appendix section on preparation of antiserum).

A significant improvement in the production of antibodies specific for plant viruses was effected by the development of phage-displayed recombinant antibody technology (McCafferty et al., 1990). The DNA from antibody-producing β-lymphocytes or hybridomas is genetically linked to the gene 3 DNA of the M13 phage. When the MI3 phage carrying the gene fusion infects *E. coli,* the proteins encoded by the antibody DNA and gene 3 DNA are coexpressed in the bacterial cells. Any antibody DNA linked to the phage DNA and any antibody proteins fused to the phage proteins will be assembled and secreted in the same manner as the phage. The phage-displayed recombinant antibody technology has been used for producing antibodies specific to several plant viruses such as potato leaf roll virus (Harper et al., 1997, 1999), potato virus Y (Boonham and Barker, 1998), African cassava mosaic virus (Ziegler et al., 1998), potato virus A (Merits et al., 1998), cucumber mosaic virus Fny strain (He, Liu, et al., 1998; Gough et al., 1999), beet necrotic yellow vein virus (Uhde et al., 2000), tobacco mosaic virus (Fischer et al., 1999), tomato spotted wilt virus (Griep et al., 2000), and citrus tristeza virus (Terrada et al., 2000).

The use of the phage-displayed recombinant technique offers distinct advantages:

1. These antibodies can be produced in large quantities at a lower cost.
2. They may be used in different ELISA formats such as PABs and MABs.
3. Strains of virus may be differentiated by selecting suitable single chain Fv antibody fragment (Boonham and Barker, 1998).
4. The phage-displayed peptides, which specifically bind to CPs of plant viruses, are useful for rapid identification and differentiation (Ziegler et al., 1998).
5. Detection of viruses such as beet necrotic yellow vein virus in stored sugar beets is possible (Uhde et al., 2000).
6. The scFv fragments may be expressed in plants such as tobacco for use in biological and medical applications (Fischer et al., 1999).

A genetic immunization system was developed by engineering scFv fragments capable of recognizing epitopes on plant virus capsids. An epichromosomal expression vector derived from potato virus X (PVX) was employed to determine the functional expression of an scFv fragment capable of recognizing an epitope of the glycoprotein G1 conserved among large numbers of tospoviruses. The gene encoding the secretory scFv was used to transform *Nicotiana benthamiana* (Franconi et al., 1999). In another study, antibodies could be expressed in plants to confer novel traits such as virus resistance. Successful intracellular expression of antibody fragments depends primarily on their amino acid sequence. A transient expression system based on PVX was used to compare different cDNA constructs for expression and stability of antibody-variable gene fragments in plants (Ziegler et al., 2000). The engineered scFvs may be useful as inexpensive tools for virus disease diagnosis and also for development of a "plantibody"-mediated resistance to plant viruses such as tospoviruses.

Another novel approach of producing antibodies against plant viruses was developed by Pal et al. (2000). Antibodies specific to the 22 kDa coat protein (CP) of barley yellow dwarf virus (BYDV-PAV) were generated by cloning a cDNA sequence encoding the CP into a mammalian expression vector (pc DNA 22 k) entrapped in liposomes followed by intramuscular injection into BALB/C mice. The antibody titers of the mouse serum were significantly increased by giving booster injection of DNA-immunized mice with purified BYDV-PAV strain. By adopting this procedure, it may be possible to produce antibodies against luteoviruses which generally reach low concentrations in host plants and are difficult to purify.

The development of monoclonal antibodies has facilitated the identification of viruses and their strains/serogroups more precisely. The MABs raised against Chinese wheat mosaic virus (CWMV) were used to identify and differentiate the wheat and oat furoviruses, CWMV, soilborne wheat mosaic virus-Oklahoma isolate (SBWMV-Okl), oat golden stripe virus (OGSV), and European wheat mosaic virus (EWMV) (Ye, Xu, et al., 2000). Grapevine can be infected by two or more viruses simultaneously. This causes difficulty in the identification of causative virus(es). By using specific MABs in ELISA, the grapevine virus D (GVD) was detected efficiently in cortical shavings from mature grapevine canes (Boscia et al., 2001).

MABs prepared by using a mixture of isolates of cucumber mosaic virus (CMV) belonging to subgroups I and II, differentiated 12 well-characterized strains of CMV and the presence of epitopes that were virus and subgroup-specific was demonstrated (Hsu et al., 2000). Two serotypes of sweet potato feathery mottle virus (SPFMV) occurring in Uganda were differentiated by a specific MAB raised against the CP of SPFMV. These two serotypes were present in different areas and they exhibited differences in ability to infect some cultivars (Karyeija et al., 2000). MABs specific for grapevine leaf roll virus-1 (GLRaV-1) were characterized by their recognition of virus CP by DAS-ELISA and Western blotting. Two MABs (IC4 and IB7) that detected 25 and 32 isolates of GLRaV-1 were used for the diagnosis of this virus, whereas one MAB (IG10) reacted specifically with both GLRaV-1 and GLRaV-3 coat proteins (Seddas et al., 2000). Strain-specific MABs reacting with sweet cherry isolate of plumpox virus (PPV-SC) in ELISA, but not with any of additional 44 PPV isolates, were developed by Myrta et al. (2000). Likewise, the development of specific MABs to differentiate tomato mosaic virus (ToMV) and tobacco mosaic virus when they occur together frequently under field conditions was essential. By employing the specific MAB (10.H1) in plate-trapped antigen (PTA)-ELISA, ToMV could be detected in infected tomato plants and there was no cross-reaction with TMV. Furthermore, this MAB recognized only the band corresponding to the coat protein of ToMV (17.5 kDa), indicating the specificity of the immunoassays in detecting and identifying the desired virus in mixed infections (Duarte et al., 2002).

An epidemiological model for the spread of serotypes of BYDV that are transmitted by different species of aphids was developed. By using specific MABs, the occurrence and spread of PAV serotype (transmitted by *Rhopalosiphum padi avenae*) and MAV serotype (transmitted by *Macrosiphum avenae*) in the field, was shown to be directly related to their respective vector species population (Quillec et al., 2000). The transmission of cucumber mosaic virus by the aphid *Myzus persicae* could be entirely prevented by mixing the purified virus preparation with the homologous MAB. Both pre- and post-acquisition feeding treatment of aphids with MABs significantly decreased CMV transmission (Gera et al., 2000). The comparison of reactivity of MABs raised against isolates of *Prunus* necrotic ring-spot virus (PNRV) infecting roses indicated that the most common

PNRV serotype in rose was different from the most prevalent serotype in *Prunus* spp. This difference was reflected in the infectivity of these serotypes. All rose isolates (27) infected *Prunus persica* seedlings, whereas three of four *Prunus* isolates were poorly infectious on *Rosa indica* (Moury et al., 2001).

The vectors of plant viruses such as insects, fungi, and nematodes have an important role in the incidence and spread of viral diseases. The presence of viruses must be detected in their vectors to restrict the spread of diseases and to assess the extent of viral infection in seeds, planting materials, and whole plants. Different immunodiagnostic techniques have detected the presence of several viruses in vectors such as tomato spotted wilt virus (Cho et al., 1988; Chamberlain et al., 1993; Medeiros et al., 2000), sugarbeet western yellows, barley yellow dwarf, and alfalfa mosaic viruses (Kastirr, 1990, Ahoonmanesh et al., 1990), chickpea chlorotic dwarf virus (Horn et al., 1994), beet mild yellowing virus (Stevens et al., 1995, 1997), pineapple wilt-associated closterovirus (Hu et al., 1996), faba bean necrotic yellows virus (Franz et al., 1998), tobacco ringspot virus (Wang and Gergerich, 1998), and tobacco rattle virus (Karanstasi et al., 2000).

Enzyme-Linked Immunosorbent Assay (ELISA)

The development of the ELISA technique by Clark and Adams (1977) represents an important advancement in the detection, identification, quantification, and differentiation of plant viruses and their strains. Among the different formats of ELISA, double antibody sandwich DAS-ELISA and indirect-ELISA have been more frequently used. Alkaline phosphatase (ALP) is quite stable and can be conveniently coupled with the antibody (IgG) using glutaraldehyde to prepare the conjugate for detecting the reaction between the antigen and antibody. Use of maleimide for conjugation was a more sensitive detection of bamboo mosaic potexvirus (BaMV) than when glutaraldehyde was used (Chen and Lu, 2000). Horseradish peroxidase (HRP) has also been employed as a marker in place of ALP. β-galactosidase (Neustroeva, et al., 1989) and penicillinase (Sudharsana and Reddy, 1989; Abraham and Albrechtsen, 2001) were used to label the antibodies against plant viruses. Penicillinase (penicillin as substrate) was found to be as effective as ALP and its enzyme and substrate are easily available and less expensive.

ELISA has been used to study different aspects of several viruses such as serological relationships between viruses, identification of different epitopes on virus coat protein (CP), epidemiology and host range of viruses, assessment of resistance/susceptibility of cultivars to viruses, and virus-vector relationships. Indexing of plant materials by ELISA is useful and indispensable. Viruses remaining latent in plant materials may spread to other countries as the world market has been considerably widened under the General Agreement on Tariffs and Trade (GATT).

Among the several viruses detected by ELISA, the following cause some of the more economically destructive crop diseases: banana bunchy top virus (Su and Wu, 1989), groundnut (peanut) mottle and clump viruses (Hobbs et al., 1987), cucumber mosaic virus (Lin et al., 1991), beet necrotic yellow vein virus (Sukhacheva et al., 1996; Uhde et al., 2000), potato virus X (Darda, 1998), apple chlorotic leafspot and apple stem grooving viruses (Wu et al., 1998), citrus tristeza virus (Bar-Joseph et al., 1979; Terrada et al., 2000), potato leaf roll virus (Harper et al., 1999), cacao swollen shoot virus (Hoffmann et al., 1999), tomato spotted wilt virus (Chamberlain et al., 1993; Griep et al., 2000); cardamom mosaic virus (Jacob and Usha, 2001), oat mosaic virus (Clover et al., 2002), *Prunus* necrotic ringspot virus (Petrzik et al., 2001; Bertozzi et al., 2002), garlic viruses (Dovas et al., 2002), and yam mosaic virus (Njukeng et al., 2002).

Several modifications of ELISA have been made to simplify the procedure, make it less expensive, and reduce the volume of antisera and labor required. The fluorogenic ultramicro-enzyme immunoassay (DAS-UM-ELISA) developed for the detection of citrus tristeza virus was as sensitive as (DAS)-ELISA. Only 10 µl of reactants are needed and the test can be completed in five hours (Peralta et al., 1997). The petri dish-agar dot immunoenzymatic assay (PADIA) developed for the detection of plant viruses is performed in a polystyrene petri dish instead of microtiter plates with wells. Circular areas outlined with a hydrophobic cryomarker pen are marked on the inner surface of the petri dish. The antiserum and ALP conjugate are added into the marked areas and incubated and washed as in standard ELISA protocol. The substrate solution mixed with warm agar is added to cover the inner surface of the petri dish. A positive reaction appears as distinct dark blue to purple dots on the agar matrix at the sites where the virus has been trapped. PADIA requires five to ten times less reagents and is as sensitive as ELISA in addition to being less expensive (four times) (Abraham and Albrechtsen, 2000). In a similar test, named petri dish-ELISA, a wax pen was used to divide the inner surface of a polystyrene petri dish into many squares or circles. Drops (50 µl) of the reagents were placed in the marked areas and the steps as in ELISA were followed. The absorbance values at 405 nm were similar or higher than the ELISA values. This test was found to be suitable for the detection of tomato mosaic virus, cowpea mosaic virus, and blackeye cowpea mosaic virus in the seeds (Abdalla and Albrechtsen, 2001).

Tissue Blot Immunoassay (TBIA)

In this technique, a tissue imprint is made on the nitrocellulose membrane by applying slight pressure on the freshly cut plant tissue surface to transfer viral antigens in the infected plant tissue. This method does not require any sample preparation and reaction with host plant protein can be avoided. Furthermore, this technique is simple and sensitive and can yield results in three hours. The imprinted membranes can be stored for long periods and transported (Cambra et al., 1994; Hsu et al., 1995). The pineapple mealybug wilt-associated virus was detected in infected plants by TBIA technique (Sether et al., 1998). By using an MAB specific to pineapple closterovirus (PCV), 2,000 samples were tested by one person in five days during a field survey, indicating its rapidity, reliability, and amenability for large-scale testing (Hu et al., 1997). Garlic virus A (GarV-A) was detected in the cloves of garlic by TBIA technique and it has been recommended for use for the certification of virus-free garlic (Helguera et al., 1997) (see Appendix secion on detection of garlic virus A). The presence of plumpox virus in different tissues of apricot was detected by the TBIA method (Dicenta et al., 2000). Barley yellow mosaic virus type 2 (BaYMV-2) was also detected in infected winter barley (Kuntze et al., 2000). A modified method named tissue print-ELISA was developed for the detection of citrus tristeza virus (CTV), using genetically engineered single-chain antibody proteins (Terrada et al., 2000).

Comparative efficacy of TBIA and ELISA techniques in detecting some plant viruses has been assessed. Tuberose mild mosaic virus (TMMV), which is widespread in Taiwan, was detected more efficiently by ELISA than by TBIA. On the other hand, for indexing tuberose bulbs stored at 5°C, TBIA was more sensitive in all (100 percent) infected bulbs, as compared to the detection of the virus by ELISA in 33 percent of the bulbs (Chen and Chang, 1999). The movement and distribution of tomato spotted wilt virus in *Capsicum* spp. were monitored by direct tissue blotting (Soler et al., 1999). Likewise, the movement of potato leaf roll virus (PLRV) from the tubers to the leaves was tracked by TBIA and ELISA methods. TBIA may be best used for detecting PLRV in foliage of plants grown from infected tubers (secondary PLRV infection), but it is less accurate for detecting primary PLRV infection (Whitworth et al., 2000). By using PABs against bacterially expressed (BE) CP of faba bean necrotic yellows virus (FBNYV), the virus was more efficiently detected by TBIA than by DAS-ELISA test (Kumari et al., 2001). The TBIA technique was used to detect beet western yellows virus (BWYV) in canola crops and in wild radish *(Raphanus raphinistrum)* which may serve as a reservoir of the viruses infecting canola (Coutts and Jones, 2000). Sugarcane yellow leaf virus (ScYLV) was detected by both TBIA and RT-PCR methods and there was excellent correlation between results obtained from these two techniques (Chatenet et al., 2001).

Dot Immunobinding Assay (DBIA)

Dot immunobinding assay (DIBA) was developed as an alternative to ELISA by replacing microtiter plates with nitrocellulose or nylon-based membranes. The nonspecific protein-binding sites present on the surface of the membranes are blocked with BSA or nonfat dry milk powder which is inexpensive and easily available. DIBA technique was found to be more sensitive than ELISA in detecting tomato spotted wilt virus and purified potyviruses (Berger et al., 1985). Several viruses have been detected by employing DIBA technique: barley stripe mosaic virus and bean common mosaic virus (Lange and Heide, 1986), tomato spot wilt virus (Vaira et al., 1993), maize dwarf potyvirus strains (Leonardon et al., 1993), cherry mottle leaf trichovirus (CMLV) (James and Mukerji, 1996), lily symptomless virus (LSV), tulip breaking virus-lily (TBV-L), and cucumber mosaic virus (CMV) (Niimi et al., 1999) (Figure 10.1). The grapevine rupestris stem-pitting-associated virus (GRSPaV) was detected by using the

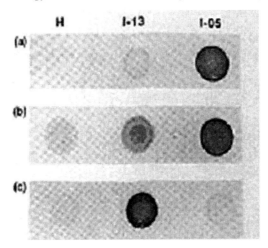

FIGURE 10.1. Detection of tospovirus isolates by dot-immunobinding assay (DIBA): reactions of tospovirus isolates—I-05 (tomato spotted wilt virus [TSWV]) and I-13 (impatiens necrotic spot virus [INSV])—using antiserum to TSWV (a), IgGs of BR-01 (b), and NL-07 (c) kits; (H) healthy sample (*Source:* Vaira et al., 1993; Reprinted with permission by Blackwell Scientific Publications, Oxford, United Kingdom.)

antiserum against bacterially expressed CP. GRSPaV was detected in leaf petioles and cortical scrapings from dormant canes during the whole vegetative season. DIBA was recommended for a wide survey of GRSPaV, since ELISA was ineffective for the same (Minafra et al., 2000). The possibility of easy storage and transport of membranes, less expensive membranes, and the requirement of extremely small volumes of plant extracts are the distinct advantages of DIBA over ELISA.

Immunosorbent Electron Microscopy (ISEM) and Gold Labeling Techniques

Electron microscopy has been a valuable tool for studying virus particle morphology and to spot check the presence of viruses in plant extracts and tissues using the leaf-dip technique (Figure 10.2). By employing antibodies specific for plant viruses, it is possible not only to visualize viruses but also to establish the relationship between viruses by immunosorbent electron microscopy (ISEM). The use of

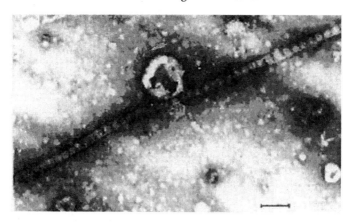

FIGURE 10.2. Detection of sunflower necrosis virus (SFNV) by leaf-dip technique (*Source:* Ramiah et al., 2001; Reprinted with permission by the Advanced Centre for Plant Virology.)

ISEM has distinct advantages over ELISA such as greater sensitivity, rapidity, absence of cross-reactivity with proteins of plant origin, and the possibility of using crude antiserum without fractionation (Figures 10.3 and 10.4) (Milne, 1993). Some of the plant viruses detected by ISEM are potato leaf roll and moptop viruses (Roberts and Harrison, 1979), and sugarcane bacilliform virus (Viswanathan and Premachandran, 1998), banana streak badnavirus (Thottappilly et al., 1998; Ndowora and Lockhart, 2000), garlic virus-A (Helguera et al., 1997), shallot yellow stripe and Welsh onion yellow stripe (WoYSV) viruses (van der Vlugt et al., 1999), pineapple mealybug wilt-associated virus (Sether et al., 1998), cymbidium mosaic potexvirus (Vejaratpimol et al., 1999), barley yellow dwarf virus (Du et al., 2000), and wheat yellow mosaic virus and barley yellow mosaic virus (Ye, Zheng, et al., 2000) (see Chapter 4, Immunoelectron Microscopy, Immunosorbent Electron Microscopy).

ISEM is considered to be the most sensitive technique for the detection of cymbidium mosaic virus (CyMV) isolates. When compared to plant indexing and electron microscopy methods that detected CyMV in 51.43 percent and 42.86 percent of orchid plants, respectively, ISEM could detect CyMV in 71.43 percent of orchid plants suspected to be infected (Vejaratpimol et al., 1999). Detection of CyMV by ISEM was shown to be more sensitive (two times) than

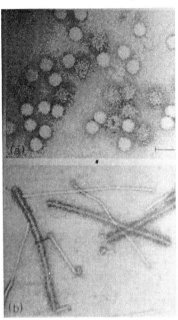

FIGURE 10.3. Immunosorbent electron microscopy (ISEM) for identification of plant viruses: (a) mixture of tomato bushy stunt virus and carnation mottle virus treated with antiserum specific to carnation mottle virus; (b) mixture of two potyviruses present in naturally infected white bryony plants treated with antiserum to one of them; note the coating of antibodies to one of them to which they are homologous (*Source:* Van Regenmortel, 1982. Reprinted from *Serology and Immunochemistry of Plant Viruses* with permission from Elsevier.)

detection by ELISA in crude sap of infected orchid leaves (Hsu et al., 1992). The serological relationship between the potyviruses infecting onion, leek, and shallot could be established by a combination of ISEM and decoration techniques (van der Vlugt et al., 1999).

Decoration of plant viruses with homologous antibody improved the visualization of viruses (Milne, 1992) as in larger viruses such as tospoviruses. Gold labels (5 to 20 nm diameter) may be conjugated either directly to the primary or coating antibody or more commonly to the protein A (PA) or a secondary goat antirabbit antibody (IgG) (see Chapter 4, Immunoelectron microscopy, Immunocytochemistry). The presence of viruses in infected tissues can be observed easily by the gold-labeling technique as in the case of alfalfa mosaic virus

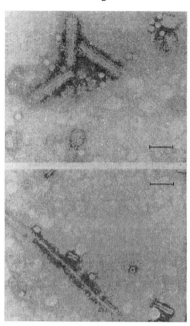

FIGURE 10.4. Identification of plant viruses by immunosorbent electron micros-copy (ISEM): (Top) mixture of raspberry ringspot virus and tobacco rattle virus incubated antiserum specific to tobacco rattle virus; (Bottom) mixture of rasp-berry ringspot virus, tobacco rattle virus, and celery mosaic virus treated with antiserum specific to celery mosaic virus; note the coating of antibodies on viruses to which they are homologous (*Source:* Van Regenmortel, 1982. Re-printed from *Serology and Immunonchemistry of Plant Viruses* with permission from Elsevier.)

(Pesic et al., 1988), pear vein yellows virus (Giunchedi and Pollini, 1992), *Festuca* leaf streak virus (Lunsgaard, 1992), tobacco vein mottling virus (Ammar et al., 1994), cucumber mosaic virus (Yang et al., 1997), potato virus Y (Arbatova et al., 1998), beet yellows closterovrius (BYV-CA) (He et al., 1998), potyviruses (Riedel et al., 1998), rice ragged stunt oryzavirus (Lu et al., 1999), tomato spotted wilt virus (TSWV) (Vaira et al., 1996; Heinze et al., 2000), and garlic virus-A (Gar V-A) (Lunello et al., 2000).

Localization of plant viruses in the tissues of their vectors has also been observed by ISEM technique. TSWV was detected in the midgut epithelium and hemocoel of larvae of *Frankliniella occiden-*

talis fed infected plants (Ullman et al., 1993; Wijkamp et al., 1993). Later studies by Nagata et al. (1999) showed that the midgut, foregut, and salivary glands of *F. occidentalis* were the only organs in which TSWV accumulated. Potato leaf roll virus (PLRV) was localized in the intestinal epithelium of aphids *M. persicae* (Garret et al., 1993). Rice ragged stunt virus could be detected by gold labeling in the tissues of the vector *Nilaparvata lugens* (Lu et al., 1999). The fungal vector, *Polymyxa graminis,* transmitting barley mild mosaic virus, showed the presence of labeled bundles of viruslike particles in the zoospores released from infected roots (Jianping et al., 1991). In another fungus, *Polymyxa betae,* vector of beet necrotic yellow vein virus (BNYVV), labeled virus particles were observed inside zoosporangia and zoospores (Peng et al., 1998). Unequivocal evidence of tobacco rattle virus particles attached to the cuticle lining of the posterior tract of the pharyngeal lumen of the nematode vector *Paratrichodorus anemones* was obtained by Karanastasi et al. (2000). Tobacco ringspot virus (TRSV) and tomato ringspot virus (TomRSV) are transmitted by the nematode *Xiphinema americanum.* Immunofluorescent labeling and electron microscopy revealed that TRSV and TomRSV were localized in different regions of the food canal of *X. americanum.* TRSV was primarily localized to the lining of the lumen of the stylet extension and the anterior esophagus, but only rarely in the triradiate lumen. On the other hand, TomRSV was localized only in the triradiate lumen (Wang et al., 2002).

Cytofluorimetric Method

The cytofluorimetric method developed by Iannelli et al. (1997) is based on the translation of biological properties into measurable fluorescence intensity. Plant virus particles in plant extracts are adsorbed onto latex particles and then fluorescence intensity is assessed by flow cytometry. Cucumber mosaic virus present in plant samples is incubated with latex particles, washed and further incubated successively with rabbit anti-CMV antibodies and antirabbit IgG labeled with fluorescein. The virus particles are detected by the laser of the cytometer, the detection limit being 10 pg/ml compared to 2.5 ng/ml detected by ELISA. By using latex particles with different sizes and differential fluorescent dyes, two or more viruses can be detected in plant extracts simultaneously. CMV, PVY, and ToMV were detected by sensitizing latex particles (3, 6, and 10 μm diameter) separately with virus-specific antibodies (Iannelli et al., 1997).

In Situ Immunoassay (ISIA)

Citrus tristeza virus (CTV) can be detected by the in situ immuno-assay (ISIA), which is simple, specific, and rapid-requires only a light microscope. Sections of stems, petioles, or leaf veins of healthy and CTV-infected citrus plants are fixed with ethanol (70 percent) and incubated with specific PAB or MAB. Enzyme-conjugated species-specific secondary antibodies are used for labeling the bound primary antibodies. The sections are then exposed to a substrate mixture consisting of nitroblue tertrazolium and 5-bromo-4-chloro-indolyl phosphate. Development of a purple color indicates the presence of CTV antigens. The ISIA test was found to be as sensitive as direct tissue blot immunoassay (DTBIA) and can be used for routine CTV diagnosis (Lin, Rundell, et al., 2000) (see Appendix section on detection of CTV). In a further study for the detection of mild and severe isolates of CTV in grapefruit trees which were earlier immunized with mild isolates, ISIA was found to be more sensitive than ELISA, using two MABs that could differentially react with CTV isolates (Lin et al., 2002).

Immunocapture-Reverse Transcription-Polymerase Chain Reaction (IC-RT-PCR)

The IC-RT-PCR technique combines the desirable features of both serological and nucleic acid-based diagnoses and provides greater sensitivity, rapidity, and reliability. This technique is reported to be 100 to 1,000 times more sensitive than ELISA, depending on the virus-host combination. The IC-RT-PCR assay was at least 100 times more sensitive than DAS-ELISA in detecting cacao swollen shoot virus (CSSV) isolate 1A (Hoffmann et al., 1997), whereas it was about 1,000 times more sensitive for the detection of lettuce mosaic virus in leaf extracts (van der Vlugt et al., 1997). This assay may be performed either in PCR tubes or microplate wells. The amenability of the IC-RT-PCR assay for large-scale application under field conditions for the detection of plumpox potyvirus (PPV) was demonstrated by Varveri and Boutsika (1998). For precise identification of new viruses and rapid differentiation of virus strains, IC-RT-PCR has been advantageously used in detection of black currant reversion-associated virus (BRAV) (Lemmetty et al., 1998). The PVY isolate PVY [NTN] could be differentiated from the PVY[N] isolate (Tomassoli et al., 1998). Grapevine leaf roll-associated viruses (GLRaV-1 and GLRaV-3) were rapidly detected and precisely identified by IC-RT-PCR assay (Acheche et al., 1999; Sefc et al., 2000). Detection of viruses in seeds is essential to prevent disease incidence and movement of viruses to other areas/countries. The IC-RT-PCR assay was demonstrated to be a very sensitive method for the detection of peanut stripe virus (PStV) and peanut mottle virus (PeMoV) in peanut seeds (Gillaspie et al., 2000) (see Appendix section on detection of peanut viruses).

VIRUS DISEASE DEVELOPMENT

Virus Replication and Synthesis of Virus-Encoded Proteins

Viruses do not have any detectable metabolic activity and physiological function. However, they direct the synthetic machinery of susceptible host plants to produce all materials required for their replication (multiplication). Viruses differ entirely from other microbial pathogens in processes of infection and replication. They cannot enter into plants through natural openings. Only a few viruses, such as TMV and PVX may enter through wounds created by rubbing of leaves against one another or through injuries caused by knives or agricultural operations. All other viruses depend on insects, mites, nematodes, and fungi which introduce the viruses into the susceptible host plant cells. Within minutes, the process of viral replication is initiated as viral nucleic acid is released from the protein coat. Further steps vary depending on the viruses and no distinct stages in pathogenesis can be recognized as in the case of fungal and bacterial pathogens. The sites of viral nucleic acid and CP synthesis and their assembly may also differ. In a recent study by Más et al. (2000), immunocytochemical techniques revealed a clear association of the CP of cherry leaf roll virus with virus-induced cytopathological structures, whereas viral RNA preferentially accumulated at the microsomal fraction of the infected tobacco leaves. However, the essential requirements for viral replication, independent synthesis of viral nucleic acids and coat proteins, and assembly of these viral components to form intact virus particles appear to be quite similar for all plant viruses. The information required for the synthesis of viral components is carried out by the viral nucleic acid. Studies on the molecular biology of viruses have laid the foundation for the clear understanding of the similarities between primitive viruses and highly evolved human beings in the mechanism of information transfer from genomic DNA through RNA (transcription) and from RNA through amino acid synthesis (translation).

Among plant viruses, the molecular genetics of TMV has been more thoroughly understood. TMV has one single-stranded, positive-sense RNA molecule which encodes four major proteins. The CP with a molecular mass of 17.5 kDa encloses the viral RNA, whereas the movement protein (MP) (30 kDa) is involved in the regulation of cell to cell movement of the virus. Two proteins (126 and 183 kDa) are required for the synthesis of RNA-dependent RNA polymerase which is needed for replication of the viral nucleic acid. The nature of the CP determines the immunological properties of the viruses and the variations in the epitopes differentiate strains and related viruses as in the case of four furoviruses analyzed by Western blotting technique (Ye, Xu, et al., 2000) (Figure 10.5). The presence of CP of banana bunchy top virus (BBTV) could be reliably detected by Western blot analysis (Manickam et al., 2001) (Figure 10.6).

The gene encoding the CP is involved in several functions: induction of hypersensitive response (HR) in TMV (Saito et al., 1989), determinant of virulence of PVX isolates (Santa Cruz and Baulcombe, 1993; Spence, 1997) and cucumber mosaic virus (Ryu et al., 1998), and symptom expression by CMV (Sugiyama et al., 2000). Movement proteins (MPs), in addition to regulation of virus movement, may exert an influence on sugar metabolism and resource allocation at sites far away from their sites of expression of MP gene of CMV. Evidence exists for the interaction between viral MPs and phloem sap proteins (PSPs) of host plants as in the case of CMV infecting squash (*Cucurbita pepo* ssp. *pepo*) (Shalitin and Wolf, 2000). The putative MPs (ORF3 products) of grapevine virusA (GVA) and grapevine virus B (GVB) were localized in the cell wall and plasmodesmata of infected cells and especially for GVA in association with cytoplasmic aggregates of virus (Saldarelli et al., 2000).

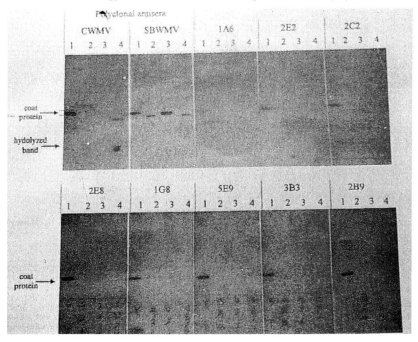

FIGURE 10.5. Western blotting of four furoviruses with two polyclonal antisera and eight monoclonal antibodies raised against Chinese wheat mosaic virus (CWMV): (1) CWMV—Chinese wheat mosaic virus; (2) EWMV—European wheat mosaic virus; (3) SBWMV—soil borne wheat mosaic virus; (4) OGSV— oat golden stripe virus (*Source:* Ye, Xu, et al., 2000; Reprinted with permission by Blackwell Wissenchafts-Verlag, Berlin, Germany.)

Other proteins are encoded by viral genomes. The viral genome linked protein (VPg) of pea seed-borne mosaic virus (PSbMV) seems to have a role in the virulence of the virus (Keller et al., 1998). The helper component (HC) protein of potyviruses has a pivotal role in the transmission by aphid vectors. Single amino acid changes in the HC protein have affected symptoms in the herbaceous hosts of plumpox potyvirus (Sáena et al., 2001). The bacterially expressed PS9 protein (38 kDa), which is one of the major proteins present in the CP of rice ragged stunt virus (RRSV), was used to generate specific antibodies. The immunogold-labeling technique revealed that the PS9 polypeptide was localized in the spikes present on the CP of RRSV (Lu et al., 1999). The 5'-terminal open reading frame (ORF)

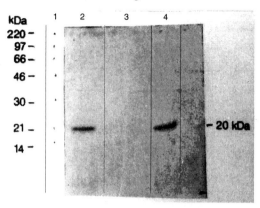

FIGURE 10.6. Detection of coat protein of banana bunchy top virus (BBTV) by Western blot technique: (Lane 1) rainbow markers (standard); (Lanes 2 and 4) BBTV-infected banana samples; (Lane 3) healthy banana sample (*Source:* Reprinted with permission from Manickam et al., 2001, American Chemical Society.)

1a of beet yellows virus (BYV) encodes a 295 kDa polyprotein with domains of papainlike cystein proteinase, methyl transferase (MT), and helicase (HEL), whereas ORF 1b encodes an RNA-dependent RNA polymerase. MABs recognizing MT and HEL detected MT-like and HEL-like proteins in BYV-infected *Tetragonia expansa* plants in immunoblots, indicating that these proteins, like other related viral enzymes, are associated with membrane compartments in host plant cells (Erokhina et al., 2000).

Characterization of virus-coded proteins and their distribution in infected plant tissues have been studied in several host-virus patho-systems using immunological techniques. By employing PABs against p24 and CP of sugarbeet yellows closterovirus (BYV) colocalization of p24 and CP in vascular petiole tissues was observed. Drastic reductions in transmission of BYV by aphids was demonstrated in infectivity neutralization tests conducted with viruliferous aphids fed on preimmune serum or antiserum to p21, another virus-coded protein (He, Harper, et al., 1998). Localization of the first triple gene block protein (TGBp1) of peanut clump virus (PCV) in infected *N. ben-thamiana* plants was studied using the immunogold-labeling technique. The TGBp1-specific labeling was most commonly associated with the plasmodesmatal collar region in the mesophyll cells. In transgenic *N. benthamiana* no TGBp1-specific immunogold labeling of plasmodesmata could be seen. The viral infection appeared to target and/or anchor the transgene TGBp1 to the plasmodemata (Erhardt et al., 1999). Four citrus tristeza virus (CTV)-coded proteins— P1, P2, P3, and P4—were detected in *Citrus excelsa* and Mexican lime *(C. aurantifolia).* The presence of these proteins in CTV-infected plants differed depending on the isolates of CTV causing infection (Lin, Xie, et al., 2000).

Although plant viruses cause systemic infection in most host plant species, pathogens are not present in all tissues of infected plants nor do they reach equal concentration in all infected tissues/organs. The virus distribution patterns in infected plants vary and influence the virus acquisition by their vectors and spread to other plants/locations. The accumulation and distribution of BYMV and CMV in *Gladiolus* plants were assessed using ELISA. BYMV was evenly distributed in different leaves, but younger leaves tended to accumulate higher concentrations. On the other hand, CMV was present in higher concentrations in lower older leaves. Hence the lowest old leaves or mixtures of young and old leaves were tested for indexing *Gladiolus* plants for BYMV and CMV infection (Chen et al., 1999). Likewise, fluctuations in the concentration of *Prunus* necrotic ringspot virus (PNRSV) in six cultivars of peach *(P. persica)* and also in almond during different seasons significantly affected virus detection by ELISA. The plants have to be tested during flowering and sprouting or flowering and dormancy to arrive at a reliable conclusion (Zotto et al., 1999; Bertozzi et al., 2002).

Viral infection of seeds generally decreases as the age of plants at time of infection increases (Narayanasamy and Jaganathan, 1975). However, TMV and ToMV are able to infect the seeds of tomato and bell pepper *(Capsicum annum),* irrespective of growth stage at the time of infection. The concentration of these viruses in seeds, as determined by ELISA, was greater when young plants were infected (Chitra et al., 1999). Two isolates of pathotypes II (AF198) and IV (AF199) of lettuce mosaic virus (LMV) differed significantly in their ability to pass through seeds of susceptible and tolerant cultivars of lettuce, as determined by PTA-ELISA. The AF 199 isolate was transmitted more easily in the susceptible cultivar (16.5 percent) as compared to AF 198 (1.33 percent). The tolerant cultivar allowed the transmission of AF 199, but not AF 198 through seeds, indicating host genotypic influence on virus seed transmission. The results indicate the importance of epidemiological consequences of distribution of virus pathotypes and the need for the precise identification of pathotypes/strains of the virus occurring in a location/country (Jadãq et al., 2002).

Localization of plant viruses has differentiated the types of tissues that support virus replication/accumulation. Sugarcane yellow leaf virus was detected by TBIA in all leaves, but the highest percentage of infected vascular bundles was found in the top young leaves (Chatenet et al., 2001). Using the immuno tissue printing procedure, localization of plumpox virus (PPV) in stem and petiole tissues of four susceptible and resistant apricot cultivars was determined. In stem tissue, PPV was localized in the pith and xylem tissues, while the virus was observed in the epidermis, cortical, and medullae parenchyma of the petiole tissues (Dicenta et al., 2000). Using ISIA, visualization of CTV antigen in stems, petioles, or leaf veins of different citrus tissues was possible (Lin, Rundell, et al., 2000). Direct TBIA technique was employed for the detection of citrus psorosis virus (CPsV) in the ovaries of infected flowers (D'Onghia et al., 2001).

The soil-borne wheat mosaic virus (SBWMV) was transmitted artificially by the fungal vector *P. graminis* through the roots. The path of transport of SBWMV from roots to leaves of inoculated plants was tracked by immunogold-labeling technique. SBWMV entered primary xylem elements before cell death occurred and then moved upward in the plant after the maturity of xylem forming hollow vessels. The virus might move laterally between xylem vessels (Verchot et al., 2001). The antiserum produced against purified pigeonpea sterility mosaic virus (PPSMV) preparation detected a virus-specific 32 kDa protein in the sap of sterility mosaic disease (SMD)-affected plants by ELISA and Western blotting. Using this antiserum, immunolabeling technique revealed specific labeling of membrane bound bodies (MBBs) and associated amorphous-electron dense material (EDM), but not fibrous inclusions (FIs) composed of randomly dispersed fibrils with electron lucent areas, present in the cytoplasm of palisade cells of pigeonpea plants (Kumar et al., 2002). Subcellular localization of brome mosaic virus (BMV) replicase-related 1a and 2a proteins and the 3a MP in infected barley leaves was studied by immunogold electron microscopy. The 1a and 2a proteins colocalized at electron dense cytoplasmic inclusions and the 3a MP protein was also associated with these inclusions which were oval or amorphous with electron lucent regions (Dohi et al., 2001).

Some plant viruses induce the formation of intracellular inclusions in the cells of certain host plant species and these inclusions are of diagnostic value. Potyviruses produce inclusions both in the cytoplasm and nuclei (intranuclear inclusion). The antiserum produced against the nonstructural proteins (NSP) of PPV specifically reacted with cytoplasmic and/or nuclear inclusions induced by 17 different potyviruses. By using immunogold labeling technique, the protein profiles of inclusions could be determined. Most of the cytoplasmic or nuclear inclusions were composed of two or more NSPs, while the cylindrical inclusions consisted of CI protein only. The crystalline nuclear inclusions induced by tobacco etch virus contained helper component proteinase (HC. Pro) in addition to proteins NIa and NIb. All cytoplasmic or nuclear inclusions of potyviruses may be considered as deposition sites of excessively produced viral NSPs (Riedel et al., 1998). MABs raised against cylindrical inclusion proteins (CIPs) of turnip mosaic virus (TuMV) and zucchini yellow mosaic virus (ZYMV) were employed to investigate the antigenic nature of CIPs. All MABs (23) against ZYMV-CIP reacted only with ZYMV-CIP, whereas some MABs (4) specific to TuMV reacted with CIPs of both viruses and many MABs (14) reacted with TuMV-CIP only. The results suggest that cross-reactive and major virus-specific epitopes were located at the N-terminal half of the respective CIPs (Kundu et al., 2000). The sites of synthesis/accumulation of NIa protein of PVY were visualized by in situ immunofluorescence method. The anti-NIa MAB accumulated both in the nucleus and cytoplasm. This MAB could detect the presence of different forms of NIa protein including precursors and cleavage products (Rouis et al., 2001). The time course and subcellular localization of MDMV and its HC. Pro in infected maize leaves were monitored. The concentrations of MDMV and its HC. Pro were higher in the upper leaves just above the inoculated leaves. The immunogold-labeling technique showed the presence of labels on pinwheel inclusions, laminated inclusion, and virus particles in the cytoplasm, suggesting that HC. Pro was present in these locations (Li et al., 2001).

Tomato golden mosaic virus (TGMV) replicates its genomic ss-DNA through ds-DNA intermediates in the nuclei of differentiated plant cells by exploiting host DNA replication machinery. In a study to determine the distribution of viral and plant DNA in the nuclei of infected leaves, a combination of indirect immunofluorescence (IF) and fluorescence in situ hybridization (FISH) techniques and antibodies against viral coat protein were employed. The TGMV virions were found associated with the viral DNA compartments. However, the CP-antibodies failed to cross-react with some large viral DNA inclusions. This may be due to encapsidation occurring after significant viral DNA accumulation. TGMV infection altered nuclear architecture and could induce plant chromatin coordination characteristic of cells arrested in early mitosis (Bass et al., 2000).

Plant viruses do not have any physiological functions. Yet they induce drastic changes in various physiological processes such as respiration, protein and nucleic acid metabolism, in addition to hormonal disturbances that affect the growth and reproduction of susceptible host plants. Virus replication requires the synthesis of virus-specific abnormal proteins. In addition, varying amounts of abnormal proteins serologically related to virus protein (TMV) are also present in infected plants. The activities of several enzymes of host plant origin such chlorophyllase, polyphenol oxidase, peroxidase, and IAA-oxidase are stimulated following virus infection. Although the changes in the activities of these enzymes do not appear to have any influence on virus replication, these changes have a role in the expression of symptoms such as mosaic, necrotic lesions, and growth abnormalities.

A novel group of proteins may be induced in virus-infected plants. This is more abundant in plants reacting with necrotic lesions than in plants infected systemically. The production of these proteins depends on the type of symptoms produced rather than on the genetic constitution of plants. These proteins, termed pathogenesis-related (PR) proteins, were first observed in tobacco producing local necrotic lesions on inoculation with TMV (Van Loon and Van Kammen, 1970). The term PR protein was coined by Antoniw et al. (1980). Bacteria and fungi are also able to induce PR proteins in various plant species, especially in incompatible combinations leading to hypersensitive necrosis. Eleven families of PR proteins (PR 1 to PR 11) have been recognized and the genes encoding them have been identified (Van Loon, 1999). The role of PR proteins in the development of resistance to microbial pathogens is discussed in Chapter 11.

Symptom Expression

The external symptoms observed in virus-infected plants appear to reflect the extent of derangement of host plant metabolism. The enhanced activities of chlorophyllase have a bearing on the mosaic or chlorosis exhibited by infected leaves (Narayanasamy and Ramakrishnan, 1965; Narayanasamy and Palaniswami, 1973). The effect of barley stripe mosaic virus (BSMV) on chlorophyll biosynthesis was assessed. The mosaic or chlorotic symptoms can be due to the inhibition of chlorophyll biosynthesis and also due to decomposition of functioning of photosynthetic apparatus formed before infection (Almási et al., 2000). The molecular basis of expression of mottling was elucidated in the cowpea-cowpea chlorotic mottle virus (CCMV) pathosystem. The type strain of CCMV-T induced bright chlorosis, whereas the attenuated variant of CCMV-T caused mild green mottle symptoms. This variation in symptoms was due to the amino acid at position 151 in the coat protein of CCMV, but not due to the nucleotide sequence of the viral genome. The amino acid residue 151 may exert subtle influences on the viral coat protein-cowpea interaction resulting in chlorotic and mild green mottle symptoms (de Assis Filho et al., 2002). The necrosis of tissues leading to the confinement of viruses within lesions (local lesions) is considered to be due to stimulated activities of oxidases, particularly polyphenol oxidase (PPO). In addition, phenylalanine-ammonia lyase (PAL) and peroxidase (PO) also show higher activities that are indicated as responses of the host plants exhibiting incompatible interaction. Dark green islands (DGIs) are observed in mosaic virus-infected leaves. The DGIs are clusters of green leaf cells that are free of virus, but surrounded by yellow virus -infected tissues. They are resistant to infection by the same virus causing mosaic symptoms. A mechanism termed post-transcriptional gene silencing seems to operate in the DGIs (Moore et al., 2001). The different growth abnormalities such as stunting, excessive tillering, stimulation of axillary buds leading to sterility, and conversion of floral parts into green leafy structures are due to disturbed hormonal metabolism in virus-infected plants (Bos, 1970).

SUMMARY

Plant viruses with simple constitutions are considered exciting biological models providing the information on the nature of genes and their functions. Most of the immunodiagnostic techniques originally developed for the detection of viruses were later modified for the detection of other microbial pathogens. Immunodiagnostic techniques are more rapid, sensitive, specific, and reliable than the conventional methods of biological indexing. Currently, they are widely used for checking seeds and planting materials. Plant viruses, being different from fungal and bacterial pathogens, require direct introduction into susceptible cells either through injured cells or by their vectors. Although viruses need protein and nucleic acid for their replication, they induce drastic changes in physiological processes which result in adverse effects on the growth and reproduction of susceptible host plants.

APPENDIX: TECHNIQUES FOR DETECTION OF VIRAL PATHOGENS

Preparation of Antiserum Against Bacterially Expressed Virus Proteins (Helguera et al., 1997)

Viral Protein Expression and Purification from Escherichia Coli *Cells*

1. Introduce the plasmid pRSET/C6CP (plasmid ligated to a fragment of CP gene of garlic virus-A) into *E. coli* cells by a standard transformation method.
2. Add IPTG (0.3 mM) to a bacterial culture grown to OD_{600} 0.5 to induce expression of the C6CP polypeptide and incubate for two hours at 37°C.
3. Suspend the bacterial cells in STE buffer consisting of NaCl (100 mM), Tris-HCl (50 mM), pH8, and EDTA (1.0 mM); disrupt the bacterial cells by sonication and centrifuge at 9,000 g_n for ten minutes.
4. Resuspend the pellet containing the viral protein in STE buffer (2 ml) supplemented with $MgCl_2$ (8 mM), DNase I (10 µg/ml) and allow the mixture to stand for 15 minutes at 4°C.

5. Centrifuge the mixture at 9,000 g_n for ten minutes and resuspend the pellet in STE buffer (1 ml).
6. Analyze the expression product (viral protein) in Western immunoblots using the antibodies specific for the virus concerned.

Generation of Antibodies Against Bacterially
Expressed Virus Proteins

1. Separate the viral protein preparation by following sodium dodecylsulfate polyacrylamide gel electrophoresis (SDS-PAGE) techniques; excise the C6CP protein band from the gel, crush and lyophilize.
2. Inject the (female California) rabbits with 0.6 mg of the protein emulsified in 1 ml of Freund's complete adjuvant (Sigma Chemical Co., St. Louis, Missouri) intradermally; provide equal amount of viral protein in Freund's incomplete adjuvant after an interval of two weeks.
3. Collect the blood twice at 15 and 21 days after the booster injection and use the antiserum at appropriate dilution (1:2000).

Detection of Garlic Virus-A (Gar V-A) by Tissue Printing
Immunoassay (Helguera et al., 1997)

1. Cut the infected cloves transversely and longitudinally using a scalpel blade and press the cut end on BA85 nitrocellulose membranes (Schleicher and Schnell, Dassel, Germany) to have the print image of the tissue.
2. Block the nonspecific sites in the membrane with blocking solution containing Tris-HCl (50 mM), pH8, NaCl (150 mM), Tween-20 (1.0 percent), and nonfat dried milk (5 percent) for one hour at room temperature and wash three times in Tris-HCl (50 mM), pH8, NaCl (150 mM), and Tween-20 (0.05 percent) (TBS/Tween).
3. Incubate the membranes in the TBS/Tween buffer supplemented with nonfat dried milk (5 percent), extract from healthy garlic plants and the antibody specific for coat protein as primary antibody for one hour at room temperature and wash three times as done earlier.
4. Incubate with the alkaline phosphatase-linked goat antirabbit antibody (Biorad, Hercules, CA) for one hour at room temperature and wash as done earlier.
5. Immerse the membrane in solution containing nitroblue tetrazolium (14 mg) and 5-bromo-4-chloro-3-indolyl phosphate (7 mg) in 40 ml substrate buffer consisting of 0.1 M Tris, 0.1 M NaCl, and 5 mM $MgCl_2$, pH9.5
6. Observe the development of purple color indicating positive reaction.

Detection of Citrus Tristeza Virus (CTV) by in Situ Immunoassay (ISIA) (Lin, Rundell, et al., 2000)

1. Cut sections (100 to 200 μm thick) of fresh stems, petioles, or veins from healthy and CTV-infected citrus plants using new razor blade and maintain 4 to 6 replicates for each sample.
2. Transfer the sections to 24-well plastic plates (Corning Glass Works, Corning, New York); fix the sections with 70 percent ethanol (300 to 700 μl) for 5 to 20 minutes at room temperature and remove the ethanol by pipetting out.
3. Incubate the sections with specific PABs or MABs raised against CTV (1 μg/ml) in PBST containing NaCl (0.15 M), sodium phosphate (0.15 M), pH7.0, Tween-20 (0.05 percent), and fetal bovine serum or BSA (3.0 percent) for 30 to 60 minutes at 37°C and wash the sections with PBST-PVP (PBS + polyvinyl pyrrolidone) (2.0 percent) for five to ten minutes.
4. Incubate the sections with alkaline phosphatase (ALP) conjugated goat antimouse Ig (for sections exposed to MABs) or ALP conjugated goat antirabbit IgG (for sections exposed to PABs) and wash with PBST-PVP for five to ten minutes followed by washing with TTBS buffer consisting of Tris (hydroxy-methyl) amino methane (20mM), NaCl (500mM), and Tween-20 (0.05 percent), pH 7.5, for five to ten minutes.
5. Incubate the sections with 300 to 500 μl of freshly prepared NBT-BCIP substrate mixture consisting of 66 μl of nitroblue tetrazolium (0.3 mg/ml) and 33 μl of 5-bromo-4-chloro-3-indolyl phosphate (0.15 mg/ml) in 10 ml of sodium carbonate buffer (0.1 M), pH 9.8, for 5 to 15 minutes.
6. Stop the reaction by removing the substrate solution from the wells and add water (500 to 1000 μl) to each well.
7. Transfer the sections to glass slides separately and observe under a light microscope (×100 magnification) for the development of purple color in the phloem tissue indicating a positive reaction.

Detection of Peanut (Groundnut) Viruses by Immuno Capture-Reverse Transcription-Polymerase Chain Reaction (IC-RT- PCR) Assay (Gillaspie et al., 2000)

Extraction of Peanut Viruses

1. Grind cotyledon tissues (0.5 g) in RNA extraction Qiagen lysis buffer containing guanidine isothiocynate, mixed with sarkosyl and heat at 70°C.

2. Process the macerate through a Qiagen mini-prep spin column (Qiagen Inc., Chatsworth, California) to get preparations containing total RNA.

Preparation of Seed Samples

1. Remove a small slice from each cotyledon distal to the radicle using a sharp razor blade; prepare 100-seed samples by selecting seeds at random from each seed lot and combine seed slices from each 100-seed sample.
2. Triturate the slices in extraction buffer (20 ml) consisting of NaCl (0.137 M), KH_2PO_4 (1 mM), NA_2HPO_4 (8 mM), KCl (3 mM), Tween-20 (0.05 percent), NaN_3 (3 mM), and water 1000 ml (PBST) and add NA_2SO_3 (0.01 M), polyvinyl pyrrolidone MW 40,000 (2 percent), NaN_3 (3 mM), powdered milk (2 percent), and Tween-20 (2 percent), pH 7.4, and centrifuge the slurry at 14,000 g_n for five minutes.

IC-RT-PCR Assay

1. Soak the PCR tubes (200 µl) in 0.1 M HCl for 15 minutes; then in NaOH (4 M) for 15 minutes; rinse in PBST; soak in ethanol (95 percent) for 15 minutes and air-dry.
2. Prepare the antiserum dilution coating buffer containing Na_2CO_3 (15 mM), $NaHCO_3$ (35 mM), NaN_3 (3 mM), pH 9.6; coat the tubes with 50 µl of a 1:500 dilution of whole antiserum specific against the virus to be detected; incubate for three hours at 37°C or overnight at 4°C; the effective dilution of antiserum has to be determined by a titration experiment using a DAS-ELISA protocol.
3. Wash the tubes three times with PBST (200 µl); add seed slice extract supernatant or a leaf extract (50 µl) and incubate for two to three hours at room temperature or overnight at 4°C.
4. Wash three times with PBST as done earlier; blot dry on a tissue; place the tubes at −70°C for at least ten minutes and thaw the contents at 94°C for two minutes to disassemble the antibody-captured virus and free the viral RNA from coat protein.
5. Add to each tube a total of 50 µl of RT-PCR mix consisting of 4 µl of 5x first-strand reverse transcription buffer (Life Technologies, Frederick, Maryland), 2 µl of 0.1 M DTT, 2.6 µl of 10 mM dNTP, 0.1 µl of RNasin RNase inhibitor (Promega Corp., Madison, Wisconsin), 0.2 µl of SuperScript RNase H⁻ Reverse Transcriptase (Life Technologies), 3 µl of 10X PCR buffer (Promega Corp.), 4.2 µl of 25 mM $MgCl_2$, 1 µl of reverse primers (primers at 100 pM/µl), 0.5 µl of forward primer,

0.2 µl of *Taq* polymerase (Promega Corp.,) and 32.75 µl of nuclease-free water.

6. Treat the tubes at 37°C for one hour, 94°C for two minutes (35 cycles of 94°C for 30 seconds, 50°C for 30 seconds, and 72°C for 60 seconds) and 72°C for ten minutes.

7. Subject the RT-PCR products to electrophoresis in 1.5 percent agarose gels in TBE buffer consisting of Trisborate (89 mM) and EDTA (2 mM), pH 8.3 and stain with ethidium bromide.

PART IV:
APPROACHES FOR PLANT HEALTH MANAGEMENT

Chapter 11

Improvement of Host Plant Resistance Through Genetic Manipulation

Plants are exposed to numerous microorganisms that have varying pathogenic potential. However, most plant species are susceptible only to a limited number of pathogens and exhibit resistance/immunity to many pathogens that may infect some other plant species. The compatible interactions between microbial pathogens and plant species may lead to development of disease (indicating the susceptibility of the plant species), whereas incompatible interactions result in slow or failure of disease development (depending on the degree of resistance of the plant species). By using immunological techniques, it has been possible to detect the presence of and quantify microbial pathogens in infected plants and to determine the changes induced in the structure and functions of diseased plants by fungal pathogens (Chapter 8), bacterial pathogens (Chapter 9), and viral pathogens (Chapter 10).

Enhancement of resistance in susceptible crop cultivars has been acknowledged as the preferred approach for plant health management. Concerted attempts have been made to identify the genes that confer resistance to one or more diseases infecting a particular crop. Several crop diseases have been effectively managed through wise and strategic use of genes for resistance, in spite of the ability of pathogens to "defeat" these genes. This situation indicates the need for constant monitoring of new races/strains of pathogens capable of infecting cultivars that were previously resistant to pathogens (Keen, 1999). The wheat-stem rust pathosystem provides an excellent example of extraordinary efforts taken to contain this devastating disease of wheat. The responses of resistant plants may be discrete or distinct from the moment at which contact between the host and pathogen is established. Immunological techniques have been used to assess the levels of resistance of test entries and visualize the changes that are induced by microbial pathogens in resistant plants.

BREEDING FOR RESISTANCE TO CROP DISEASES

Assessment of Levels of Resistance to Fungal Diseases

Immunological techniques have been useful in detecting and quantifying fungal antigens in cultivars/genotypes following inoculation with fungal pathogens to assess levels of resistance. A sensitive and accurate immunoassay was developed for quantitative measurement of *Alternaria cassiae* infection of *Cassia obtusifolia*. The radioimmunosorbent assay (RISA) employed for quantification of fungal mycelium. RISA could detect as little as 1.6ng/ml dry weight equivalent of fungal mycelium by using the antiserum raised against the homogenate of *A. cassiae* (Sharon et al., 1993) (see Appendix section on quantification of *A. cassia* infection). Roots of soybean lines susceptible and resistant to *Aphanomyces euteiches* (root rot disease) were exposed to 100 zoospores/ml and the development of the pathogen was assessed by using the specific antiserum in ELISA. The buildup of the pathogen was significantly slower in resistant lines. There was a positive correlation between the rate of lesion development on the roots and pathogen population as measured by ELISA, suggesting that the resistance to root rot disease may be associated with reduction in oospore production, pathogen multiplication, and zoospore germination (Kraft and Boge, 1994, 1996) (see Appendix section on quantification of *A. euteiches*). Different ELISA formats have been useful in quantifying fungal pathogen populations as in the case of *C. fulvum* in tomato leaves (Kaprovich-Tate et al., 1998), *M. grisea* in rice leaves (Hasegawa et al., 1999), *S. tritici* in wheat leaves (Jenny et al., 1999), *B. cinerea* in pear stems (Meyer et al., 2000), and *Peronospora parasitica* in tobacco (Carzaniga et al., 2001).

Assessment of resistance by using serodiagnostic tests is particularly advantageous in the case of perennial crops which require a very long incubation period for the expression of symptoms of infection. For example, the symptoms of eastern filbert blight disease of hazelnut *(Corylus avellana)* are visible at 13 to 22 months after inoculation with the pathogen *Anisogramma anomala*. In contrast, by using indirect ELISA, infected plants could be identified at 3 to 5 months after inoculation when no recognizable symptom was seen. By employing the ELISA technique in a breeding program, a hazelnut progeny with a gene conferring a high level of resistance could be selected (Coyne et al., 1996). Likewise, different grades of resistance (susceptible, resistant, and highly resistant), were differentiated in narrow-leaved lupins *(Lupinus angustifolius)* against *Diaporthe toxica,* causing latent stem infection (Shanker et al., 1998). To identify sugarcane varieties exhibiting resistance to red rot disease *(C. falcatum),* ELISA technique was employed to detect the pathogen in different tissues of inoculated sugarcane cultivars whose reactions to the disease had been already determined by conventional method. The ELISA technique was shown to provide reliable results much earlier than the conventional method (Viswanathan et al., 2000).

Molecular Basis of Host Resistance to Fungal Diseases

Plants defend themselves against pathogen invasion through different resistance mechanisms. Nonhost plant pathogens show general resistance as reflected by the development of the hypersensitive response (HR). This kind of general resistance may be due to the passive or preexisting defense mechanisms involving structural barriers such as waxy cuticle or strategically positioned pools of antimicrobial compounds that may prevent the development of the pathogen(s). On the other hand, the specific resistance of cultivars of a plant species may be due to the presence of resistance (R) genes. The expression of R genes leads to the synthesis of new compounds that can interact with the pathogen directly or inactivate toxic metabolites. The specific resistance may be induced by the products of avirulence genes of the pathogen or nonpathogens, chemicals, or wounding. These active defense mechanisms depend on significant changes in the host plant metabolism. Microbial pathogens, fungi, bacteria, and viruses may induce active defense responses in their hosts when the structural barriers have been breached.

Plant cells possess the ability to sense pathogen invasion by recognizing either the endogenous signal molecules formed following degradation of plant cell wall components or exogenous molecules of pathogen origin. These signal molecules (compounds), known as elicitors, can be proteins, glycoproteins, oligosaccharides, or lipids. The elicitors may be either race specific or non-race specific and they can induce a complete set of defense responses. A novel protein elicitor (PaNie$_{234}$) from *Pythium aphanidermatum* contains a putative eukaryotic secretion signal with a proteinase cleavage site. The heterologously expressed elicitor protein, without the secretions signal of 21 amino acids (PaNie$_{213}$), induced programmed cell death (PCD) and de novo formation of 4, hydroxybenzoic acid in cultured carrot cells. When PaNie$_{213}$ was infiltrated into the intercellular spaces of leaves of *Arabidopsis thaliana,* necrosis and deposition of callose on the cell walls of spongy parenchyma cells surrounding the necrotic mesophyll cells were observed. Likewise, tobacco and tomato leaves also reacted with necroses. Maize and oat did not show any such reaction following infiltration of PaNie$_{213}$, indicating that monocotyledons were unable to perceive the signal (Veit et al., 2001). Race-specific elicitors are frequently products of avirulence *(avr)* genes encoded by pathogens and they are specifically recognized by the products of plant resistance (R) genes. This interaction between gene products represents the molecular basis of race-/cultivar-specific host resistance (gene-for-gene interaction). In contrast, non-race specific elicitors may activate defense mechanisms not dependent on the resistance genes of the host plants (Benhamou, 1996; Baker et al., 1997). The R gene products can be grouped into five classes based on their structural characteristics: (I) intracellular protein kinases; (II) transmembrane receptor-like proteins with extracellular leucine-rich repeats (LRR) and cytoplasmic protein kinase domain; (III) intracellular receptor-like proteins with LRR domains and nucleotide-binding sites (NBS); (IV) intracellular receptor-like proteins with LRR domains, NBS and region of homology to *Drosophila* Toll and the mammalian interleukin-1 receptor; and (V) transmembrane receptor-like proteins with extracellular LRR domains (Sessa and Martin, 2000).

Three kinds of active defense responses have been differentiated: (1) The primary responses are observed in cells that remain in contact with the pathogen or infected by the pathogen (as in the case of viruses); these cells recognize the specific signal molecules of pathogen origin; the outcome of this primary response leads to programmed cell death (PCD), resulting in the development of visible necrotic lesions. (2) The secondary responses are exhibited by the adjacent cells surrounding the initial infection site in response to diffusible signal molecules (elicitors) formed following primary interaction; the elicitors initiate the activation of plant defense response genes in cultivars carrying the matching or complementary disease resistance gene; several of these genes contain leucine-rich-repeat (LRR) domains which convey the specificity for elicitor recognition. (3) The third kind of active defense responses are the systemic acquired resistance (SAR) observed in organs/tissues far away from the site of induction of resistance. The SAR is hormonally induced throughout the plant.

Studies on the molecular basis for active defense responses of plants have indicated that the elicitation of these responses may be due to recognition events frequently involving protein-protein interactions. Plants are able to recognize and respond to one or more stimuli produced by an invading pathogen during the early stages of pathogenesis. Plant cells posses effective surveillance systems that are involved in the recognition of stimuli arising from the pathogen and elicitation of specific responses to arrest pathogen development. These surveillance systems are encoded by the host-resistance genes whose products are the receptor proteins. Many of the avirulence determinants of the pathogen eliciting host-resistance gene-dependent responses are known to be proteins with no detectable enzymic activity. The physical interaction between elicitor proteins (of the pathogen) and receptor proteins (of host plants) has been demonstrated in several pathosystems. Several diverse defense responses are activated when the presence of the pathogen is recognized by the plant. In incompatible interactions, the appearance of necrotic lesions is observed at the primary site of infection, due to rapid localized cell death. This response is called hypersensitive response (HR) which activates a signal transduction pathway resulting in PCD. The molecular responses associated with HR are the production of reactive oxygen species, transient opening of ion channels, cell wall fortifications, production of antimicrobial compounds (phytoalexins), and induction of pathogenesis-related (PR) proteins such as glucanases, chitinases, and thaumatin-like proteins (TLPs) (Somssich and Hahlbrock, 1998; Hutcheson, 1998; Dempsey et al., 1998; Van Loon, 1999; Jia, 2000).

The resistance of nonhost plants may be due to the presence of antimicrobial peptides as in the case of cowpea seeds. Two cysteine-rich peptides (6.8 and 10 kDa) present in cowpea seeds inhibited the development of *F. solani* and *F. oxysporum*. Sequence analysis of these peptides indicated the presence of a defensin and a lipid transfer protein (LTP) with high degrees of homology to other antifungal peptides derived from plants. Localization of this LTP in the cell wall and cytosolic compartments was observed using immunofluoresence assays (Carvalho et al., 2001). Resistance of tomato to *C. fulva* causing leaf mold disease, is based on the specific recognition of extracellular fungal protein resulting in HR. Five proteins secreted by *C. fulvum* were purified. Different tomato breeding lines and accessions of *Lycopersicon pimpenellifolium* were evaluated for their recognitional specificity by injecting the purified fungal proteins. Recognition of one or more of these proteins was inferred from the development of HR. In addition, the race-specific elicitors AVR4 and AVR9 proteins of *C. fulvum* were also recognized by the resistant lines. This ability to recognize was governed by a single dominant gene. Furthermore, *Nicotiana paniculata,* a nonhost of *C. fulvum* could also recognize one of the extracellular proteins of *C. fulvum,* indicating that plants have a very efficient surveillance system (Laugé et al., 2000).

The direct interaction between the products of host plant resistance genes and pathogen avirulence genes was demonstrated in the rice-blast pathosystem. Rice plants expressing *Pi-ta* genes are resistant to strains of rice blast pathogen *M. grisea. Pi-ta* encodes a putative cytoplasmic receptor with a leucine-rich domain (LRD) and AVR-Pi-ta protein of *M. grisea* is predicted to encode a metalloprotease with an N-terminal-secreting signal. The AVR-Pi-ta protein was shown to bind specifically to the LRD of the Pi-ta protein inside the plant cell to initiate a *Pi-ta*-mediated defense response (Jia et al. 2000).

The accumulation of defense-related proteins in the stem tissues of carnation plants inoculated with virulent and avirulent races of *F. oxysporum* f. sp. *dianthi* was monitored. Constitutive expression of chitinase activity in the intercellular fluids (IFs) occurred in healthy plants. However, the total chitinase activity increased progressively with increasing time intervals after inoculation. By using the antiserum raised against β-(1,3)-glucanase P3 of tomato, two bands were detected on immunoblots in the IFs of pathogen-inoculated stem tissue and these bands were absent in stem tissues treated with water (Pelt-Heerschap and Smit-Bakker, 1999). The induction of defense-related ultrastructural modifications, as exemplified by the formation of appositions on the outer host cell wall surface, occlusion of intercellular spaces, and formation of papillae was discernible in maize seedlings infected by *F. moniliforme (Gibberella zeae)*. Immunolocalization technique revealed that the pathogenesis-related maize seed (PRms) proteins accumulated at very high levels in those cell types that represented the first barrier for fungal penetration such as the aleurone layer of the germinating seed, as well as the subcuteller epithelial cells of isolated germinating embryos. The presence of a large number of abnormal fungal cells showing PRms-specific labeling seems to indicate a function for PRms in plant defense response (Murillo et al., 1999; Chen et al., 2001).

Plants constitutively express proteins inhibiting fungal endopolygalacturonases (endo PGs) and these proteins (polygalacturonase-inhibitor protein [PGIP]) are localized on cell walls of most plant species. The presence of the endo PG in infected tissues was detected by immunological techniques. Production of PG was observed in soybean seedlings infected by *Sclerotinia sclerotiorum,* but not in plants infected by *Diaporthe phaseolorum* var. *caulivora* and *P. sojae,* indicating the importance of PG-PGIP interaction leading to the resistance response of soybean only in certain host-pathogen interactions (Favaron et al., 2000). The different isoforms of PGIP from leek *(Allium porrum)* inhibited the endo PGs of *S. sclerotiorum* and *Botrytis aclada (B. allii)* to varying degrees (Favaron, 2001). A PGIP was purified from the leaves of potato cv. Spunta and was inhibitory to several fungi grown in liquid cultures. In the incompatible interaction with *P. infestans,* the levels of the PGIP increased significantly as assessed by using heterologus antibodies in Western blots. Similar increases in PGIP levels occurred following wounding and treatment with salicylic acid (SA) which is a known inducer of resistance to several pathogens. The synthesis of PGIP may occur due to elicitation of the active defense mechanisms of potato plants by signal molecules known to induce plant defense genes (Machinandiarena et al., 2001).

The activation of other enzymes and proteins in plants following inoculation of fungal pathogens may contribute to the development of resistance as in the case of the pearl millet downy mildew pathosystem. The H^+-ATPase was activated in all pearl millet cultivars showing resistance to the fungal pathogen *S. graminicola*. A positive correlation between the levels of resistance of pearl millet cultivars and extent of activation of H^+-ATPase was observed. Upregulation of H^+-ATPase activity was revealed specifically in the coleoptile region, which is the susceptible site for infection by *S. graminicola.* By using a monoclonal antibody MAb 46E 5B 11F6 specific for maize H⁺-ATPase, the upregulation of H^+-ATPase activity only in resistant cultivars was demonstrated by ELISA and immunoblot. In addition, a 100 kDa band was specifically increased by threefold in the resistant cultivars (Madhu et al., 2001). In barley leaf epidermis inoculated with *E. graminis* f. sp. *hordei,* the 14-3-3 proteins with a key role in many eukaryotic signalling networks were induced. These proteins can bind with several proteins including a 100 kDa membrane protein probably plasma membrane and H^+-ATPase and they appear to be involved in an epidermis-specific response to this pathogen present only in this tissue (Finnie et al., 2002). Two novel cell wall structural proteins were rapidly deposited in the cell wall matrix in the spring wheat *(T. aestivum)* line DH 1015 following the application of *Fusarium graminearum*-derived elicitor (El-Gendy et al., 2001).

Maize genotypes resistant or susceptible to aflatoxin production (by *A. flavus*) or contamination were analyzed for differences in both constitutive and inducible proteins. Five additional constitutive proteins were shown to be associated with resistance of eight of ten genotypes examined. Among the new proteins globulin-1 and globulin-2 with MWs of 58 kDa and 46 kDa, respectively, were identified. Induction of specific antifungal proteins such as zeamatin (22 kDa) in resistant genotypes reaching high levels was also observed. Both constitutive and inducible proteins seem to be essential for kernel resistance, since the embryo-killed kernels, which are unable to synthesize new proteins, support the high level of aflatoxin production. The results suggest that the synthesis of new proteins by the embryo has an important role in the development of resistance to *A. flavus* (Huynh et al., 1992).

Microbial pathogens—fungi, bacteria, and viruses—are able to induce the synthesis of new protein compounds in different plant species, particularly in incompatible interactions, resulting in hypersensitive necrosis. These proteins are host plant-coded proteins and have been classified into eleven families. They are known as pathogenesis-related (PR) proteins and defined as "proteins coded for by the host plant but induced only in pathological or related situations" (Antoniw et al., 1980, p. 3). PR proteins have a role in the resistance response, particularly in the development of acquired resistance following the formation of necrotic lesions. Induction of PR proteins is associated with the development of SAR in tissues/organs distant from the site of inoculation (Ward et al., 1991). PR proteins may be considered as defense proteins, as they are likely to limit the multiplication/spread of pathogens in noninfected distant leaves. The genes involved in the production of PR proteins have been identified and gene symbols *(ypr)* have been assigned.

PR proteins have been grouped based on amino acid sequences, serological relationships, and/or enzymatic or biological activity (Table 11.1). The major acidic PR proteins (PR-1 to PR-5) are localized in the intercellular space of leaf tissues where it is possible for them to come in contact with invading fungal and bacterial pathogens. In addition, a few of the inducible acidic PR proteins associated with SAR have been shown to exhibit significant antipathogenic properties. PR-1 protein accumulated in tomatoes infected with *C. fulvum* (leaf mold disease) earlier in incompatible interactions than in compatible inter-

TABLE 11.1. Classification of pathogenesis-related (PR) proteins into families

Family of PR proteins	Type member	Properties
PR-1	Tobacco PR-1 a	Not determined
PR-2	Tobacco PR -2	β–(1,3)-glucanase
PR-3	Tobacco P, Q	Chitinase type I, II, IV, V, VI, VII
PR-4	Tobacco R	Chitinase type I, II
PR-5	Tobacco S	Thaumatin-like
PR-6	Tomato inhibitor I	Proteinase inhibitor
PR-7	Tomato P_{69}	Endoproteinase
PR-8	Cucumber chitinase	Chitinase type III
PR-9	Tobacco "lignin-forming peroxidase"	Peroxidase
PR-10	Parsley "PR1"	Ribonuclease-like
PR-11	Tobacco class V chitinase	Chitinase type I

Source: Van Loon, 1999.

actions. However, the PR-1 content reached even higher levels in the compatible interaction eventually, although the rate of synthesis was less, indicating that this PR-1 protein may not be required for the development of resistance in this pathosystem.

Wheat leaf rust disease development and β-(1,3)-glucanase (PR-2) expression in resistant and susceptible near-isogenic wheat seedlings were monitored. The resistance to leaf rust disease caused by *Puccinia recondita* f. sp. *tritici* is conferred by *Lr29* and *Lr34* genes. The levels of activity and immunological profiles of PR-2 proteins were determined. Induction of PR-2 protein activity was noted in all resistant lines and susceptible control concurrent with the formation of substomatal vesicles by the leaf rust pathogen. The resistance offered by *Lr29* and *Lr34* genes did not appear to be related to higher levels of β-(1,3)-glucanases reached in the resistant lines (Kemp et al., 1999). On the other hand, the adult plant resistance of wheat *(T. aestivum)* conferred by *Lr35* to leaf rust disease was found to be associated with high constitutive levels of β-(1,3)-glucanase (EC 3.2.1.39). However, four intercellular proteins (35, 33, 32, and 31 kDa) serologically related to β-(1,3)-glucanase were present in both resistant and susceptible genotypes during all stages of plant growth (Anguelova-Merhar et al., 1999). In a subsequent study, the resistance in wheat to the pathotype UVPrt 9 was shown to be associated with high constitutive expression of chitinase and induction of β-(1,3)-glucanase activities. At 168 hours postinoculation, several infection sites in the resistant isogenic line (Karee/*Lr35*) displayed a necrosis-confirming HR, while the susceptible line exhibited HR at a limited number of infection sites (Anguelova-Merhar et al., 2001).

The temporal expression patterns for the defense response genes encoding PR-1, PR-2 [β-(1,3)-glucanase], PR-3 (chitinase), PR-4, PR-5 (TLPs), and peroxidase (PR-9) in wheat spikes of resistant and susceptible cultivars inoculated with *F. graminearum (G. zeae)* were determined. No differences in the accumulation of transcripts for the six defense response genes during early stages (6 to 12 hours after inoculation) could be observed. The timing of defense-response gene induction in the pathosystem was not related to the pathogen infection (Pritsch et al., 2000). However, a more rapid accumulation of two acidic isoforms of β-(1,3)-glucanases, class IV and class VII chitinases was recorded in the resistant cultivar (Suman 3) than in the susceptible mutant during the first 24 hours after inoculation with *F. gram-*

inearum (Li et al., 2001). The role of PR-2b-β-(1,3)-glucanase in the development of resistance in chickpea *(Cicer arietinum)* to *Ascochyta rabiei* is unclear, although a strong accumulation of intracellular and extracellular β-(1,3)-glucanase activity was observed during a six-day infection period. No inhibitory effect of the purified PR-2b protein on *A. rabiei* was evident. Moreover, in the time-course studies no qualitative or quantitative difference in the accumulation of the PR-2b protein in the resistant and susceptible chickpea cultivars could be noted (Hanselle and Barz, 2001).

Localization of chitinase (CA Chi2) in the compatible and incompatible interactions of pepper stems with *P. capsici* was visualized by employing immunogold labeling technique. In the compatible interaction, the quantity of gold particles in the fungal cell surface was limited, but there was even distribution over the entire fungal wall which was in intimate contact with the host cell wall. In contrast, most of the gold particles were seen over the cell wall area of the pathogen interacting with resistant tissues. The hyphae spreading intercellularly showed homogenous labeling over their cell walls. Degradation of the fungal cell wall was evident at the hyphal tips which had dense depositions of gold particles indicating that the activity of the host chitinase on the pathogen is limited to the cell wall, since the cytoplasm of fungal cells was nearly free of gold labels (Lee, et al., 2000).

The defense gene responses of grapevine inoculated with *B. cinerea* were assessed by immunodetection methods. Among the several newly synthesized proteins, four acidic proteins that cross-reacted with a tobacco PR-2a [β-(1,3)-glucanase] antiserum, were detected. One of these four isoforms showed strong similarities with β-(1,3)-glucanases from many other plant species. The β-(1,3)-glucanase proteins appeared at day 3 and increased up to day 7 after inoculation with *B. cinerea* (Renault et al., 2000).

Detection and immunolocalization of PR proteins in the seminal root tissues of barley and wheat infected with *Bipolaris sorokiniana* was studied employing respective antisera by Western blot analysis and immunohistological techniques. Accumulation of PR-1 and PR-5 proteins in the roots of both wheat and barley as revealed by Western blot analysis, was observed (Figure 11.1). Root inoculation with *B. sorokiniana* induced PR-protein accumulation in the leaves of barley but not in wheat as revealed by ELISA format. Enhanced accumu-

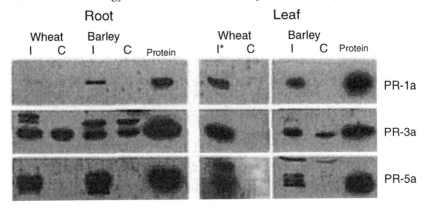

FIGURE 11.1. Western blot analysis of protein extracts from barley roots and leaves inoculated with *Bipolaris sorokiniana:* (I) infected; (C) healthy control; (*) leaf inoculation (*Source:* Liljeroth et al., 2001; Reprinted with permission by Blackwell Wissenchafts-Verlag, Berlin, Germany.)

lation of PR proteins (PR-1a, Pr-2c, PR-3a, and PR-5a) in response to infection was noted only when the roots were inoculated close to the root tip (Figure 11.2). Immunohistological labeling with antiserum against PR-3a protein showed a strong signal in the cortex tissue of uninoculated plants indicating the constitutive expression of this PR protein (Figure 11.3). When plants were inoculated close to the root tips, a significant increase in the signal in the stele and endodermis was discernible. In the case of PR-1a, activation of accumulation was noticed in all tissues of barley roots, but it was less evident in wheat. Accumulation of PR proteins in leaves of barley but not in wheat following inoculation of roots indicated systemic induction of PR genes (Liljeroth et al., 2001) (see Appendix section on immunological techniques).

In barley, two novel proteins belonging to the newly designated PR-17 family were detected following inoculation with *Blumeria (Erysiphe) graminis* f. sp. *hordei*. By using the antisera raised against these *Hv*PR-17a and *Hv*PR-17b proteins, their accumulation in the mesophyll apoplast as well as in leaf epidermis (the only tissue invaded by this pathogen) was observed. These PR proteins showed strong similarity to NtPRp 27 from tobacco and WCI-5 from wheat formed in response to viral and fungal infections, respectively (Christensen et al., 2002). Inoculation of rose shoots with *Diplocarpon*

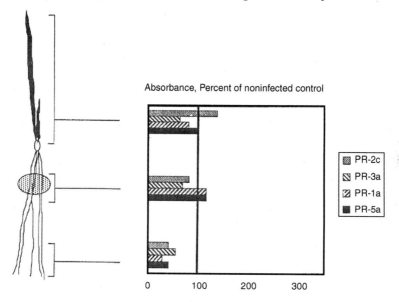

FIGURE 11.2. Acid soluble proteins of barley seedlings grown in filter paper rolls after inoculation with *Biploaris sorokiniana* determined by ELISA (*Note:* The absorbance values are expressed as percentages relative to healthy controls. *Source:* Liljeroth et al., 2001; Reprinted with permission by Blackwell Wissenchafts-Verlag, Berlin, Germany.)

rosae (black spot disease) induced accumulation of PR proteins in the intercellular spaces of leaves. Western blot analysis showed that the rose PR proteins were serologically related to tobacco PR proteins. Strong accumulation of PR-2, PR-3 and PR-5 proteins occurred in the early stages, while PR-1 protein was induced later (Suo and Leung, 2002).

The direct involvement of PR proteins in determining resistance to fungal pathogens seems unlikely in some pathosystems. The PR-2 [β-(1,3)-glucanases] and PR-3 (chitinases) proteins have the ability to hydrolyze the cell wall components [β-(1,3)-glucan and chitin] of some fungal pathogens. The pathogens belonging to Oomycetes do not contain chitin and some pathogens, such as *Pythium aphanidermatum, P. ultimum,* and *Phytophthora cactorum,* are insensitive to a mixture of chitinase and β-(1,3)-glucanase (Mauch et al., 1988). However, the extracellular PR proteins may be involved in pathogen

Uninfected   Infected

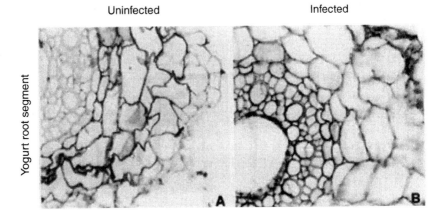

FIGURE 11.3. Immunological labeling of root tissues with antiserum against barley Pr-3a protein: (A) young healthy root segment; (B) young root segment infected by *Bipolaris sorokiniana* (*Source:* Liljeroth et al., 2001; Reprinted with permission by Blackwell Wissenchafts-Verlag, Berlin, Germany.)

recognition processes releasing defense-activating signal molecules from the walls of invading pathogens. The chitinases and glucanases may be able to liberate elicitor-type carbohydrate molecules from pathogen cell walls, as demonstrated in the soybean-*Phytophthora* pathosystem. β-(1,3)-glucanase induced in seedlings infected by *P. megasperma* f. sp. *glycinea* was shown to release elicitor-active fragments from cell wall preparations of this fungal pathogen (Ham et al., 1991, 1995). Fungal pathogens can also secrete metabolites that are capable of inhibiting specific plant enzymes (Ham et al., 1997). The plants can, in turn, produce a combination of several PR proteins that may arrest the development of the pathogen resulting in reduction or prevention of disease initiation/development depending on the degree of resistance of the plant to the pathogen in question. The role of PR proteins in the development of resistance to fungal pathogens infecting certain crop species may be insignificant. The maize mutant *rhm1* is resistant to *Bipolaris maydis* causing Southern corn blight disease. The levels of PR proteins chitinase, PR-1, and peroxidase did not show any difference in *rhm1* mutant and the wild type irrespective of infection by *B. maydis*. It seems that *rhm1* resistance may operate via a mechanism not dependent on the induction of PR genes (Simmons et al., 2001).

Assessment of Levels of Resistance to Bacterial Diseases

Immunological techniques have been used for the assessment of grades of resistance of cultivars/genotypes to bacterial pathogens only to a limited extent. However, immunodiagnostic methods have been employed to detect latent infection of *Ralstonia solanacearum* in peanut (Shan et al., 1997) and potato seed tubers (Priou et al., 1999). Different cultivars of sugarcane were tested for the presence of *Clavibacter xyli* ssp. *xyli* causing ratoon stunting disease (RSD). The susceptible cultivars used as indicator hosts such as CO421, CO997, and Q28 allowed the pathogen to reach high titres. On the other hand, cultivars COC86062, CO7704, CO8021, and COC671 had low concentrations of the pathogen, as indicated by the absorbance values in ELISA (Viswanathan, 1997). Tissue blot immunoassay (TBIA) was employed to detect colonized vascular bundles (CVBs) in sugarcane infected by RSD. Stained imprints of CVBs on nitrocellulose membranes were obtained for sugarcane cultivars which were susceptible, highly susceptible, or resistant to RSD. A high negative correlation was found between the percentages of CVBs determined by TBIA technique and the levels of resistance of sugarcane cultivars. The use of TBIA for screening sugarcane genotypes for resistance to RSD has been suggested (Giglioti et al., 1999).

The evaporative binding (EB)-ELISA was employed to compare the relative populations of *Leifsonia (Clavibacter) xyli* ssp. *xyli* in 25 cultivars of sugarcane. The test cultivars were inoculated by planting them through a contaminated mechanical planter. Highly significant differences in the populations of the bacterial pathogen in the xylem extracts of cultivars were indicated by (EB)-ELISA method. This is a simple method that can be used for rating sugarcane cultivars/genotypes for resistance to ratoon stunting disease (Croft, 2002). An ELISA on nitrocellulose membrane (NCM-ELISA) was developed to detect latent infection of *R. solanacearum* in potato clones showing resistance or moderate resistance to bacterial wilt disease. The frequency of latent infection in the test clones determined by NCM-ELISA formed the criterion for selection of clones. Fifteen clones of potato were selected as resistant to wilt based on this criterion (Priou et al., 2001).

Molecular Basis of Host Resistance to Bacterial Diseases

The nature of interaction between a host plant species and a bacterial pathogen depends on the presence of preformed antibacterial compounds and/or production of inhibiting compounds in response to infection by the pathogen in question. The presence of cross-reactive antigenic proteins between bacterial pathogens and their respective host plants and their involvement in the determination of degree of susceptibility (compatibility) or resistance (incompatibility) have been studied in some host-pathogen interactions. The pathotype A of *Xanthomonas axonopodis* pv. *citri (XacA)* can infect all citrus varieties, while the serologically distinct pathotype C *(XacC)* is able to infect only the Mexican lime *(C. aurantifolia)*. Crude preparations from *C. aurantifolia* contained cross-reactive antigens (proteins) that could be detected by using antisera raised against both pathotypes *XacA* and *XacC* by DAS-ELISA format, since *C. aurantifolia* was susceptible to both pathotypes. On the other hand, the presence of significant amounts cross-reactive antigens in the crude preparations from *C. sinensis* was revealed by the antiserum specific to pathotype *XacA,* but not to *XacC,* which was not pathogenic to *C. sinensis.* The antigenic disparity between *C. aurantifolia* and *C. sinensis* as established by the antiserum to pathotype *XacC* indicates the involvement of cross-reactive antigens in determining the nature of interaction between *Xac* and *Citrus* spp. resulting in the development of canker disease (Bach and Alba, 1993).

The interaction between the host plant and the bacterial pathogen becomes incompatible when the pathogen expressing specific avirulence *(avr)* genes interacts with plants carrying corresponding resistance (R) genes. The products of over 40 *avr* genes from *P. syringae* and *Xanthomonas* strains have been shown to be proteins with molecular mass ranging from 18 to 10 kDa. *Pseudomonas syringae* pv. *tomato* strains carrying *avrD* gene produced syringolides both in culture and *in planta,* in addition to eliciting a response in soybean lines having the *Rsg4* resistance gene (Keen, 1999). Soybean contains an allergenic protein Gly m Bd 30K and its presence in leaves could be detected by immunoblotting with an anti-Gly m Bd 30K monoclonal antibody (MAb F5). Treatment of leaves of soybean carrying *Rpg4(+)* resistance gene elicited HR (Ogawa et al., 2000).

Bacterial avirulence gene function is determined by its interactions with **h**ypersensitive **r**esponse and **p**athogenicity *(hrp)* genes of several bacterial pathogens. The *hrp*-encoded protein translocation complex is considered to be involved in the translocation of *avr* gene products. The *avrXv3* gene of *X. campestris* pv. *vesicatoria* was found to be plant inducible and controlled by the *hrp*-regulatory system. A direct role for AvrXv3 protein in the induction of HR in tomato near-isogenic line (NIL) 216 was demonstrated by Astua-Monge et al. (2000). The molecular mechanism of host specificity in *P. avenae,* which has a wide host range among monocots, was elucidated by using a rice-incompatible strain N1141 of *P. avenae*. This strain induced a resistance response of rice cells. A strain-specific antiserum was raised against N1141 bacterial cells and then absorbed with H8301 cells (compatible strain). This antiserum contained antibodies against N1141 flagellin protein which induced hypersensitive death in cultured rice cells within six hours of treatment, whereas the flagellin of H8301 strain (compatible) did not induce HR (Che et al., 2000).

A resistance gene-dependent cell death program in soybean plants and cell cultures was activated after inoculation with *P. syringae* pv. *glycinea* carrying the avirulence gene (*avrA*) or zoospores of *P. sojae* race 1. A new gene that was specifically activated within two hours after inoculation with *P. sojae* zoospores was identified. This gene, encoding for a deduced protein of 368 amino acids with a very high content of aspargine, was designated N-rich protein (NRP). The antibody specific to NRP recognized a protein (42 kDa) located in the cell wall, as indicated by cell fractionation studies. The NRP gene appeared to be a marker that was activated in soybean early in the development of disease resistance (Ludwig and Tenhaken, 2001). The incompatible interaction between grapevine and *P. syringae* pv. *pisi* race 2 (strain 1) was examined at molecular levels. Two genes, stilbene synthase (SS) and a PR-10 gene encoding putative defense proteins were analyzed. The expression of *SS* gene resulted in the accumulation of the phytoalexin resveratrol, whereas the genomic clone (Vv PR10-1) coding for PR-10 protein was responsible for the synthesis of PR-10 protein. The polypeptide encoded by the corresponding mRNA was detected by immunoblotting at 24 to 96 hours postinoculation. The interaction of nonhost grapevine leaves (infiltrated with bacterial suspension) resulting in the development of necrosis has to be considered as an HR-like response (Robert et al., 2001).

The differential multiplication, colonization, and production of the wilt-inducing virulence factor exopolysaccharide I (EPS I) by *R. solani* in one susceptible and two resistant tomato cultivars were studied. Viable bacterial cell counts in the midstems indicated that the vascular tissues of susceptible cultivars were colonized by this bacterial pathogen at a faster rate compared to resistant cultivars. By using ELISA to quantify the EPS I produced in infected plants, greater amounts of EPS I (per plant) in susceptible cultivars than in resistant cultivars was revealed. Furthermore, immunofluorescence (IF) microscopic observations using the antibodies against either EPS I or *R. solani* cells, showed that the bacterial pathogen and EPS I were present throughout the vascular bundles and intercellular spaces of the pith in the susceptible cultivars, while pathogen cells and EPS I were confined to the vascular tissues of the resistant cultivars (Mc Garvey et al., 1999).

The interaction between rice vascular tissues and *X. oryzae* pv. *oryzae (Xoo),* causing bacterial leaf blight has been studied. In rice seedlings, the exposed pit membrane separating the xylem lumen from associated parenchyma cells permits contact with pathogen cells. The resistance response was characterized by the thickening of secondary walls of xylem within 48 hours, in addition to narrowing of pit diameter, effectively reducing the area of pit membrane exposed to access. The activity and accumulation of a secreted cationic peroxidase, PO-C1, increased in xylem vessel walls and lumen. By employing peptide-specific antibodies and immunogold labeling, production of PO-C1 in the xylem parenchyma and secretion to the xylem lumen and walls were demonstrated. The results suggest a role for vessel secondary wall thickening and PO-C1 accumulation in the defense response in this pathosystem (Hilaire et al., 2001).

Assessment of Levels of Resistance to Viral Diseases

Sources of resistance to viral diseases may be available in the same or related species of host plants. Immunological methods have been particularly useful not only for the detection, but also for the quantification of virus concentrations that reflect the grades of resistance of test entries. Different species of *Arachis* were tested for their resistance to peanut (groundnut) stripe virus by ELISA test. Some lines of *A. diogi, A. halodes,* and *Arachis* sp. did not allow the virus to reach detectable concentrations thus indicating their high levels of resistance to the virus (Culver et al., 1987). Rice tungro disease (RTD) is caused by a complex of two viruses—rice tungro bacilliform virus (RTBV) and rice tungro spherical virus (RTSV). The relative concentrations of RTBV and RTSV in five rice cultivars were determined by (DAS)-ELISA format. The rice cultivars TN1, ASD8, and FK135 (susceptible) had higher virus titers as compared to cultivars Balimau Putih and Utri Merah (tolerant) (Sta. Cruz et al., 1993) (Figure 11.4). The resistance of plum germplasm entries to plum pox potyvirus, apple mosaic virus, and *Prunus* necrotic ringspot virus was assessed by testing flowers, leaves, and bark of clones by ELISA and resistant entries were identified (Karešová and Paprštein, 1998). Twenty-five root stock-scion combinations of plum were tested for their resistance to tobacco ringspot virus (ToRSV) using root and bark samples in ELISA. The presence of ToRSV was detected in all 25 combinations of five root stocks and five scions in the field plots (Kommineni et al., 1998).

Using ELISA, three cultivars of rice *(O. sativa)* and four cultivars of *O. glaberrima* were evaluated for their resistance to rice yellow mottle virus (RYMV). The cultivars Tog 5681 and Gigante were found to be highly resistant to RYMV and no symptoms of infection were observed in these cultivars. Appearance of symptoms was associated with an increase in the absorbance values (A_{405}nm) of ELISA test. The results indicated the presence of a single recessive gene common to both resistant cultivars (Ndjiondjop et al., 1999). The usefulness of ELISA for the identification of resistant cultivars/genotypes was demonstrated in tobacco-potato virus Y pathosystem. When tobacco cultivars (8) were tested at four weeks after inoculation with PVY^O and PVY ^{NTN} strains, two cultivars (Pollagi 4 and Hevesi 12) were found to be immune to both strains. These two cultivars were recommended as good sources of resistance for use in breeding programs (Takács et al., 2000). By performing ELISA tests for the detection of five serotypes (MAV, PAV, RPV, RMV, and SGV) of barley yellow dwarf virus (BYDV) and wheat dwarf virus, ten breeding lines of winter barley were found to have high levels of resistance to these viruses, indicating their suitability for breeding for resistance programs (Poscai and Muranyi, 2000).

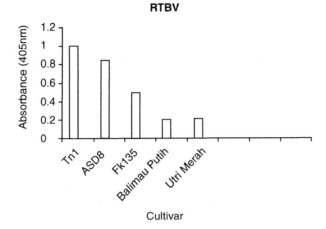

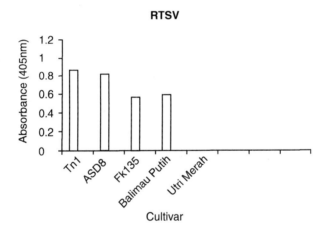

FIGURE 11.4. Determination of relative concentrations of RTBV and RTSV in rice clutivars by (DAS)-ELISA (*Source:* Sta Cruz et al., 1993; Reprinted with permission by Blackwell Wissenchafts-Verlag, Berlin, Germany.)

To facilitate selection of reliable resistance sources the levels of resistance/susceptibility of germplasm genotypes may be assessed by determining the multiplication/concentration of the virus attained in them. Germplasm entries were tested for resistance to citrus tristeza virus (CTV). All cultivars of *Poncirus trifoliata, Severenia luxifolia, Atalantia ceylanica,* and *Fortunella crassifolia* were resistant to CTV.

Because the fruits of *F. crassifolia* are edible, the use of this species for the improvement of CTV resistance may be preferable (Asíns et al., 1999). A tomato line (Y118) derived from a cross with *Lycopersicon chiliense* (LA 1938) showed resistance to tomato spotted wilt virus (TSWV). Among the F3 population, one line (Fla 925-2) showed field resistance to TSWV, as determined by ELISA. The plants of this line recovered as the virus titer was reduced from initially high levels in the plant tissues to subsequent levels below detection with ELISA (Canady et al., 2001).

Assessment of the levels of susceptibility/resistance using immunological methods may be useful when the symptom expression is poor or even absent. The infection of vegetatively propagated chrysanthemum cultivars by TSWV and their susceptibility levels were determined by ELISA. All fifteen chrysanthemum cultivars tested were, however, found to be susceptible to TSWV, indicating the need for further research for the sources of resistance (Van Wetering et al., 1999).

In pineapple mealybug wilt-associated closter viruses (PMWa V-1 and PMWa V-2), none of the pineapple accessions maintained by the United States Department of Agriculture/Agricultural Research Service National Germplasm System was found to be resistant to PMWaV-1 and PMWaV-2 based on the results of TBIAs (Sether et al., 2001). The intensity of symptoms observed may not be reflected in the virus titer determined by ELISA as in the case of gladiolus infected by bean yellow mosaic virus (Selvarajan et al., 1998) (Table 11.2). Such a combination of two criteria viz. symptom expression and virus titer determined by immunological techniques

TABLE 11.2. Relationship between symptom intensity induced by bean yellow mosaic virus in gladiolus cultivars and ELISA absorbance values

Number of cultivars	Symptom intensity	Absorbance at A_{405}
35	Mild mosaic	High
42	Mild mosaic	Low
2	Mild mosaic	< Detectable level
10	No visible symptom	Low
1	No visible symptom	High

Source: Selvarajan et al., 1998.

may be effective in the identification of dependable sources of resistance to viruses.

The influence of disease incidence on varietal mixtures consisting of susceptible and resistant cultivars was assessed in the case of soil-borne wheat mosaic virus (SBWMV). Winter wheat *(T. aestivum)* was raised as mixtures in the proportions of 1:1 and 1:3 (suscepti-ble:resistant cultivars) and also as pure stands of the varieties in the mixture. By using DAS-ELISA, about 88 percent of the plants of susceptible cultivars were found to be infected by SBWMV. The disease reduction relative to the pure stand was 32.2 percent in the 1:1 mixture and 39.8 percent in the 1:3 mixture. In addition, the concentrations of SBWMV as determined by OD values of ELISA in the infected plants of the two mixtures were consistently lower than infected plants in pure stands. By growing mixtures of resistant and susceptible cultivars, the disease incidence is reduced and the virus concentration is less compared to control plants (Hariri et al., 2001).

Molecular Basis of Host Resistance to Viral Diseases

Virus infections are considered as derangements in the metabolism of proteins and nucleic acids that are essential for viral replication. Susceptible plants allow viruses to direct synthetic machinery to provide the necessary compounds to satisfy the requirements of the viruses. In contrast, resistant plants activate the mechanisms involved in the production of compounds that may adversely affect viral nucleic acid synthesis translation and movement to other cells or tissues, thus restricting the progress of disease development. Viral proteins may provide stimuli that may be recognized as foreign in the host cell, thus eliciting a defense response in incompatible interactions. Plant cells possess surveillance systems that may detect the stimuli generated by the viral proteins.

The tobacco mosaic virus (TMV) produces at least three known proteins that may function as avirulence determinants involved in gene-for-gene interactions (Culver, 1996). The TMV replicase protein (Padgett and Beachy, 1993), coat protein (Culver and Dawson, 1991), and movement protein (Weber et al., 1994) are avirulence determinants inducing resistance to TMV. The N protein product of tobacco *N* gene is postulated to be a cytoplasmic receptor for the viral replicase protein which is an avirulence determinant inducing HR in tobacco lines with N-resistance gene (Jones and Jones, 1997).

Some host-virus combinations have shown involvement of defense-related compounds in the development of resistance to viruses. The presence of nine induced proteins was observed in potato infected by a highly virulent isolate (potato virus Y^{NTN}) of PVY^N. Tolerant cultivars showed greater expression of these proteins. However, the potato cv. Sante, which is highly resistant to PVY^{NTN}, did not exhibit any change in protein expression. However, a constitutive protein with a MW of 43 kDa accumulated and reached high concentrations, following inoculation with this virus isolate. The presence of a similar protein in comparable concentrations in resistant cultivar Carlingford indicates the putative resistance-related function for this constitutive protein (Gruden et al., 2000).

The replication and movement of CMV in the susceptible melon *(Cucumis melo)* and resistant *C. figarei* were analyzed. The numbers of infected protoplasts isolated from CMV-inoculated *C. figarei* leaves were significantly lower than those from infected melon leaves. The amount of CMV antigen determined by ELISA or Western blotting in infected leaves of *C. figarei* was approximately tenfold lower than that in infected melon leaves. Results suggest that the resistance to CMV in *C. figarei* may be partly due to the interference in virus replication, but the inhibition of virus movement via both cell-to-cell and long-distance pathways may be the primary mechanism restricting disease development (Saiga et al., 1998).

In another study, the resistance to tomato spotted wilt virus (TSWV) in *Capsicum* spp. was shown to depend on the restriction of viral cell-to-cell movement. The resistance of *C. chinense* (accessions PI 15225 and PI 159236-resistant to TSWV) and *C. annuum* cv. Negral (susceptible) were evaluated by employing DAS-ELISA and DTBIA techniques. The DTBIA revealed restriction of and slower viral movement within the inoculated leaves of resistant plants. The weak ELISA positive reaction of inoculated and adjacent areas in resistant plants may have been due to low foliar area infected but not due to reduced viral replication (Soler et al., 1999). Grapefruit trees that were cross-protected by inoculation with mild isolates of CTV exhibited resistance to the citrus tristeza disease. However, the presence of severe isolates of CTV could be detected in cross-protected trees by using ISIA technique (Lin, Rundall, et al., 2002).

The distribution of RYMV in the partially resistant cv. Azucena *(Oryza sativa japonica)* and susceptible cv. IR 64 *(Oryza sativa indica)* was studied using ELISA and immunoprinting techniques. Immunoprinting tests revealed the presence of RYMV in both Azucena and IR 64 at the same intensity with similar distribution patterns, though symptom severity was less pronounced or symptoms were not visible in cv. Azucena. The virus titers, as determined by ELISA, were less in the partially resistant Azucena compared to susceptible IR 64. The highly resistant cv. Gigante had significantly lower titers of RYMV compared with Azucena. The decline of partial resistance of Azucena was clearly associated with an increase in virus content at 14 to 21 days after inoculation. Using immunofluorescence microscopy, localization of RYMV in the vascular tissues, mesophylls, and the bundle sheaths of systemically infected leaves of IR 64 was observed (Ioannidou et al., 2000).

A clone of the diploid potato species *Solanum phureja* cv. Egg Yolk (Clone 5010) showed strong resistance to potato leaf roll virus (PLRV). The resistance to PLRV was found to be very strongly expressed in leaf tissue, since no virus could be detected by the quantitative ELISA, although other tissues were infected. The petiole and stem tissues were comparatively less resistant to the virus infection. The virus titer in the leaves of a resistant cultivar was as little as approximately 2 percent or less of the titer in the susceptible *S. tuberosum* cv. Maris Piper (Franco-Lara and Barker, 1999). The distribution of tobacco rattle virus (TRV) in the tubers of 59 potato cultivars or breeding lines was traced by ELISA and reverse transcription polymerase chain reaction (RT-PCR). About 2 percent of plants grown from either symptomatic or asymptomatic tubers showed the presence of TRV. Systemic infection of TRV was detected more frequently in the foliage of susceptible genotypes (Crosslin et al., 1999).

The PR proteins are induced in several host-virus interactions, more prominently in the hosts reacting with HR. Tobacco cultivars with N gene reacted to TMV infection by producing necrotic local lesions and the appearance of four new proteins (b1, b2, b3, and b4) was associated with the HR. When tobacco was inoculated with different viruses, the same proteins were induced, whereas infection of *Nicotiana glutinosa* by TMV resulted in the production of different proteins, indicating that the induction of new proteins was associated with the type of symptoms (necrotic local lesions), rather than with the genetic constitution of the host plants (Van Loon and Van Kamenen, 1970; Gianinazzi et al., 1977; Antoniw et al., 1980). The discovery of newly induced proteins in tobacco that reacted hypersensitively following inoculation with other viruses has prompted research into various aspects of the nature and functions of these proteins. Moreover, similar proteins were induced by fungal and bacterial pathogens in incompatible interactions and also by abiotic stresses.

PR proteins have a role in resistance response and more specifically in the development of SAR following the formation of necrotic local lesions in inoculated leaves. The involvement of PR protein expressed constitutively in the hybrids of *N. glutinosa* and *N. debneyi* showing high resistance to TMV infection was strongly indicated (Ahl and Gianinazzi, 1982). An association between the presence of PR proteins and reduced TMV replication and spread in inoculated leaves of tobacco cv. Samsun reacting with systemic symptoms was reported by Van Loon and Antoniw (1982). PR proteins are not likely to function as general defense proteins. As viruses spread through the symplastic connects, these extracellular proteins cannot selectively affect intracellular viral replication and transport. The PR proteins may make cells at distance less conducive for virus infection, but no convincing evidence is available yet for such a possibility.

The role of PR proteins in the development of resistance to plant viruses is not clearly established. The synthesis of PR proteins in tobacco is preceded by the operations of resistance reactions. Furthermore, no quantitative relationship between the amount of PR-1 proteins present and level of resistance observed in inoculated primary leaves has been established. In addition, even before the accumulation of PR proteins, SAR seems to develop in distant leaves (Fraser, 1982). Constitutive expression of PR-1 or PR-5 in tobacco did not improve the level of resistance of transformed tobacco to TMV (Linthorst et al.,1989). These results indicate that PR proteins alone may not be effective in protecting plants against viral infection and they may act in combination with other factors to enhance the level of resistance of plants.

Induction of PR proteins by virus infection may result in the development of resistance to other microbial pathogens. For example, following inoculation of cucumber plants with tobacco necrosis virus (TNV), chitinase (PR-2) synthesis was induced thus making the plants resistant to the bacterial pathogen *Pseudomonas lachrymans*

(Metraux et al., 1991). Systemic induction of PR proteins may be due to marked increases in endogenous levels of SA. Parallel increases in contents of SA and PR-1 protein occur in TMV-resistant Xanthi nc (NN) tobacco but not in susceptible (nn) tobacco. SA may act as an endogenous signal that triggers local and systemic induction of PR-1 proteins and possibly some other components of SAR also in NN tobacco (Yalpani et al., 1991).

ENGINEERING RESISTANCE TO CROP DISEASES

Breeding crops for resistance to diseases is undoubtedly the best long-term disease management strategy. Though conventional methods of breeding for disease resistance have been effective in some pathosystems, they are time-consuming and expensive. Another major limiting factor is the nonavailability of dependable sources of resistance in cultivars or wild relatives and, even if available, it is difficult to transfer the new characters (resistance) from one species or variety to another without the loss of desirable agronomical attributes. In this context, genetic engineering, which can overcome some of the drawbacks of conventional breeding procedures, has become an attractive alternative to obtain crop varieties with resistance to disease(s). Genetic engineering techniques offer numerous options for selecting desired resistance gene(s) from a wide range of sources available not only in the plant kingdom, but also in the animal kingdom. Thus, the resistance genes from diverse sources such as bacteriophage (T4 lysozyme), hen (egg white lysozyme), or frog skin magainins can be expressed in suitable plant species to improve their resistance to diseases caused by microbial pathogens (Ward et al., 1994; Sticher et al., 1997).

Development of Transgenic Plants Resistant to Viral Diseases

Among the diseases caused by microbial plant pathogens, management of viral diseases presents more difficulties since crop varieties need to be resistant to both viral and vector species that transmit viruses from plant to plant. Hence, additional measures have to be taken to restrict vector populations.

Pathogen-Derived Resistance (PDR)

The possibility of engineering resistance to viruses by genetically modifying a susceptible variety with genes derived from the pathogen itself was suggested by Sanford and Johnson (1985). Expression of viral gene products in the transformed plants was expected to adversely affect the ability of the virus to sustain infection. The functional nature of this concept was first established by Powell et al. (1986) who demonstrated that tobacco plants expressing the CP gene of TMV were resistant to infection by this virus. Since this novel approach, several attempts have been made to examine the suitability of this strategy for development of cultivars resistant to several viruses (Table 11.3). In addition to the CP gene of viruses, the nonstructural virus genes such as replicase, movement protein (MP), and protease genes have also been transferred to susceptible plants to improve their resistance (Table 11.4).

Immunological techniques have been employed to quantify the amounts of coat proteins or the concentrations of virus in transgenic plants. Plants accumulating the maximum levels of transgenic coat protein displayed the highest level of resistance to TMV. Plants without symptoms had low virus titer or the virus could not be detected by ELISA (Powell et al., 1989, 1990; Beachy et al., 1990). The extent of protection offered by CP gene-mediated resistance seems to be positively related to the level of CP expression or amount of CP present in the transgenic plants as in the case of PVX (Hoekema et al., 1989), alfalfa mosaic virus (Hill et al., 1991), and rice stripe virus (Hayakawa et al., 1992). A significant decrease in the incidence of watermelon mosaic and zucchini yellow mosaic viruses in transgenic lines of squash and cantaloupe expressing CP genes was evident when ELISA tests were performed (Clough and Hamm, 1995). Potato clones transformed with CP genes of potato virus X showed high levels of resistance to the virus under field conditions, as determined by the optical density in ELISA tests. Good correlation was found between the resistance to PVX and expression of CP gene of PVX (Doreste et al., 2002). Soybean plants were transformed with CP genes of soybean mosaic virus (SMV) and their resistance to SMV was evaluated by quantifying the temporal and spatial spread of SMV released from the point source in the field using ELISA. Two SMV-CP transformed lines showed significantly lower infection rates and

TABLE 11.3. Transgenic crops/plants transformed with viral coat protein (CP) genes

Virus	Transgenic crops /plants	References
Alfalfa mosaic virus (AlMV)	Alfalfa	Hill et al., 1991
	Tobacco cv. Xanthi nc	Van Dun et al., 1987, 1988
	Tomato	Tumer et al., 1987
Arabis mosaic virus (ArMV)	Tobacco	Bertioli et al., 1992
	Nicotiana benthamiana	Spielmann, Douet-Orhant, et al., 2000
	N. benthamiana and grapevine	Spielmann, Krastanova, et al., 2000
Cucumber mosaic virus (CMV)	Cucumber	Gonsalves et al., 1992
	Tobacco	Cuozzo et al., 1988
	Tomato	Fuchs et al., 1996 Kaniewski et al., 1999 Tomassoli et al., 1999
Grapevine chrome mosaic virus (GCMV)	Tobacco	Brault et al., 1993
Grapevine fanleaf virus (GFLV)	Grapevine	Xue et al., 1999
Grapevine leafroll-associated virus (GLRaV-3)	Grapevine	Xue et al., 1999
Odontoglossum ringspot virus (ORSV)	Eggplant (brinjal)	Dardick and Culver, 1997
Papaya ringspot virus (PRSV)	Papaya	Fitch et al., 1992 Yeh et al., 1998 Ferreira et al., 2002
Pepper severe mosaic virus (PeSMV)	Tobacco	Rabinowicz et al., 1998
Plum pox virus (PPV)	Plum	Ravelonandro et al., 1997
Potato aucuba mosaic virus (PAMV)	Tobacco	Leclerc and Abou Haider, 1995
Potato leafroll virus (PLRV)	Potato	Kawchuk et al., 1990 Zhang et al., 1995 Thomas et al., 1997
Potato mop-top virus (PMTV)	Potato	Barker et al., 1998
Potato virus (PVS)	Potato	Mackenzie et al., 1991

Virus	Transgenic crops /plants	References
Potato virus X (PVX)	Potato	Hoekema et al., 1989
		Doreste et al., 2002
	Tobacco	Hemenway et al., 1988
		Lawson et al., 1990
Potato virus Y (PVY)	Potato	Lawson et al., 1990
		Maiti et al., 1999
PVY NTN	Potato	Racman et al., 2001
Rice stripe virus (RStV)	Rice	Hayakawa et al., 1992
Soybean mosaic virus (SMV)	Tobacco	Stark and Beachy, 1989
Sweet potato feathery mottle virus (SPFMV)	Sweet potato	Okada et al., 2001
Tobacco mosaic virus (TMV)	Tobacco	Powell et al., 1986
		Beachy et al., 1990
	Tomato	Nelson et al., 1988
	Eggplant	Dardick and Culver, 1997
Tobacco rattle virus (TRV)	Tobacco	Van Dun and Bol, 1988
		Angenent et al., 1990
Tobacco streak virus (TSV)	Tobacco	Van Dun et al., 1988
Tomato mosaic virus (ToMV)	Tomato	Sanders et al., 1992
Tomato spotted wilt virus (TSWV)	Tobacco	Mackenzie, and Ellis, 1992
		Pang et al., 1992
Turnip mosaic virus (TuMV)	Chinese cabbage	Zhu et al., 2001
Vanilla necrosis virus (VNV)	*Nicotiana benthamiana*	Wang et al., 1997
Watermelon mosaic virus (WMV)	*Cucumis melo* *Cucurbita pepo* var. *melopepo*	Clough and Hamm, 1995
Wheat streak mosaic virus (WSMV)	Wheat	Sivamani et al., 2002
Zucchini yellow mosaic virus (ZYMV)	*C. melo* and *C. pepo* var. *melopepo*	

TABLE 11.4 Transgenic crop/plants transformed with noncoat viral genes

Virus/genes or sequences	Transgenic crop plants	References
African cassava mosaic virus (ACMV)/replication-associated AC1 coding sequence	Tobacco proto-plasts and *N. benthamiana*	Hong and Stanley, 1996
Apple chlorotic leafspot virus (ACLsV)/MP gene	*Chenopodium quinoa*	Yoshikawa et al., 1999
Cucumber mosaic virus (CMV)/replicase gene	Tomato	Gal-On et al., 1998, 1999
Cymbidium ringspot virus/ replicase gene	*N. benthamiana*	Rubino et al., 1993
Groundnut ringspot virus (GRSV) and tomato chlorotic spot virus (TCSV)/sequences	Tobacco	Mackenzie and Ellis, 1992
Pea early browning virus (PEBV)/replicase gene	Pea	MacFarlane and Davies, 1992
Potato moptop virus (PMTV)/modified triple gene block gene	Potato	Melander et al., 2001
Potato virus X (PVX)/ ORF1 gene (replicase)	Tobacco	Braun and Hemenway, 1992; Longstaff et al., 1993; Vardi et al., 1993
Potato virus Y (PVY)/NIa, NIb and CP genes	Potato	Maiti et al., 1999
PVY° strain/P1 gene	Potato	Mäki-Valkama, Pehu, et al., 2000, 2001
Tobacco etch virus (TEV)/NIa gene	Tobacco	Fellers et al., 1998
Tobacco vein mottling virus (TVMV)/NIa gene	Tobacco	Fellers et al., 1998
Tobacco mosaic virus (TMV)/ replicase gene sequence (50kDa)	Tobacco	Golemboski et al., 1990
Tomato mottle virus (ToMoV)/ MP genes (*BC1* and *BV1*)	Tobacco	Duan et al., 1997
Tomato spotted wilt virus (TSWV)/nucleoprotein (N) gene sequences	Tobacco	Prins et al., 1995, 1996
Tomato yellow leaf curl virus (TYLCV)/Rep protein C1 gene	Tomato cv. Money Maker	Brunetti et al., 1997

lesser final SMV incidence values and they yielded better compared with nontransformed control plants. This appears to be the first field study demonstrating the effectiveness of PDR on temporal and spatial dynamics of virus spread in soybean (Steinlage et al., 2002).

A transgenic clone (C-5) of transgenic plum trees *(Prunus domestica)* expressing CP genes of plum pox virus (PPV) was found to be resistant by testing the clone inoculated by viruliferous aphids or chip budding using DAS-ELISA format (Ravelonandro et al., 1997). The tomato lines transformed with CMV-CP gene contained lower concentrations of the virus, as determined by ELISA test and also exhibited only mild symptoms of the disease (Murphy et al., 1998). The resistance of homozygous lines of tomato variety UC 82B transformed with four different CP genes of CMV was demonstrated under field conditions in Italy by Tomassoli et al., (1999). The P-1 sequence of potato virus (PVY^O) in antisense orientation was shown to provide effective, strain-specific resistance in transgenic potato cv. Pito plants whose level of resistance was determined by ELISA based on the virus titer (Mäki-Valkama, Pehu, et al., 2000). A subsequent study to analyze the mechanisms and specificity of resistance indicate that the resistance was based on posttranscriptional gene silencing (PTGS). The resistance provided by P-1 transgene was specific to PVY^O isolates, the PVY strain group from which the P-1 transgene was derived (Mäki-Valkama, Valkonen, et al., 2000). Field evaluation of the transformed lines showed that the transgenic lines were fully protected from infection by PVY^O transmitted by aphids (Mäki-Valkama et al., 2001). Potato cv. Igor became unfit for cultivation because of its high susceptibility to PVY^{NTN} strain which caused potato tuber necrotic ringspot disease (PTNRD). Transgenic lines of this potato cultivar expressing CP gene sequence of PVY^{NTN} strain, showed high resistance. No symptoms of infection on foliage and tubers were seen and the virus could not be detected by ELISA or infectivity assay (Racman et al., 2001).

In the transgenic lines of sweet potato expressing CP gene of sweet potato feathery mottle virus (SPFMV) causing russet crack disease, virus accumulation at three months after inoculation was suppressed, as indicated by ELISA. These transgenic lines showed high resistance not only to primary but also to secondary infection by SPFMV-S strain (Okada et al., 2001). The effectiveness of transgenic resistance in protecting papaya against papaya ringspot virus was demonstrated under field conditions. The yields of transgenic cultivars (SunUp and Rainbow) were threefold greater than nontransformed controls (Ferreira et al., 2002).

In developing grapevine plants resistant to viruses transmitted by nematodes, transgenic *Nicotiana benthamiana* plants expressing arabis mosaic nepovirus (ArMV)-CP gene were generated. Accumulation of ArMV-CP in the form of virion-like isometric particles (VLPs) could be observed by employing ELISA and ISEM. The VLPs reacted with the antiserum specific to ArMV. Resistance to ArMV was expressed as delay in infection and a reduction in percentage of transgenic plants infected. However, transgenic grapevine *(Vitis rupestris)* did not accumulate the ArMV-CP at levels detectable by ELISA and no VLPs could be visualized by ISEM as in *N. benthamiana* (Spielmann, Krastanova, et al., 2000). In another study, transgenic lines of *N. benthamiana* and grapevine *(V. rupestris)* expressing CP genes of ArMV or grapevine fanleaf nepovirus (GFLV) were developed and the accumulation of CP was analyzed by ELISA. Resistance to both ArMV and GFLV could be demonstrated only in *N. benthamiana,* but not in grapevine. No transgenic CP of ArMV and GFLV was detectable in grapevine (Spielmann, Douet-Orhant, et al., 2000). In contrast, various levels of expression of GFLV-CP in transgenic plants of rootstock Coudere 3309 were detected by Xue et al. (1999). However, level of resistance to transgenic lines to GFLV has not been determined.

Among 11 independently transformed wheat *(T. aestivum)* lines transformed with CP gene of wheat streak mosaic virus (WSMV), one line showed high resistance to WSMV strains. This line exhibited milder symptoms and had lower virus titer as determined by immunological tests. However, no CP expression could be detected, though a high level of degraded CP mRNA expression was revealed by northern hybridization (Sivamani et al., 2002).

Tobacco plants were transformed with nucleocapsid (N) gene of TSWV-D (dahlia isolate). The highly resistant Burley 21 transgenic lines were resistant to only 44 percent of heterologous TSWV isolates tested. Symptomless plants of Burley 21 transgenic lines accumulated appreciable amounts of nucleocapsid protein, as determined by ELISA. This transgenic resistance against TSWV was effective in reducing disease incidence under field conditions (Herrero et al., 2000). The expression of N gene of TSWV in *Osteospermum eck-*

lonis was estimated by DAS-ELISA and Western and Northern blotting. The development of symptoms in transgenic lines, after inoculation of TSWV mechanically or by the thrips *(Frankliniella occidentalis),* was monitored by triple antibody sandwich enzyme-linked immunosorbent assay (TAS-ELISA). One clone with multiple transgene integration was identified. However, there was no improvement in the level of resistance of transgenic lines to TSWV (Vaira et al., 2000).

Transgenic plants expressing nonstructural proteins of plant viruses have been evaluated for resistance to viruses. In addition, the mechanism of development of resistance has been studied using immunological techniques. Immunogold electron microscopy was employed to visualize in situ localization of the 50 kDa protein encoded by ORF-2 of apple chlorotic leafspot virus (ACLSV) genome in infected *Chenopodium quinoa.* The 50 kDa protein (considered as the movement protein of ACLSV) was localized in the plasmodesmata and nearby cytoplasm. In transgenic *Nicotiana occidentalis* leaves expressing the 50 kDa protein fused to enhanced green fluorescent protein (EGFP), green fluorescence was seen as spots on the cell wall of several types of epidermal, palisade, and spongy mesophyll and collenchyma cells. The viral protein-EGFP fusion accumulated in sieve elements and formed an extensive interconnecting network of threadlike structures. The ACLSV 50 kDa protein may target plasmodesmata and traffic into sieve elements (Yoshikawa et al., 1999). Barley cv. Golden Promise expressing hairpin (hp) RNA containing barley yellow dwarf virus-PAV (BYDV-PAV) sequences exhibited immunity to the virus. BYDV-PAV could neither be detected by ELISA nor recovered from resistant plants by the aphid vector *Rhopalosiphum padi.* However, these plants were susceptible to a related luteovirus, European cereal yellow dwarf virus-RPV, indicating that the immunity was virus specific and not broken down by the presence of the related virus under field conditions (Wang et al., 2001).

It may be possible to enhance the level of resistance by the expression of multiple genes from viral pathogens which may exhibit additive action. Tomato plants transformed with a combination of two CP genes from isolates of CMV belonging to two different subgroups of CMV showed extremely high resistance to the virus. The transformants exhibited broad resistance against strains of both groups at high frequency, indicating the practical value of the resistance that can be exploited advantageously (Kaniewski et al., 1999).

In contrast, such an approach may be undesirable in potyviruses. Different sequences of potyviral genomes such as NIa, NIb, and CP coding regions have been used to obtain transgenic plants with resistance to PVY. Tobacco plants transformed with CP or NIa + NIb + CP (NNC) genes of TVMV were evaluated for resistance to TVMV. Plants expressing NNC genes in combination were less resistant than plants with CP gene alone. Furthermore, lines with NNC genes showed no resistance to tobacco etch virus (TEV), another potyvirus to which lines with CP gene alone were resistant. Thus, use of more than one pathogen-derived resistance (PDR)-determinant may be undesirable for the development of virus-resistant plants in some pathosystems (Maiti et al., 1999). On the other hand, coexpression of potato PVY^O CP and the codon-modified *cryV-Bt* gene from *Bacillus thuringiensis* ssp. *kurstaki* (toxic to Lepidopteran and Coleopteran insects) in potato cv. Spunta resulted in the development of resistance to both the virus and insect pests. A similar approach was made to develop a potato cultivar resistant to potato leaf roll virus (PLRV) and Colorado potato beetle *(Leptinotarsa decemlineata)* by expressing the unmodified full-length PLRV replicase gene and *Cry 3A* gene from *B. tenebrionis* in potato cv. Russet Burbank. No virus could be detected in symptomless potato plants. The resistance was high under natural conditions and against plant-to-plant transmission of PLRV by the aphid vector *(Myzus persicae)*. Three resistant lines have been released for commercial cultivation (Thomas et al., 2000). Hence, by selecting suitable genes of resistance, transgenic plants showing resistance to more than one biotic stress may be generated.

Developing virus-resistant plants by expression of antibodies in plants has been the subject of much research. The problem of expression of entire antibody was overcome by expressing only a single chain (scFv) of the antibody molecules. Tavladoraki et al. (1993) were able to demonstrate that transgenic tobacco lines expressing an antibody chain directed against artichoke crinkle mottle virus (ACMV) were highly resistant to the virus. The resistance was due to specific binding of the antibody chain to ACMV particles in protoplasts. Two approaches, as extensions of this concept, appear to be promising: targeting double-stranded RNA, which is an intermediate generated during replication of RNA viruses, and using mammalian defense systems in plants through the expression of oligoadenylate synthetase. This enzyme is activated by the presence of ds-RNA and polymerizes ATP into an oligomeric form that activates a latent ribonuclease which can degrade viral and cellular RNAs. Truve et al. (1993) successfully generated potato plants expressing oligoadenylate synthetase and demonstrated that the transgenic potato plants were resistant to PVX under field conditions

A monoclonal antibody (MAb 24) specific to TMV was expressed in tobacco cv. Petite Havana SR1 suspension culture derived from a stably transformed transgenic plant. This full-size recombinant antibody was retained by the plant cell wall and was not present in the culture medium. SDS-PAGE and immunoblot analyses showed that highly purified Mab 24 could be obtained by suspension culture and affinity purification thus indicating the potential of plant cell suspension cultures as bioreactors for the production of recombinant antibodies (Fischer, Liao, et al., 1999; Fischer, Schumann, et al., 1999). Transgenic tobacco plants expressing an scFv region of antibody fragment derived from a broad spectrum MAb 3-17 that was raised against Johnson grass mosaic potyvirus were generated. This MAb reacted strongly with 14 potyvirus species. The expressed recombinant proteins were targeted either to apoplasm or cytoplasm. The transgenic plants were protected against both local and systemic infections when challenged with PVY strain D and clover yellow vein virus (ClYVV) strain 300 (Xiao et al., 2000).

Membrane-anchored antiviral antibodies have been postulated to confer resistance in transgenic plants against viruses. The TMV-specific scFv24 was fused with human platelet-derived growth factor receptor (PDGFR) transmembrane domain or the T-cell receptor β-domain (TcRB) transmembrane domain to target expression of scFv24 as an extracellularly facing plasma membrane protein. The functional expression of the recombinant fusion scFv-PDGFR and scFv24-TcRB in transgenic tobacco suspension cultures and transgenic plants was confirmed by ELISA and Western blot analyses. Immunofluorescence (IF) and electron microscopy techniques revealed that the TcRB transmembrane domain targeted scFv24 to the tobacco plasma membrane. The transgenic tobacco plants expressing scFv24-TcRB exhibited resistance to infection by TMV, indicating that the membrane-anchored antiviral antibody fragments are functional and can be targeted to the plasma membrane *in planta* for developing virus-resistant plants (Schillberg et al., 2000).

Development of Transgenic Plants Resistant to Fungal Diseases

Because the fungal and bacterial pathogens are more complex than viral pathogens different kinds of approaches have to be adopted. The transformation of plants may be attempted (1) for the synthesis of novel protein(s) that may inhibit pathogen development; (2) to activate synthesis of compounds such as PR proteins involved in the development of host-plant resistance; and (3) to produce compounds that may inactivate enzymes or detoxify the toxins of pathogen origin. Among the approaches, the introduction of genes encoding potential antimicrobial PR proteins with chitinase or β-(1,3)-glucanase has been attempted for containing fungal pathogens. In addition, many biotic and abiotic inducers of resistance activate the synthesis of chitinases and glucanases in response to pathogen infection thus leading to development of SAR (see Chapter 12). Since single genes are involved in the production of these proteins/enzymes, the possibility of achieving success in developing resistant cultivars is greater than in the case of resistance governed by multiple genes.

A significant correlation between rapid and high-level expression of one or more PR proteins and host-plant resistance has been observed in some pathosystems. Hence, overexpression of genes encoding specific or multiple PR proteins is likely to improve the resistance level of host plants to fungal pathogens. This expectation has formed the basis for attempts to generate transgenic plants which have high-level expression of combinations of PR proteins with different modes of action against different pathogens leading to broad-spectrum, durable resistance. Chitin and β-(1,3)-glucanase form the major constituents of cell walls of several fungal pathogens. The chitinase gene from bean *(Phaseolus vulgaris)* was used to transform tobacco and oilseed rape *(Brassica napus)* plants that exhibited higher levels of chitinase activity resulting in significant increases in the resistance of transformed plants to *R. solani* (Broglie et al., 1991). Immunocytochemical observations revealed the damaged hyphal tips of *R. solani* in which greater vacuolization and cell lysis were evident (Benhamou et al., 1993). Likewise, enhanced levels of resistance of transformed *B. napus* var. *oleifera* with tomato endochitinase gene to fungal pathogens *Phoma lingam, S. sclerotiorum,* and *Cylindrosporium concentricum* was reported by Grison et al. (1996). The use of rice chitinase genes for improving the resistance of cucumber to *B. cinerea* (Tabei et al., 1997) and rose to *D. rosae* has been demonstrated. The constitutive expression of chitinase gene was positively correlated with the level of resistance to fungal pathogens. Expression of chitinase and β-(1,3)-glucanase genes simultaneously in tomato increased resistance to *F. oxysporum* f. sp. *lycopersici,* whereas expression of individual genes did not have any effect on the resistance level of transformed tomato (Jongedijk et al., 1995). A rice Class I chitinase gene *(chi-11)* normally expressed in seeds was introduced into the indica rice cultivar. The presence of two proteins (about 30 and 35 kDa) that reacted with the chitinase antibody could be detected by Western blot analyses in the transgenic rice lines. These transformed rice plants showed resistance to *R. solani* (Lin et al., 1995). A positive correlation between the level of chitinase expression and grades of resistance was observed in the transgenic lines of rice. In another study, the rice plants expressing high levels of thaumatin-like-protein (TLP) (PR-5) were protected against *R. solani* infection (Datta, Valazhahan, et al., 1999). These studies indicate that multigenic combinations of PR protein genes may provide more effective protection against fungal pathogens (Datta, Mutukrishnan, et al., 1999).

The effectiveness of chitinase genes from microbes that possess antimicrobial activity has been tested against certain fungal pathogens. The bacterial chitinase gene from *Serratia marcescens* was used to transform tobacco. Transgenic tobacco cells accumulated high levels of chitinase protein (eightfold) and the plants exhibited enhanced levels of resistance to *R. solani* (Jach et al., 1992).

The genes encoding antifungal proteins endo-and exochitinases from *Trichoderma atroviride (T. harzianum),* which has been employed as a biocontrol agent, were used to transform apple cv. Marshall McIntosh. Resistance of the transformed apple plants to *Venturia inaequalis* causing scab disease, was positively correlated with the levels of expression of either enzyme (protein). Plants expressing both enzymes simultaneously were more resistant than plants expressing either single enzyme at the same level. Both enzymes acted synergistically resulting in greater protection to transformants against the scab disease (Bolar et al., 2001).

An antibody specific against the purified Valencia orange basic chitinase was employed to screen cDNA expression library. The identified Valencia flavedo chitinase cDNA was designated *chi1* and encoded a predicted polypeptide of 231 amino acid. The CHI1 protein shares 60, 58, and 56 percent identity with basic chitinase proteins of rice, grape, and maize, respectively. Expression of *chi1* gene induced resistance in orange fruits against *Penicillium digitatum* causing green mold disease. Similar resistance was induced by other treatments such as UV-irradiation, hot water brushing, and application of β-aminobutyric acid and the biocontrol agent *Candida oleophila* (Porat et al., 2001). Broccoli (*Brassica oleracea* var. *italica*) plants expressing the endochitinase gene of *T. harzianum* showed significant increase in resistance to *A. brassicola* and this resistance was correlated to the level of expression of endochitinase activity (Mora and Earle, 2001).

Transgenic plants with increased resistance to fungal pathogens have been developed by transferring genes governing production of antifungal proteins, genes controlling the functions such as RIPs, and genes that are involved in the modification of host metabolism such as synthesis of phytoalexins (Narayanasamy, 2002). Puroindolines (PINs) from wheat have antimicrobial properties. Transgenic rice (cvM202) plants expressing the PIN genes *pinA* and/or *pinB* throughout the plants were generated. PIN extracts from the leaves of transformed plants reduced the growth of pathogens *M. grisea* and *R. solani*. Significant increase in the tolerance of the transgenic rice lines to diseases (blast and sheath blight) was also observed (Krishnamurthy et al., 2001). The gene encoding trichosanthin, a type I ribosome inactivating protein (RIP) isolated from the tuber of *Trichosanthes kirilowii* was used to transform rice (*O. japonica* cv. Zhonghua 8) callus cells by bombardment. Transgenic plants expressing trichosanthin were generated and tested for their resistance to rice blast fungus *M. grisea (Pyricularia oryzae)* by inoculating the leaves with the spores of the pathogen. The transgenic plants exhibited resistance to the blast disease, indicating the protective role of trichosanthin against the fungal pathogen (Yuan et al., 2002). The garden rose cultivars were transformed with different combinations of antifungal defense genes. The secretion of RIPs into extracellular space resulted in the reduction in the susceptibility of transformants to black spot disease caused by *Diplocarpon rosae* (Dohm et al., 2002).

The hemolymph of the giant silk moth *Hyalophora cecropia* contains three classes of proteins/peptides: cecropins, attacins, and lysozymes (Bonman and Hultmark, 1987). Cecropins belonging to the group of cytolytic pore-forming peptides possess broad-spectrum antimicrobial activity. A cecropin A-based peptide effectively inhibited the germination of *Colletotrichum coccodes* causing postharvest rot disease in tomato fruits. The yeast *Saccharomyces cerevisiae* was transformed with the DNA sequence encoding cecropin A peptide and the transformants inhibited the growth of germinated *C. coccodes* conidia and subsequently inhibited the development of decay in tomato fruits cv. Roma caused by *C. coccodes*. This study opens up new avenues for the biological control of postharvest diseases of vegetables and fruits (Jones and Prusky, 2002).

Development of Transgenic Plants Resistant to Bacterial Diseases

Transgenic plants showing resistance to bacterial pathogens have been generated by introducing alien genes of plants, genes encoding for antibacterial proteins from diverse sources such as nonhost plants, bacteriophage lysozyme, cytolytic pore-forming peptides, and human lactoferrin (Narayanasamy, 2002). The *Xa21* gene from the wild rice *Oryza longistaminata*, which is highly resistant to the bacterial leaf blight (BLB) disease caused by *X. oryzae* pv. *oryzae (Xoo)*, was used to transform rice cultivar IR24. The *Xa21* gene coding for a receptor kinase-like protein was inherited as a single gene in transgenic rice progenies and the transgenic rice plants were resistant to various isolates of *Xoo* (Wang et al., 1996; Ronald, 1997). The combination of *Xa21* with *Xa4*, which was already present in rice cv. IR 72, further enhanced the level of resistance of this rice cultivar (Tu et al., 1998). The cecropins (from *H. cecropia*) placed in a group of cytolytic pore-forming peptides exhibited broad-spectrum antibacterial activity against both Gram-negative and Gram-positive bacteria. Expression of cecropin gene in tobacco provided effective protection to transformed tobacco plants against wild fire, a serious disease caused by *P. syringae* pv. *tabaci* (Huang et al., 1997).

Another novel approach for generating transgenic plants resistant to *P. syringae* pv. *tabaci* uses the pathogen gene itself. This pathogen produces a toxin called tabtoxin which induces systemic necrosis in susceptible plants. Because the bacterial pathogen produces an enzyme acetyl transferase that acetylates tabtoxin to an inactive form, the pathogen is not affected by its own toxin. Hence the bacterial tabtoxin resistance gene *(ttr)* was used to transform tobacco plants. The transgenic tobacco plants were resistant to wild fire disease under field conditions (Batchvarova et al., 1998).

Attacins are another group of antibacterial proteins present in the hemolymph of *H. cecropia.* Attacins are small lytic peptides active against several Gram-negative bacteria. The expression of the attacin E in pear cv. Passe-Crassane (susceptible to fire blight pathogen *E. amylovora*) resulted in enhanced levels of resistance compared to nontransformed control plants. Clones showing higher levels of synthesis of attacin as determined by Western blot analysis were selected (Reynoird et al., 1999). Potato clones were transformed with either the gene encoding the acidic attacin from *H. cecropia* or the gene encoding the cecropin analogue peptide SB-37. Some of the clones exhibited resistance to potato black leg and soft rot diseases caused by *E. carotovora* ssp. *atroseptica* (Arce et al., 1999). Chicken lysozyme gene *(chly)* was used to transform potato to improve resistance to *E. carotovora* ssp. *atroseptica.* The level of transgene expression in transformed potato lines correlated with the degree of resistance of transformants (Serrano et al., 2000).

SUMMARY

Development of resistant cultivars is an important long-term strategy for effective management of plant health. To achieve this objective, two approaches have been made. Breeding for resistance to crop diseases results in the transfer of disease-resistant genes from wild relatives of crop plants. However, transfer of resistance genes through conventional methods has been difficult in many plant species. Engineering resistance has helped to overcome some of the limitations of the earlier approach and uses resistance genes from both plant and animal kingdoms. The studies on the molecular biology of host-plant resistance have provided an insight into the mechanisms of development of resistance to diseases. Various immunological techniques have been used to assess the levels of resistance of parents and progenies of crosses/engineered plants to determine the nature of products of pathogens and host plants. Localization and distribution of defense-related compounds such as PR proteins have also been visualized using appropriate immunological methods.

APPENDIX: QUANTIFICATION OF INFECTION AND DETECTION BY IMMUNOLOGICAL TECHNIQUES

Quantification of Infection of Alternaria cassiae by Radioimmunosorbent Assay (RISA) (Sharon et al., 1993)

Preparation of Antiserum

1. Culture the fungal pathogen on appropriate medium; harvest the mycelium by gently using a rubber policeman; wash the mycelia three times with 100 ml of PBS containing mixed sodium phosphates (10 mM), NaCl (150 mM), and KCl (3 mM), pH 7.4, and subject to centrifugation at 10,000g for 5 min.
2. Separate the mycelium from the supernatant; suspend the mycelium (100 mg) in PBS (10 ml) and homogenize at 0°C with a glass homogenizer.
3. Inject the mycelial homogenate into a female New Zealand White rabbit subcutaneously; provide five booster injections at ten-day intervals commencing at 21 days after the first injection; collect the blood by incising marginal ear vein at ten days after last booster injection; separate the antiserum after allowing the blood to stand for two hours at room temperature, followed by centrifugation at 5,000 g_n for ten minutes; store in one ml aliquots at –20°C and dilute the antiserum (1:200) before use.

Radioimmunosorbent Assay (RISA)

1. Dispense the dilutions of fungal homogenates in PBS at 100 μl per well of 99-well polyvinyl chloride microtiter plates and incubate overnight at 4°C.
2. Wash the wells three times (for one minute each) with 200 μl of PBS, containing bovine serum albumin (1 percent) (blocking buffer) for blocking nonspecific binding sites; incubate for two hours at 37°C and wash the wells three times as done earlier.

3. Transfer the antiserum (at 1:200 dilution) (75 µl/well); incubate for two hours at 37°C and wash again three times as done earlier.
4. Add ^{125}I protein A (0.025 µCi) in blocking buffer to each well; incubate for two hours at 37°C and wash the wells again three times as done earlier.
5. Air-dry; cut the wells from microtiter plates with scissors; place them in glass tubes and count in a gamma counter (Gammamatic, Kontron, Zurich, Switzerland).

Quantification of Aphanomyces euteiches *in the Pea Lines with Different Levels of Resistance by ELISA (Kraft and Boge, 1994)*

Preparation of Antiserum

1. Maintain the culture of *A. euteiches* in cornmeal agar and multiply in oatmeal broth in the dark at 22°C for 21 days; wash the mycelial mat three times in sterile distilled water free of residual nutrients and grind the mycelial mat submerged in distilled water in a mortar with a pestle.
2. Sonify the triturate at the highest power setting in an ice bath to disrupt the oospores; air-dry the triturate; grind into powder and store at –4°C until it is used.
3. Inject eight-week-old New Zealand white female rabbits intramuscularly with a mixture consisting of the fungus preparation (1ml) and Freund's complete adjuvant (1 ml); provide booster injection at 11, 16, and 56 days after the first injection and bleed the animal at 18 days after last injection and continue bleeding at weekly intervals for six weeks.
4. Test the antiserum collected after blood clotting and centrifugation against the powdered fungus preparation (already prepared) after diluting to desired concentrations (0.25, 0.10, 0.04, and 0.02 mg/ml) in the CEP buffer containing Na_2CO_3 (1.59 g), $NaHCO_3$ (2.93 g), NaN_3 (0.2 g), polyvinyl pyrrolidone (10 g), and egg albumin (1 g) per liter of distilled water, pH 9.6.

Quantification of A. euteiches *by ELISA*

1. Prepare dilutions of all fungal preparations in CEP buffer; dispense the test samples to microtiter plates at 0.1 ml/well and incubate in a moist chamber for one hour.
2. Wash the wells three times (for five minutes each) with Tris-Tween containing Tris-base (1.2 g), NaCl (8.0 g), KCl (0.2 g), Tween-20 (0.5 ml) in water (1 l).

3. Dilute the antiserum to 1:1000 in Tris-EP consisting of Tris-Tween (500 ml), polyvinyl pyrrolidone 40 (10 g), and egg albumin (1 g); dispense 0.1 ml to each well and incubate for one hour and wash the plates as done earlier.
4. Dilute the goat antirabbit IgG-conjugated with alkaline phosphatase to 1:1000 in Tris-EP; transfer 0.1 ml of diluted conjugate to each well; incubate for one hour and wash three times as done earlier.
5. Prepare the substrate (Sigma 104) in diethanolamine (97 ml) + H_2O_2 (800 ml) + NaN_3 (0.2 g) (at 1 mg/ml) and transfer to each well and incubate for one hour.
6. Record the absorbance at 405 nm using the ELISA microplate reader and calculate the average of readings of at least three wells for each sample.

Assay of Pathogen Content in Root Tissues

1. Plant seeds of test pea lines after surface sterilization with NaOCl (0.5 percent) for one minute.
2. Inoculate the seedlings by applying 50 ml of zoospore suspension directly to the base of each seedling and incubate for five, seven, and nine days under greenhouse conditions, maintaining the moisture level at a saturated condition.
3. Collect the taproot tissue (0.3 g/sample) and grind with CEP buffer (3.5 ml) in mortar with a pestle.
4. Store overnight at 4°C and perform the ELISA tests as previously and repeat the tests twice.

Immunological Techniques for Detection and Immunolocalization of PR Proteins (Liljeroth et al., 2001)

PR Protein Extraction

1. Freeze the plant tissues in liquid nitrogen and store at –80°C until analysis.
2. Homogenize samples (5 g) in liquid nitrogen using a Waring blender and extract with citrate buffer, pH 2.8, containing mercaptoethanol (14 mM).
3. Centrifuge the homogenate at 5,000 rpm for 15 minutes; filter the supernatant through Whatman 1F paper and dialyze against Tris buffer (1 mM), pH 6.8.
4. Concentrate the samples by dialysis against polyethylene glycol (PEG) (20 M) and determine the protein concentration of the samples.

Enzyme-Linked Immunosorbent Assay (ELISA)

1. Dispense 50 µl (2 µg protein) aliquots of samples to microtiter plate wells; incubate at 37°C overnight to evaporate all water and block the nonspecific sites in the wells with bovine serum albumin (BSA) (0.1 percent).
2. Add antisera diluted in TBST consisting of TBS buffer, pH 7.5, and 0.1 percent Tween-20 to the wells.
3. Add antirabbit IgG horseradish peroxidase conjugate diluted to 1:10,000 in TBST as secondary antibody.
4. Develop color by adding 2,2'-azino-bis (3-ethylbenzthiazoline-6, sulfonic acid (ABTS) as substrate (Vector Lab Burlingama, California) according to the manufacturer's instructions and determine the absorbance at 405 nm after an incubation for 30 minutes at room temperature.

Note: After each step (1), (2), and (3) wash the wells three times with TBST (100 µl/well)

Immunolabeling Technique

1. Fix the tissue (roots) in paraformaldehyde (2 percent), glutaraldehyde (0.1 percent) in phosphate buffer, pH 7.2, for three hours; dehydrate through an ethanol series (25, 50, and 75 percent v/v) for 20 minutes at each concentration and overnight at 100 percent concentration and pass through a xylene series (25, 50, 75, and 100 percent in ethanol) for one hour each.
2. Embed the root tissue in paraplast by a graded series of 25, 50, and 75 percent (v/v) paraplast in xylene for three hours each and finally incubate in 100 percent paraplast overnight.
3. Cut transverse sections (10 µm) using Leitz microtome and mount the sections on Superfrost plus slides (Histolab products, AB, Göteborg, Sweden).
4. Incubate the slides at 45°C overnight and then at 60°C for 30 minutes.
5. Deparaffinize the sections as follows: 2 × 5 minutes in xylene; 2 × 3 in 100, 75, 50 percent ethanol in distilled water and 2 × 3 in distilled water and wash the slides with phosphate-buffered saline (PBS), pH 7.2, for five minutes.
6. Block the slides with undiluted blocking solution from the BM Chemiluminiscence Western Blotting Kit, Roche Diagnostics, Mannheim, Germany and rinse the slides in PBS-BSA (0.1 percent).
7. Add antiserum against PR protein diluted in PBS-BSA (150 µl/slide); incubate in a moist chamber for 2.5 hours at room temperature and

rinse in PBS-BSA 3 × 5 min, maintaining normal rabbit serum as control.

8. Incubate the slides with alkaline phosphatase labeled secondary antibody (Bio-Rad Laboratories, Hercules, California) for one hour and rinse in PBS-BSA 3 × 5 minutes.

9. Stain the slides with bromo-chloro-indolyl posphate/nitroblue tetrazolium (BCIP/NBT) for 20 minutes at room temperature.

Chapter 12

Induction of Disease Resistance and Production of Disease-Free Plants

Development of cultivars resistant to microbial pathogens either by breeding for resistance or engineering resistance (Chapter 11) is undoubtedly the most desirable strategy for the management of crop diseases. However, the limitations of these approaches have necessitated the search for alternative methods for containing diseases caused by microbial pathogens. Induction of resistance in existing high-yielding but susceptible cultivars by using biotic or abiotic inducers of resistance is considered a feasible approach that can be applied to all crops. Furthermore, neither time-consuming breeding procedures nor expensive techniques are necessary for inducing resistance to diseases. Use of disease-free propagative materials and plants is a basic disease management strategy that is effective in avoiding adverse effects of microbial plant pathogens on seedlings. The usefulness of immunological techniques in achieving these objectives is discussed in this chapter.

INDUCTION OF RESISTANCE TO MICROBIAL PLANT PATHOGENS

Induction of pathogen resistance in plants has been differently named as physiological acquired immunity (Chester, 1933), acquired immunity (Price, 1936), and immunization (Tuzun and Kuć, 1991; Kuć, 1991). The use of biocontrol agents against plant pathogens has also been designated as acquired immunity by Cook and Baker (1983). Resistance that has developed in inoculated leaves against tobacco mosaic virus is called localized acquired resistance, whereas resistance induced in uninoculated leaves far away from inoculated leaves is termed systemic acquired resistance (SAR) Ross (1961 a,b). The development of SAR frequently depends on the formation of necrotic lesions (HR). Such resistance can be induced by viruses, bacteria, and fungi (biotic inducers), as well as by abiotic inducers such as salicylic acid (SA), 2,6-dichloro-isonicotinic acid, chitosan (Métraux et al., 1991; Ward et al., 1991; El-Ghaouth et al., 1994), and plant products such as antiviral principles (AVPs) (Narayanasamy, 1990). The effectiveness of protection by SAR depends on the agent employed and the extent of development of HR. Plant growth-promoting rhizobacteria (PGPR) offer effective protection to crop plants by direct action on pathogens and production of antibiotics and also by indirect action by inducing resistance (Narayanasamy, 2002). The resistance induced by PGPR differs from SAR in that development of resistance is not dependent on the formation of HR and hence it is referred to as induced systemic resistance (ISR) (Pieterse et al., 1996). PGPR-mediated ISR against plant pathogens and insect pests has been critically reviewed by Ramamoorthy et al. (2001).

Induction of Resistance by Viruses

Two types of resistance—localized and SAR—have been recognized in tobacco cv. Samsun NN inoculated with TMV. The resistance induced was nonspecific and the inoculated plants developed resistance not only to unrelated viruses such as tobacco necrosis virus (TNV), tobacco ringspot virus (TRSV), and turnip mosaic virus (TuMV), but also to the fungal pathogen *Thielaviopsis basicola* (Ross, 1966). The possibility of inducing resistance to several diseases infecting various crops has been indicated (Table 12.1). The time required for development of SAR depends on the crop plant and the nature of the inducer of resistance. Following the application of inducers certain families of genes collectively known as SAR genes are activated. Irrespective of the nature of the inducer (biotic or abiotic) the same spectrum of SAR gene expression occurs and it leads to the development of resistance in treated plants. The proteins encoded by SAR genes have been shown to possess antimicrobial properties and they include PR proteins and enzymes related to the development of resistance to diseases caused by microbial pathogens.

Immunological techniques have been used to detect and quantify defense proteins and to detect microbial pathogens. The products of

TABLE 12.1. Induction of resistance to crop diseases by using biotic inducers

Inducer	Crop/Pathogen	References
A. Viruses as inducers		
Tobacco mosaic virus (TMV)	Tobacco/TMV	Ross, 1961a,b Yalpani et al., 1993
	Tobacco/tobacco necrosis virus (TNV), tobacco ringspot virus (TRSV), turnip mosaic virus (TuMV), and *Thielaviopsis basicola*	Ross, 1966
	Tobacco/*Peronospora tabacina*	Xie and Kuć, 1997
	Tobacco/*Erysiphe cichoracearum*	Raggi, 1998
Tobacco necrosis virus (TNV)	Cucumber/*Pseudomonas syringae* pv. *lachrymans*	Métraux et al., 1988
	Tomato/*Phytophthora infestans*	Anfoka and Buchenauer, 1997
	Cucumber/*Sphaerotheca fuliginea*	Sticher et al., 1997
TNV/tobacco rattle virus (TRV)	Asparagus bean/TNV	Pennazio and Roggero, 1991
B. Fungi as inducers	Pathogens	
Colletotrichum lagenarium (Cl)	Cucumber, muskmelon, watermelon/Cl	Kuć, 1987, 1990
C. lindemuthianum	French bean/Cl	Dann et al., 1996
Drechslera teres (Dt)	Barley/Dt	Reiss and Bryngelsson, 1996
Phytophthora infestans (Pi)	Tomato/Pi	Christ and Mösinger, 1989
	Potato/Pi	Enkerli et al., 1993
P. tabacina (Pt)	Tobacco/Pt	Ye et al., 1990
	Nonpathogens	
Penicillium oxalicum	Tomato/*Fusarium oxysporum* f. sp. *lycopersici*	De Cal et al., 1997
Phytophthroa cryptogea	Potato/*Pi*	Quintanilla and Brishammar 1998
(Binucleate) *Rhizoctonia* (BNR)	Bean/*Rhizoctonia solani* and *Colletotrichum lindemuthianum*	Xue et al., 1998

Inducer	Crop/Pathogen	References
C. Bacteria as inducers		
Pseudomonas aeruginosa	Bean/*Botrytis cincerea*	Meyer and Höfte, 1997
	Tomato/*Pythium splendens*	Buysens et al., 1996
P. fluorescens	Radish/*Fusarium oxysporum* f. sp. *raphani* and *Alternaria brassicola*	Leeman et al., 1995 Hoffland et al., 1996
P. fluorescens	Rice/rice tungro bacilliform (RTBV) and rice tungro spherical virus (RTSV)	Narayanasamy, 1995
	Bean/*P. syringae* pv. *phaseolicola*	Alström, 1991
	Rice/*Rhizoctonia solani* and *Pyricularia oryzae*	Vidhyasekaran and Muthamilan, 1999 Vidhyasekaran et al., 1997
	Sugarcane/*Colletotrichum falcatum*	Viswanathan and Samiyappan, 2002
P. putida and *Serratia marcescens*	Cucumber/*P. syringae* pv. *lachrymans*	Liu et al., 1996b Wei et al., 1996
S. marcescnes	Cucumber/*Fusarium oxysporum* f. sp. *cucumerinum*	Liu et al., 1995a.

defense genes include several enzymes such as peroxidase (PO) and polyphenol oxidase (PPO) which catalyze the synthesis of lignin and phenylalanine-ammonia-lyase (PAL) required for the production of phytoalexins and phenolics, which are considered to have an important role in the development of disease resistance in plants. PR proteins such as β-(1,3)-glucanases (PR-2) and chitinases (PR-3) may degrade cell walls of fungal pathogens resulting in lysis of the cells. Thaumatin-like proteins (TLPs) (PR-5) have antifungal properties and enhance resistance in plants. Transgenic plants (rice and wheat) expressing TLPs exhibit higher levels of resistance to fungal pathogens (Chen et al., 1999; Datta et al., 1999).

Resistance to viral diseases in plants may be induced by viruses, bacteria, and fungi that react by producing necrotic local lesions. Production of PR proteins in tobacco leaves and cucumber inoculated with TMV was demonstrated (Gianinazzi, 1983; Métraux et al., 1988). Constitutive expression of PR-1 protein and high levels of resistance to TMV in a hybrid of *Nicotiana glutinosa* × *N. debneyi* were observed (Ahl and Gianinazzi, 1982). PR-3 proteins could be detected by employing suitable immunological techniques. Tomato plants inoculated with TNV showed resistance to *P. infestans* causing late blight disease. Accumulation of PR proteins in the tomato plants exhibiting SAR was noted. By using PR protein-specific-antisera, four PR proteins were identified as P-14, AP24, chitinase, and β-(1,3)-glucanase. *Phytophthora infestans* was inhibited by the basic fractions of these PR proteins (Anfoka and Buchenauer, 1997). The development of SAR in tobacco inoculated with TMV against the powdery mildew pathogen, *Erysiphe cichoracearum,* appears to depend on the accumulation of hydroxyproline-rich glycoprotein (HRPG), since cell wall hydroxyproline (HyP) contents of TMV-protected plants increased significantly (Raggi, 1998).

Induction of Resistance by Fungi

Plants showing SAR to fungal pathogens accumulate many PR proteins including PR-1, β-(1,3)-glucanases (PR-2), chitinases (PR-3), PR-4, and thaumatin-like protein (PR-5) (Table 12.1). Following inoculation of barley leaves with *Drechslera teres* (blotch disease), *E. graminis* f. sp. *hordei* (powdery mildew) or *P. hordei* (rust), chitinases, β-(1,3)-glucanases, peroxidases, PR1a and PR1-b proteins, and thaumatin-like proteins accumulated. However, the role of these PR proteins in the development of resistance is not clearly understood (Reiss and Bryngelsson, 1996). Possibly, PR-2 and PR-3 proteins may be involved in the degradation of cell walls of fungal pathogens, while PR-5 proteins are likely to inhibit the development of fungal pathogens.

Induction of Resistance by Bacteria

Plant growth-promoting rhizobacteria (PGPR) act on plant pathogens directly or by inducing systemic resistance indirectly. Direct adverse effects are caused by antibiotics, siderophores, hydrogen cyanide produced, and competition for available nutrients. The effects of *P. fluorescens* on the development of *R. solani,* causing damping-off disease in sugar beet seedlings, have been studied in soil microcosms. The results of ELISA and direct microscopy have revealed the presence of PGPR on inoculated seeds and its inhibitory effect on the development of pathogen-formed mycelium biomass and sclerotia (Thrane et al., 2001). During the expression of ISR, different defense mechanisms are activated resulting in enhancement of activities of chitinases, β-(1,3)-glucanases, peroxidases and synthesis of PR proteins, accumulation of phytoalexins, and formation of protective biopolymers such as lignin, callose, and HPRGs. These compounds may either directly inhibit pathogen development or reinforce the natural barriers of the host plants leading to the restriction of pathogen invasion (Table 12.1). In some pathosystems, development of PGPR-mediated ISR appears to depend on the systemic accumulation of PR proteins (Maurhofer et al., 1994). The O-antigenic chain of outer membrane of lipopolysaccharides (LPS) from *P. fluorescens* may be the bacterial determinant required for the development of ISR in radish against *F. oxysporum* f. sp. *raphani* causing wilt disease (Leeman et al., 1995).

In rice plants (cv. IR 58) infected by *R. solani* causing sheath blight disease, the presence of two TLPs (25 and 24kDa) were detected using the antiserum raised against a Pinto bean TLP as the primary antibody in Western blots. The rice TLPs are encoded by a family of at least three genes which may be differentially expressed in responses to bacterial or fungal pathogens (Velazhahan et al., 1998). Transgenic rice plants expressing TLP gene and resistance to *R. solani* produced the expected 23 kDa TLP as confirmed by Western blot analysis. The antiserum specific to 23 kDa bean TLP was employed using horseradish peroxidase (HRP) conjugated with goat antirabbit IgG as the secondary antibody and made visible with HRP color reagent (Datta et al., 1999).

Treatment of sugarcane setts with *P. fluorescens* and *P. putida* induced accumulation of chitinase in germinating settling, whereas soil application of these PGPR strains induced chitinase activity systemically in sugarcane stalk tissues. The enhanced chitinase activity was related to suppression of development of red rot disease caused by *C. falcatum*. By using barley chitinase as the primary antibody, the induction of 18 kDa chitinase in sugarcane stalk tissues in response to PGPR treatment or inoculation with *C. falcatum* was observed using Western blot analysis (Viswanathan and Samiyappan, 2001) (Figure 12.1). Following treatment of chickpea *(Cicer arietinum)* with *P. fluorescens,* ISR against charcoal rot disease caused by *Macrophomina*

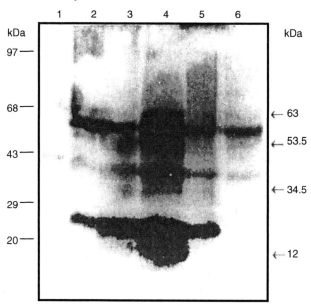

FIGURE 12.1.Detection of chitinase in sugarcane stalk tissues by barley chitinase antiserum by Western blot analysis: (Lane 1) marker; (Lane 2) *Colletotrichum falcatum* (Cf); (Lane 3) *Pseudomonas fluorescens* + Cf; (Lane 4) *P. putida* + Cf; (Lane 5) *P. putida;* (Lane 6) control (*Source:* Viswanathan and Samiyappan, 2001; Reprinted with permission by R. Samiyappan.)

phaseolina was observed. The level of chitinases and glucanases increased by six- to sevenfold up to four days postinoculation (Srivastava et al., 2001). Defense proteins and enzymes were induced following treatment of tomato plant with *P. fluorescens* and inoculation with *F. oxysproum* f. sp. *lycopersici* (wilt disease pathogen). β-(1,3)-glucanase, chitinase, and TLPs accumulated reaching high levels at three to five days after pathogen inoculation of bacterized plants. Western blot analysis using primary antibodies specific to barley chitinase and bean TLP, revealed the formation of chitinase isoform (46 kDa) and TLP isoform (33 kDa) in tomato roots treated with *P. fluorescens* (Ramamoorthy et al., 2002) (Figures 12.2 and 12.3).

The PGPR strains of *P. fluorescens* and *Serratia marscescens* have also induced systemic resistance to viral pathogens. The ISR activity could be inferred in cucumber-CMV pathosystems. Treatment of

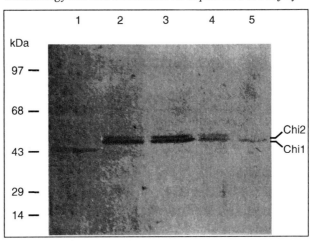

FIGURE 12.2. Western blot analysis of chitinase isoforms induced by *Pseudomonas fluorescens* (Pf) in tomato challenged with *Fusarium oxysporum* f. sp. *lycopersici* (FOL): (Lane 1) marker; (Lane 2) plants treated with Pf; (Lane 3) plants treated with Pf and challenged with FOL; (Lane 4) plants inoculated with FOL; (Lane 5) control (uninoculated) (*Source:* Courtesy of Ramamoorthy et al., 2002; Reprinted with kind permission of Kluwer Academic Publishers, Netherlands.)

seeds with PGPR strains has led to consistent reduction in mean numbers of symptomatic plants and delay in symptom expression. No viral antigen could be detected in asymptomatic plants by ELISA (Raupach et al., 1996). The strains Pf1 and CHAO strains of *P. fluorescens* were able to induce systemic resistance in rice against rice tungro virus disease when these strains were applied as seed treatment, root dipping, or foliar spray (Narayanasamy, 1995; Muthulakshmi, 1997). Tomato plants were protected by treatment with *P. fluorescens* against tomato spotted wilt virus (TSWV) (Kandan et al., 2002). Enhancement of the activities of defense-related enzymes such as peroxidase and phenylalanine-ammonia lyase was observed in rice and tomato plants treated with *P. fluorescens.*

Although PGPR have been shown to induce resistance in several pathosystems, the peanut late leafspot (LLS) disease pathosystem seems to be an exception. Application of seven PGPR strains including *P. fluorescens* or chemical elicitors such as salicylic acid failed to induce resistance to LLS disease caused by *Phaeoisariopsis per-*

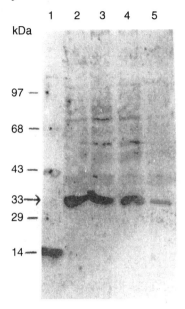

FIGURE 12.3. Western blot analysis of thaumatin-like protein (TLP) isoforms induced by *Pseudomonas fluorescens* (Pf) in tomato challenged with *Fusarium oxysporum* f. sp. *lycopersici* (FOL): (Lane 1) marker; (Lane 2) plants treated with Pf; (Lane 3) plants treated with Pf and challenged with FOL; (Lane 4) plants inoculated with FOL; (Lane 5) control (uninoculated) (*Source:* Ramamoorthy et al., 2002; Reprinted with kind permission of Kluwer Academic Publishers, Netherlands.)

sonata. LLS resistance in peanut is not systemically inducible in the same manner as other pathosystems (Zhang et al., 2001).

Induction of Resistance by Abiotic Inducers

Plant Products

The presence of antimicrobial proteins in many plant species has been detected by immunological techniques (Chapter 6). The genes encoding antimicrobial proteins have been used to transform crop plants for improved resistance to various crop diseases (Chapter 11). Many abiotic agents have been evaluated for their ability to induce resistance to microbial plant pathogens in crop plants (Table 12.2).

TABLE 12.2. Induction of resistance to crop diseases by using abiotic inducers

Inducer	Crop/Pathogen	References
A. Plant products		
Antiviral principles (AVPs) from sorghum and coconut	Peanut/bud necrosis virus (BNV)	Narayanasamy and Ramiah, 1983
AVPs from green gram and black gram	Rice/rice tungro viruses	Narayanasamy, 1995 Muthulakshmi and Narayanasamy, 2000
Phytolacca plant anti-viral protein (PAP)	Tobacco/potato virus X potato virus Y and CMV	Lodge et al., 1993
B. Chemicals		
1. Organic chemicals		
Salicylic acid (SA)	Tobacco/PVX and CMV	Métraux et al., 1991 Ward et al., 1991
	Tobacco/CMV	Naylor et al., 1998
	Sunflower/*Plasmopara helianthi*	Tosi et al., 2000
2,6-dichloroiso-nicotinic acid (INA)	French bean/ *Colletotrichum lindemuthianum*	Dann et al., 1996
Chitosan	Cucumber/*Pythium aphanidermatum*	El-Ghaouth et al., 1994
	Tomato/*F. oxysporum* f. sp. *lycopersici*	Benhamou et al., 1994
	Tomato/*P. infestans*	Oh et al., 1998
DL-β-amino-n-butyic acid (BABA) benzol-(1,2,3)-thiadiole-7-carbothioic acid *S*-methyl ester (BTH)	Sunflower/*Plasmopara helianthi*	Tosi et al., 2000
	Cowpea/*Colletotrichum destructivum*	Latunde-Dada and Lucas, 2001
2. Inorganic chemicals		
Potassium or so-dium/phosphate	Cucumber/*Colletotrichum lagenarium*	Gottstein and Kuć, 1989
	Vicia faba/*Uromyces* sp.	Walters and Murray, 1992
	Maize/*Puccinia sorghi*	Reuveni et al., 1994
3. Synthetic elicitors		
BION	Tomato/*F. oxysporum* f. sp. *lycopersici*	Takács and Dolej, 1998
Actigard [benzol-(1,2,3)-thiadiazole-7-carbothioic acid-*S*-methyl ester]	Cucumber/*C. lagenarium* and *Pseudomonas syringae* pv. *lachrymans*	Raupach and Kloepper, 1998

Abiotic inducers activate the same mechanisms of resistance that operate in resistant plants. Potential AVPs that are capable of inducing resistance to viruses in different crops such as peanut (groundnut) and rice have been identified (Narayanasamy, 1990). AVPs from sorghum and coconut leaves induced resistance in peanut plants to bud necrosis virus (BNV) (Narayanasamy and Ramiah, 1983). The AVP from coconut induced the formation of three new proteins (81 kDa, 52 kDa, and 44 kDa). These proteins correlated with the development of resistance to BNV (Narayanasamy and Ganapathy, 1986). Likewise, the protein (34 kDa) from *Clerodendrum aculeatum* induced resistance to TMV (Verma et al., 1996; Shelly Praveen et al., 2001).

Resistance to rice tungro disease caused by RTBV and RTSV was induced in susceptible rice cultivars (Co 43) using AVPs from the seed sprouts of pigeonpea and mungbean. The higher level of resistance was indicated by the low titers of RTBV and RTSV as determined by ELISA (Muthulakshmi and Narayanasamy, 1996). The changes in the levels of PR proteins following application of the AVPs on rice plants was monitored by employing specific antisera in ELISA. PR 1a protein attained a maximum level at 25 days after application of AVPs and challenged by RTV inoculation using viruliferous leafhoppers. Levels of PR-2 proteins increased. The reduction in the percentage of RTV infection and low virus titer were correlated with the induction of defense-related compounds by AVPs (Muthulakshmi and Narayanasamy 2000) (Figure 12.4).

Several plant species contain antimicrobial proteins that inhibit the development of plant pathogens and the diseases caused by them (Chapter 6). Transgenic plants expressing antimicrobial proteins from nonhost plants have been shown to be resistant to certain crop diseases (Chapter 11). The pokeweed *(Phytolacca americana)* antiviral protein (PAP) protected tobaco and pepper plants against infection by TMV and PVY (Chen et al., 1991). The mechanisms of action of antimicrobial plant proteins have not been clearly established. A putative α-glucosidase (97 kDa) was purified from coconut leaves *(Cocos nucifera),* nonhost of *R. solani,* by Western blot analysis. The α-glucosidase was able to degrade the toxin produced by *R. solani,* resulting in a significant reduction in sheath blight symptoms, indicating a role for the α-glucosidase of a nonhost in the protection of plants against fungal pathogens (Shanmugam, Sriram, et al., 2001) (see Appendix). A basic β-(1,3)-endoglucanase was purified from

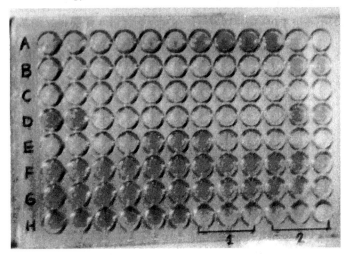

FIGURE 12.4. Determination of titers of RTBV by ELISA following treatment with antiviral principles (AVPs) and inoculation with RTV: (Well rows ABC) PAVP-treated + inoculation with RTV; (DEF) MAVP-treated + inoculation with RTV; (GH) RTV inoculation; (1) healthy (control); (2) blank (*Source:* P. Muthulakshmi, 1997; Reprinted with permission by Tamil Nadu Agricultural University, Coimbatore, India.)

C. sinensis cv. Valencia orange callus. Specific antibodies raised against this basic protein were used to screen callus cDNA expression library and to isolate its corresponding cDNA named gns1. The expression of the *gns1* gene was markedly induced by inoculation with *Penicillium digitatum* infecting orange fruits and also by biocontrol agent *Candida oleophila* which induced resistance to the fungal pathogen *P. digitatum.* The gene *gns1* appears to be part of the molecular mechanisms involved in pathogen defense responses in citrus fruit (Porat et al., 2002).

Chemicals

Organic compounds such as SA, 2,6-dichloroisonicotinic acid (INA), and acibenzolar S-methyl (BTH) have been tested for their ability to induce SAR. A positive correlation was shown between induction of SAR by SA and INA and systemic induction of several SAR genes (Dempsey et al., 1993; Mauch-Mani and Slusarenko,

1996). The activities of chitinase and β-(1,3)-glucanase showed significant increase in green bean plants treated with INA and accumulation of SA was also observed in treated plants (Dann et al., 1996). Development of resistance to CMV following application of SA was not due to inhibition of viral replication, but rather due to reduced systemic movement of CMV in SA-treated plants (Naylor et al., 1998). An MAB specific to SA was produced using an immunogen in which 5-aminosalicylic acid (5-ASA) was conjugated to KLH (hemocyanin from keyhole limpet) through its (C_5)-NH_2 group. The antibody was highly specific to SA and 5-ASA with no cross-reactivity with other structurally related compounds. The direct ELISA using this MAB had a detection range of 0.0195 to 20 nmol of SA. The fluctuations in the endogenous SA levels in cucumber following infection with *P. syringae* pv. *syringae* were monitored (Wang et al., 2002). Chitosan-treated cucumber plants were protected against *P. aphanidermatum*. In the treated plants several host-defense responses such as formation of structural barriers in root tissues and antifungal hydrolases such as chitinase chitosanase and β-(1,3)-glucanase were activated resulting in restriction of pathogen development (El-Ghaouth et al., 1994). In tomato, treatment with chitosan suppressed the development of wilt disease (*F. oxysporum* f. sp. *lycopersici*) and late blight disease (*P. infestans*). Chitosan seems to inhibit the pathogen development directly and also by inducing expression of defense-related SAR genes (Oh et al., 1998). A multicomponent defense response induced in grapevine tissues by both biotic and abiotic elicitors was studied using a cell culture system. SA, chitosan, methyl jasmonate, and elicitor released from cell walls of *B. cinerea* induced accumulation of PR proteins and key enzymes of phenyl-propanoid pathway to a great extent. The expression kinetics of four defense-related genes (*PR-1, PR-6, PAL,* and *CHI*) varied and depended on the nature of elicitor employed (Repka, 2001).

The efficiency of plant activators such as DL-β-amino-n-butyric acid (BABA) and benzol-(1,2,3)-thiadiazole-7-carbothioic acid *S*-methyl ester (BTH) in inducing production of PR proteins was assessed. In sunflower treated with SA and BTH, at least eight proteins, including two PR proteins (PR-1 and PR-5) serologically related to tobacco PR proteins, were induced in sunflower leaf discs. BTH, when applied as soil drench, also induced the production of a protein serologically related to the tobacco PR-5 protein, with or without inoculation of sunflower with *Plasmopara helianthi* (downy mildew). On the other hand, no additional proteins appeared in sunflower plants treated with BABA (Tosi et al., 2000). Induction of SAR in sugarcane following application of acibenzolar *S*-methyl as soil drench was demonstrated. Accumulation of PR-2, PR-3, and PR-5 proteins was observed in the treated plants which were protected against *C. falcatum* causing red rot disease (Sundar et al., 2001). Rose (cv. Iris Gee) treated with BTH exhibited resistance to *D. rosae* causing black spot disease. Induction and accumulation of a set of extracellular proteins were observed in the treated plants. Some of these proteins were identified as PR-1, PR-2, PR-3, and PR-5 proteins by immunoblot analysis probed with antisera raised against tobacco PR-1C, PR-N, PR-Q, and PR-S proteins. The accumulation of PR proteins was much more pronounced in BTH-pretreated leaves than in water-pretreated leaves upon inoculation with *D. rosae,* especially the 15 kDa PR-1, 36 and 37 kDa PR-2 proteins, which are likely to be more important in the expression of resistance to *D. rosae* (Suo and Leung, 2002a,b).

Among the inorganic chemicals tested, phosphates (sodium or potassium) have been found to be efficient inducers of SAR. Cucumber plants treated with phosphate salts exhibited resistance to anthracnose disease caused by *Colletotrichum lagenarium* (Gottstein and Kuć, 1989). SAR was induced by the foliar application of phosphate salts on maize against the rust pathogen *(P. sorghi).* Phosphates may generate an endogenous SAR signal because of calcium sequestration at points of phosphate application (Reuveni et al., 1994). Cucumber plants could be protected against powdery mildew caused by *Sphaerotheca fuliginea* by applying dipotassium phosphate (0.1 M). The activities of peroxidase and β-(1,3)-glucanase were markedly increased in cucumber plants treated with phosphate (Reuveni et al., 1997). These studies show the possibility of inducing resistance in plants using inexpensive chemicals such as phosphates without any adverse effects on crop plants.

CROSS PROTECTION

Protection of plants using mild or avirulent strains of a pathogen against severe strains is a practical disease management strategy, especially in perennial crops that suffer from several viral diseases. A mild strain of a virus induces a state of resistance in plants that become resistant to severe strains of the same virus or related viruses. The mechanism of development of resistance in cross-protected plants and transgenic plants expressing viral genes appears to be similar. The development of cross protection in plants inoculated with mild strains of virus may be studied using immunological techniques.

A monoclonal antibody capable of differentiating the severe strain of zucchini yellows mosaic virus (ZYMV-TW) from mild strains was identified by plate-trapped antigen (PTA)-ELISA, (DAS)-ELISA, and immunosorbent electron microscopy (ISEM) techniques. ELISA tests were employed to quantify the concentration of protecting (mild) and challenging (severe) strains. In cross-protected zucchini squash *(C. pepo)* plants that did not develop severe symptoms, the concentrations of ZYMV-TW strain was low as revealed by ELISA. In contrast, the plants showing severe symptoms, ZYMV-TW strain reached high titer, indicating the absence of cross-protection (Wang and Gonsalves, 1999). An engineered strain, ZYMV-AG, was obtained by a point mutation resulting in three amino acid changes in conserved motifs FAK-related nonkinase (FRNK) in the 455 amino acid sequence of the helper component-protease (HC-Pro) of ZYMV. This point mutation caused a dramatic change in symptoms from severe to mild in squash and to a symptom-free condition in cucumber, melon, and watermelon. The cucurbit plants inoculated with this engineered ZYMV strain were effectively protected against the wild (severe) strain (Gal-On 2000).

The efficacy of three mild isolates of citrus tristeza virus (CTV) in protecting grapefruit cv. Ruby Red on sour orange rootstock was assessed by visual observations and ELISA tests at 16 years after inoculation with mild isolates. Infection varied from 10 to 14 percent depending on the mild isolate used for cross protection. On the other hand, a high percentage (67 percent) of control plants (unprotected) showed symptoms of infection by severe strains of CTV as determined by ELISA tests. The health of grapefruits protected by particular mild isolate (DD102) was found to be significantly better (Powell et al.,1999). Cross-protected citrus cultivars/species were indexed by employing virus- and strain-specific polyclonal and monoclonal antibodies. Indirect DAS-ELISA was found to be more efficient in detecting CTV than DAS-ELISA. Incidence of severe strains of CTV seemed to be the cause for the decline of Kagzilime trees that were protected earlier by mild strain (Chakraborty and Ahlawat, 2001). Grapefruit trees inoculated with mild isolates of CTV declined after 18 years. When these trees were assayed by ELISA and ISIA, the presence of both mild and severe isolates of CTV was detected regardless of symptom expression (Lin et al., 2002). These studies indicate that constant monitoring for the incidence of severe strains/isolates is essential to maintain required production levels.

Avirulent strains of fungal and bacterial pathogens have been evaluated for their ability to protect plants against infection by virulent strains. However, no large-scale application of cross protection as a disease management strategy has been demonstrated to be effective under field conditions (Narayanasamy, 2002).

PRODUCTION OF DISEASE-FREE PLANTS

Use of disease-free seeds and propagative plant materials such as tubers, setts, and bud wood materials is vital and basic crop disease management strategy, since infected seeds and planting materials form the primary sources of infection. The plants growing from them introduce the pathogen(s) into the field at the early stages of crop cultivation and serve as potential sources of infection from which diseases can spread to other healthy plants. Various immunodiagnostic methods for detection of microbial pathogens in seeds and planting materials have been described in Chapters 8, 9, and 10. The rate of seed transmission may be affected by both host genotype and viral pathotype or strain as in lettuce mosaic virus (LMV). The isolate AF 198 (pathotype II) was less frequently (1.33 percent) transmitted through a susceptible cultivar than the isolate AF 199 (pathotype IV) which had a transmission rate of 16.5 percent, as determined by PAT-ELISA tests. The tolerant cultivar allowed seed transmission of AF 199 (11.9 percent). In contrast, AF 198 was not transmitted through the seeds of LMV-tolerant cultivars thus indicating the influence of host genotype on virus seed transmission (Jadăq et al., 2002). This study highlights the importance of monitoring the extent of seed infection by viruses to contain serious seed-borne viruses such as LMV. Different approaches have been followed to obtain virus-free plants from infected mother plants.

Meristem Tip Culture

Because the apical meristems of virus-infected plants are generally free of the infecting virus, removing the apical meristem aseptically and growing it in media such as Murashige-Skoog (MS) medium may produce virus-free plants. In some cases, exposure of source plants to higher temperatures (thermotherapy) or treatment with antiviral chemicals (chemotherapy) or auxins (to accelerate the elongation of apical portions of infected plants) may be required before the excision of meristems.

Different immunological techniques are used to check levels of virus in plants regenerated from apical meristems. Following thermotherapy, grapevine fanleaf virus (GFLV) was eliminated and verified using ELISA (Maekawa et al., 1995; Zhang et al., 1998). A more sensitive ultramicro-ELISA (UM-ELISA) technique was used for testing the presence of PVX and even low virus titers could be recognized using this technique (Hernandez et al., 1995). The successful elimination of the potato virus S (PVS), requiring several subcultures, was confirmed by testing the regenerants (Park et al., 1994; Kim et al., 1996). The thin section culture technique, using a modified Vacin and Went medium, was shown to eliminate cymbidium mosaic virus (CymMV) and *Odontoglossum* ringspot virus (ORSV) from orchid plants. The regenerants were free of these two viruses (Lim et al., 1993). Explants of *Prunus cerasifera* treated with antiviral chemicals for eradication of *Prunus* necrotic ringspot virus (PNRSV) were tested by ELISA technique. The highest frequency of healthy explants was obtained by treatment with 2,4-dioxohexahydro-1,3,5-triazine (DHT) (Triolo et al., 1999).

Viruses infecting garlic have limited attempts to increase production levels. Production of virus-free plants is considered an important tactic to overcome this obstacle. Virus-free plants of 87 accessions of garlic were obtained by meristem culture. The virus-free status of cultured plants in respect of onion yellow dwarf potyvirus (OYDV), leek yellow stripe potyvirus (LYSV), garlic common latent carlavirus (GCLV), shallot latent carlavirus (SLV), and mite-borne filamentous virus (MbFV) was ensured by using polyclonal and monoclonal antibodies specific to these viruses in ELISA tests (Senula et al., 2000). *Alstroemeria* mosaic potyvirus (AlMV) could be eliminated from infected *Alstroemeria* plants by planting excised apical meristems from stem or rhizomes in modified MS medium containing gibberellic acid and 6-benzyl aminopurine (benzyl adenine) (BA). The rate of virus elimination was 73.7 and 14.7 for plantlets developing from explants measuring 0.7 mm and 2 mm, respectively (Chiari and Bridgen, 2002).

Levels of viral concentration in meristem culture-derived plants have been tested by other immunological techniques. Tuberose mild mosaic virus (TMMV) is widespread in Taiwan. The sensitivity of ELISA and DTBIA was compared when indexing tuberose leaves for the presence of TMMV. ELISA was more sensitive, since 94 of 100 leaf samples were ELISA positive and DTBIA assay detected TMMV in only 89 leaf samples. However, the DTBIA assay was more sensitive than ELISA in detecting TMMV in tuberose bulbs (Chen and Chang, 1999).

Immunological techniques are effective and efficient for providing rapid and reliable results for plant quarantine certification. The microbial plant pathogens recognized as quarantined pathogens are grouped into three categories:

List A   includes dangerous quarantined organisms; this is further divied in two categories:
  A1—pathogens absent in all member countries (exotic pathogens)
  A2—pathogens present only in some countries in the region

List B   pathogens included by individual countries under regulations; comparatively less stringent measures are enforced by the country concerned

The number of pathogens and other pests included in these lists may vary depending on the seriousness of pathogens. The necessity for inclusion of additional pathogens or deletion of pathogens from the current list is continuously reviewed and modified when required. For further details refer to Narayanasamy (2002).

Apical meristem culture and apical bud culture procedures were applied for producing sugarcane plants free of ScYLV, which is included in the list of quarantine pathogens. Tissue blot immunoassay (TBIA) and RT-PCR were found to be equally effective in detecting the ScYLV in all sugarcane leaves. Apical meristem culture was more effective than apical bud culture for producing ScYLV-free plants (Chatenet et al., 2001). The effectiveness of the TBIA technique in detecting ScYLV was confirmed by Fitch et al. (2001). The meristem tip-culture technique was employed to produce ScYLV-free plants in six susceptible sugarcane cultivars. The regenerants were tested by TBIA to confirm their virus-free nature. These sugarcane lines remained virus free over a period of four years when they were grown either in isolated fields or in greenhouses. Using two different MABs specific for pineapple mealybug wilt-associated virus-1 (PMWaV-1) or PMWaV-2 TBIA showed that both closeroviruses were widely distributed throughout the pineapple-growing areas of the world. PMWaV-specific RT-PCR assays and TBIAs were employed to screen the pineapple accessions maintained at the USDA-ARS National Germplasm Repository for PMWaV infection and no resistant entry could be identified. Hence, the PMWaV infection was eliminated through axillary and apical bud propagation from infected crowns (Sether et al., 2001).

In addition to meristem culture, another culture technique was successfully used to eliminate viruses. ELISA tests on lily plantlets produced by anther culture procedure showed that 41 percent of the plantlets were free of lily symptomless virus (LSV), CMV, and/or tulip breaking virus (TBV) (Niimi et al., 2001).

Stem-disc dome culture was developed for producing virus-free garlic plants. The dome-shaped tissue structures that form on stem-disc explants from a single garlic clove were excised and maintained on phytohormone-free Linsmaier and Skoog medium. The regenerated plants were free from infection by severe mosaic and yellow streak viruses as revealed by TBIA and RT-PCR techniques (Ayabe and Sumi, 2001).

Many viruses are transmitted through true seeds and use of virus-free seeds must be ensured to realize the yield potential of the cultivar. Seeds of gourd *(Lagenaria siceraria)* were subjected to heat treatment at 75°C for three days to inactivate the contaminating cucumber green mottle mosaic virus (CGMMV). High-density latex particle agglutination test and ELISA techniques were applied to check the extent of virus elimination. Dry-heat treatment was found to be more effective than treating the seeds with chemicals such as Na_3PO_4, K_3PO_4, or NaOCl (Kim and Lee, 2000).

SUMMARY

Incorporating resistance genes either by conventional breeding methods or by engineering resistance has been possible in a limited number of crops. Inducing resistance in crop cultivars susceptible to disease is desirable. Resistance to diseases caused by microbial pathogens can be achieved by using both biotic and abiotic inducers. Development of SAR following treatment with inducers depends on the formation of hypersensitive (necrotic) local lesions. In contrast, the ISR that develops after treatment with PGPR is not dependent on the hypersensitive response (HR). Various immunological techniques have been employed to monitor the physiological alterations in plants showing SAR, such as production of PR proteins, and enzymes that are considered to be related to the development of resistance.

APPENDIX: PROTOCOL FOR POLYCLONAL ANTISERUM PRODUCTION

Production of Polyclonal Antiserum Against Coconut Leaf α-Glucosidase (Shanmugam, Raguchander, et al., 2001).

1. Purify the putative α-glucosidase of coconut leaves by SDS-PAGE technique.
2. Prepare a mixture of purified α-glucosidase (1 ml containing 100 μg protein) and Freund's complete adjuvant (1 ml) and emulsify using a Cyclomixer.
3. Immunize a New Zealand white rabbit by intramuscular injection; provide two booster injections with the mixture containing α-glucosidase and Freund's incomplete adjuvant at four and six weeks after the first injection.
4. Collect the blood in sterile glass vials and transfer the supernatant antiserum after clotting of blood cells.
5. Centrifuge the antiserum three times at 8,000 rpm, 4°C for ten minutes to remove the suspended blood cells.
6. Store the antiserum in sterile microfuge tubes at -70°C.

PART V:
ASSESSMENT OF FOOD SAFETY

Chapter 13

Mycotoxins and Food Safety

Microbial pathogens infect all agricultural produce such as grains, vegetables, and fruits during pre- and postharvest stages in the field and during transit and storage. Infection by microbial pathogens may result in poor quality of seeds that are to be used for sowing and in grains that are to be used as food or feed. Affected vegetables and fruits become unfit for consumption due to rotting and food products may contain harmful compounds produced by the pathogens.

Fungal pathogens are the most important because of their effect on the quality of food and feed, although infection by bacterial and viral pathogens may affect seed germination. Among fungi, different species of *Aspergillus, Fusarium, Alternaria,* and *Penicillium* produce mycotoxins affecting the quality of various food and feed materials. The toxins—pathotoxins, and phytotoxins—produced by fungal pathogens *in planta* are not known to cause any adverse effects on humans and animals.

The term mycotoxin applies to the fungal secondary metabolites that occur naturally as contaminants of agricultural and other consumable products, resulting in expression of toxicity symptoms in animals via a natural route of administration. The diseases induced following the ingestion of foods and feeds contaminated with mycotoxins are termed mycotoxicoses. Fungi may gain access to the seeds either in the field or during storage. Some fungi may be able to colonize seeds under both environments. Field fungi may be able to invade seeds either during development or after maturity, but before harvest. Storage fungi are adapted for development even without free water. Storage fungi have a major role in seed deterioration and cause many mycotoxicoses because of the presence of mycotoxins in the deteriorated seeds.

MYCOTOXIN CONTAMINATION

Fungi may produce a large number of mycotoxins (over 300) that may affect the quality of agricultural produce. Research has focused on the important mycotoxins such as aflatoxin, trichothecenes, fumonisins, and the secondary groups such as citrinin, cyclopiazonic acid, ochratoxins, patulin, and zearalenone. The chemical nature and the effects of some mycotoxins on animals have been studied in detail.

Aflatoxins (AF)

Aspergillus flavus and *A. parasiticus* produce different aflatoxins that are most common in several crops such as maize (Anderson et al., 1975; de Côrtes et al., 2000), peanut (Diener et al., 1987), cotton (Ashworth Jr. et al., 1969), mustard (Bilgrami et al., 1992), rice and wheat (Tsai and Yu, 1999). Significant levels of contamination with aflatoxins have been observed in spices such as red peppers *(C. annum)*, nutmeg and ginger (Vrabcheva, 2000; Reddy et al., 2001). Dried yam chips were contaminated with aflatoxin (Bassa et al., 2001). *Aspergillus flavus* produces aflatoxin B_1 and B_2, while *A. parasiticus* elaborates aflatoxin G_1, G_2, M_1 in addition to B_1 and B_2. Aflatoxins are basically difuranocoumarin compounds. Aflatoxin B_1 is the most important among the naturally occurring aflatoxins in agricultural commodities. Aflatoxins are generally formed after harvest (when there is rain) and during storage under high humidity and seed moisture conditions. *Aspergillus flavus* is normally a saprophyte and may become a weak parasite under favorable conditions causing aflaroot disease in peanut (groundnut).

Dairy animals consuming aflatoxin-contaminated feeds may secret the aflatoxin M_1 in milk. M_1 is a toxic metabolite derived from B_1 (Coker et al., 1984; Bhat, 1991). Human consumption of heavily aflatoxin-infected maize and rice can be fatal. Hepatitis and hepatocarcinoma are human diseases induced by aflatoxins (Krogh, 1987). Human exposure to high levels of aflatoxin B_1 (AFB_1) from food materials is an important risk factor for the development of liver cancer (Wogan, 1991). The carcinogenic properties of aflatoxin in other animal systems have also been determined (Busby and Wogan, 1984).

Aflatoxin contamination in cereals and oilseeds has been responsible for significant economic crop losses. Maize (corn) cobs may be infected even in standing crops to a great extent. Senesced silks in immature cobs provide a suitable medium for rapid growth of *A. flavus* leading to entry into the ears. In peanuts, infection under field conditions is a major source of rapid development of *A. flavus* in storage. In standing peanut crops, *A. flavus* present in soil can infect the flowers, penetrate through the pegs, and establish in the developing pods (Lindsey, 1970). In cotton, infection may occur even before bolls open. Aflatoxin was present in unopened immature bolls at 10 to 20 days after inoculation with *A. flavus* (Lee et al., 1986). Such information on the conditions favoring the development of aflatoxin-producing strains may be useful for avoiding contamination.

Fumonisins

Fusarium moniliforme (Gibberella fujikuroi) and other species associated with the *G. fujikuroi* species complex produce fumonisins. The prevalence of this fungal species complex on worldwide maize production is of great concern. Fumonisins are high polar diesters of propane-(1,2,3)-tricarboxylic acid (TCA) and various 2-amino-(12,16)-dimethyl-polyhydroxyeicosanes in which the C_{14} and C_{15} hydroxyl groups are esterified with terminal carboxyl group of TCA (Wilson et al., 1998). Biochemical analyses suggest that fumonisins may be products of either polyketide or fatty acid biosynthesis. A type I putative polyketide synthase (PKS) gene *(FUM5)* involved in fumonisin biosynthesis has been isolated and characterized (Proctor et al., 1999). *Fusarium proliferatum* causes ear rot disease in maize. After artificial inoculation of maize ears with *F. proliferatum* under field conditions, the ear rot severity of nine maize hybrids and the accumulation of fumonisin B_1 (FB_1), fumonisin B_2 (FB_2), beauvericin (BEA), and fusaproliferin (FP) were assessed. A significant correlation between mycotoxin contamination and ear rot index was established (Pascale et al., 1999).

Evidence suggests that the human food poisoning that occurred in Guangxi province, China, in 1989 and the gastrointestinal disorders affecting 130,000 people in Anhui province, China, in 1991 may have been due to the mycotoxin contamination of corn, wheat, and barley. Fumonisins, in addition to trichothecenes, deoxynivalenol (DON), and nivalenol (NIV) were present in the cereal samples (Li et al., 1999). The fumonisins B_1 and B_2 isolated and characterized from cultures of *F. moniliforme* (MRC 86) cultured on corn were shown to have cancer-promoting activity (Gelderblom et al., 1988). The natural occurrence of fumonisins in corn has been reported from several countries such as the United States (Plattner et al., 1990), Poland (Lew et al., 1991), South Africa (Sydenham et al., 1990), Taiwan (Tseng et al., 1995), and China (Li et al., 1999).

Trichothecenes

Several fungal species elaborate trichothecenes to different levels. Trichothecenes produced by different *Fusarium* species have been characterized (Table 13.1). Trichothecenes may also be produced by other fungi such as *Cephalosporium* sp., *Cylindrocarpon* sp., *Myrothecium* sp., *Phomopsis* sp., *Trichothecium* sp., and *Verticimonosporium* sp. (Scott, 1989). Seeds of cereals and animal feeds are frequently contaminated by trichothecenes in most countries. Deoxynivalenol, nivalenol, and T-2 toxin are common in the cereal samples. Feed refusal, skin irritation, hemorrhagic lesions in the gut, esophageal mucosal necrosis, and carcinogenicity are some of the toxicoses caused by trichothecenes (Scott, 1989).

Deoxynivalenol (DON), nivalenol (NIV), and their ester zearalenone (ZEA) are the trichothecenes that contaminated maize samples from Chinese villages where about 130,000 people were affected by gastrointestinal disorders in 1991 (Li et al., 1999). When one central

TABLE 13.1. Trichothecenes produced by *Fusarium* spp.

Fungus	Trichothecene formed	Reference
F. avenaceum (Gibberella avenacea)	Neosolaniol, T-2 toxin, diacetoxyscirpenol, deoxynivalenol (vomitoxin) (DON)	Agarwal and Sinclair, 1996
F. culmorum	Deoxynivalenol (DON), zearalenone and ergosterol and T-2 toxin	Perkowski et al., 1995 Bilgrami and Choudhary, 1998 Kang et al., 2001
F. graminearum (G. zeae)	Deoxynivalenol (DON), zearalenone and T-2 toxin, fusarenon X, nivalenol	Bilgrami and Choudhry, 1998 Agarwal and Sinclair, 1996 Savard et al., 2000 Ngoko et al., 2001
F. poae F. sporotrichioides	T-2 toxin, diacetoxyscirpenol (DAS), and HT-2 toxin	Bilgrami and Choudhary, 1998 Ichinoe and Kurata, 1983

spikelet of spring wheat *(T. aestivum)* was inoculated with *F. graminearum,* which causes fusarium head blight (FHB) disease, high concentrations of DON accumulated in the spikelets below the inoculation point and the corresponding internodes of the rachis also had high levels of DON, indicating rapid downward movement of the mycotoxin (Savard et al., 2000). A significant correlation between the incidence of *F. graminearum* and contamination of maize with DON was reported by Ngoko et al. (2001). The mutant strain of *F. graminearum* obtained by disruption of trichodiene synthase gene *(Tri 5)* could infect only the inoculated spikelet, but not other spikelets in the same spike. This strain did not produce DON in the infected spikelet thus indicating that DON production plays a significant role in the spread of FHB disease (Bai et al., 2002). Localization of DON in the cell walls, cytoplasm, mitochondria, and vacuoles of *F. culmorum* indicated the multipoint synthesis of the mycotoxin by the pathogen. Toxins were present in the xylem vessels, phloem sieve tubes, paratracheal parenchyma cells, and parenchyma cells outside the vascular bundles in the rachis of spikelets inoculated with *F. culmorum* (Kang and Buchenauer, 2000a,b; Kang et al., 2001). Mechanical damage to ears of maize predisposes them to infection by *Fusarium* spp. and little or no infection occurs in ears without injury. Injury caused by insects and birds increases infection by *F. graminearum* significantly (Sutton et al., 1980).

Ochratoxins

Several species of *Aspergillus* and *Penicillium* produce ochratoxin A, including *A. ochraceus, A. alliaceus, A. melleus, P. verrucosum, P. aurantiogriseum, P. chrysogenum,* and *P. purpurescens.* Barley grown in temperate zones infected by *P. verrucosum* seems to be the major source of ochratoxin A (Marquardt and Frohlich, 1990). High percentages of maize cultivars (25 to 36 percent) showed ochratoxin A and B contamination. Three isolates of *A. ochraceus* obtained from maize samples produced high concentrations of ochratoxin B (Shivendra Kumar and Roy, 2000). The pork from pigs fed with infected barley may contain high concentrations of ochratoxin A, thus posing a serious health hazard for humans. Ochratoxin may act as a carcinogen, hepatotoxin, nephrotoxin, and immunosuppressive agent. Kidney degeneration, which leads to death in extreme cases, may be due to higher intake of ochratoxin A by humans (Pitt, 1991; Marquardt and Frohlich, 1990). The cereal samples from Bulgarian villages with a history of Balkan endemic nephropathy (BEN) were contaminated with ochratoxin A and citrinin. Highest toxin levels were present in wheat, wheat bran (feed), and oats (seed) (Vrabcheva et al., 2000).

Ergot Alkaloids

Ergot sclerotia produced by *Claviceps* spp. may contain both useful pharmaceutical compounds and harmful alkaloids known to cause disorders in humans and animals. *C. purpurea* infecting rye *(Secale cereale)* produces alkaloids such as ergocristine and ergotamine in high concentrations and other alkaloids such as ergosine, ergocornine, ergokryptine, and ergonovine at lower levels (Scott et al., 1992). The adverse effects of ergot alkaloids on human beings who suffered from ergotism have been well-documented. The graphic account of Fuller (1968) describes a tragedy that occurred in France in 1951. Bread contaminated with ergot alkaloids was considered as the principal cause for the death of several people and irreversible injuries and psychotic episodes in many more persons. The contamination of animal feeds is more common. Nervous ergotism causing paralysis of the posterior limbs and gangrenous ergotism inducing contraction of smooth muscle resulting in reduced blood flow and ultimate development of gangrene are the most dreadful effects of ergotism in animals (Shelby, 1999). Because the ergot alkaloids are resistant to heat, they are able to survive heat denaturation by baking and cooking processes (Scott et al., 1992; Fajardo et al., 1995).

Alternaria *Toxins*

Several species of *Alternaria* are known to cause diseases in many crops. They produce a wide range of metabolites involved in pathogenesis and some of them are able to induce adverse effects in humans and animals when ingested through contaminated grains, fruits, or vegetables. Important mycotoxins are tenuazonic acid (TeA), altertoxin-I (ATX-I), alternariol (AOH), alternariol methyl ether (AME), and altenuene (ALT) (Jewers and John, 1990). *Alternaria alternata (A. kikuchiana)* produces a host-specific toxin which induces disease symptoms as the fungal pathogen itself. In addition, it can also produce other mycotoxins viz AOH, AME, ALT, TeA, ATX I, II, and III. The contamination of sorghum grains with *Alternaria* toxins is frequent (Seitz et al., 1975; Sauer et al., 1978; Ansari and Shrivastava, 1990). Natural occurrences of *Alternaria* toxins in ragi, barley, oats, wheat, apples, pepper, and sunflower have also been reported (Bilgrami and Choudhary, 1998). The abnormal feeding behavior in chickens and possibly in other animals is attributed to tenuazonic acid (Pitt, 1992).

DETECTION OF MYCOTOXINS

Mycotoxin contamination of agricultural commodities is an unavoidable menace because of prevailing uncontrollable environmental conditions prior to and during harvest, transit, and storage. The dangers posed by mycotoxins to human and animal health have been well demonstrated. Efforts must be made to minimize mycotoxin content in foods and feeds by monitoring, managing, and controlling quality of products from farms to markets, establishing regulatory limits or guidelines, and implementing decontamination procedures effectively. Surveillance programs to monitor incidence of microbial pathogens in agricultural products have to be organized to reduce the risk of consumption by humans and animals. Surveillance systems rely on sensitive methods to detect mycotoxins in different ecological systems. Intensive research efforts have culminated in the availability of various analytical methods such as thin layer chromatography (TLC), gas-liquid chromatography (GLC), high performance liquid chromatography (HPLC), and immunological assays (IAs).

Immunological assays are the preferred methods for detecting mycotoxins because of their simplicity, sensitivity, specificity, reliability, and versatility. Immunoassays have replaced conventional methods that may not be useful for the quantitation of low-molecular-weight compounds such as microbial toxins and pesticides. The pioneering investigations of Landsteiner (1945) on the antigenicity of small molecules laid the foundation for the development of hapten technology. Further progress has led to simple protocols that allow accurate detection of low concentrations of mycotoxins in foods, feeds, and biological fluids. Commercial kits for mycotoxin analysis have also been developed (Morgan, 1989; Wilson et al., 1998; Sinha,

2000). The usefulness of various immunoassays for the detection and quantification of mycotoxins is highlighted in this chapter.

Preparation of Antiserum

Mycotoxins are low-molecular-weight secondary metabolites produced by toxigenic fungi and as such they are not immunogenic. Hence, mycotoxins are conjugated to a protein or polypeptide carrier prior to immunization of test animals. Antibodies against several mycotoxins have been obtained following this approach. A mycotoxin marker is employed to identify the specific antibodies. Conjugation of the mycotoxin to a protein molecule such as BSA, ovalbumin, or keyhole limpet hemocynanin occurs by chemical reaction between functional groups present on the mycotoxin and the protein molecule. Toxins such as ochratoxin A (OA) contain reactive groups with which the reactive groups of protein molecules can conjugate. However, most mycotoxins like aflatoxin and trichothecenes do not have any reactive group for direct coupling. Therefore, a reactive group such as a carboxyl (-COOH) has to be introduced first by chemical synthesis. Both polyclonal and monoclonal antibodies have been raised against several mycotoxins. Polyclonal antibodies can be produced using the mycotoxin-protein conjugate as the immunogen. Monoclonal antibodies are secreted by fusions of immunized mouse spleen cells with a myeloma cell line (Tseng, 1995; Wilson et al., 1998).

Immunological Techniques

Several immunoassays have been performed for the detection of mycotoxins in agricultural commodities. Of these, ELISA, RIA, and immunoaffinity column assay have been frequently employed. Alkaline phosphatase and horseradish peroxidase are the marker enzymes generally employed, though other enzyme markers such as β-D-galactosidase and glucoamylase have also been used in different formats of ELISA.

Aflatoxins

Enzyme-linked immunosorbent assay (ELISA). Polyclonal antibodies with specificity toward the dihydrofuran portion or the cyclopentanone ring moiety of the aflatoxin molecule are generated depending on the site of conjugation (Chu, 1994). Antibodies specific to AFB_1 or AFB_1 and B_2 or for AFB_1, B_2, G_1, and G_2 are available. In the direct ELISA system, the primary antibody to the antigen is immobilized at the bottom of the wells of the ELISA plates and then the secondary antibody with affinity for the primary antibody is added. The enzyme linked to the secondary antibody reacts with a chromogenic substrate and the color that develops is proportional to the amounts of both antibodies and the antigen (see Appendix section on detection of AFB_1 by ELISA). The comparative sensitivities of different immunoassays are presented in Table 13.2.

The extent of contamination of aflatoxin (AFB_1) in maize, peanut butter, and wheat was accurately estimated by ELISA test which was shown to be more precise and cost effective compared with analytical

TABLE 13.2. Sensitivity of immunoassays for detection of aflatoxins

Mycotoxin	Immuno-assay	Commodity	Applicable range (pg/assay)/ Detection limits (μg/kg)	References
Aflatoxin AFB_1	Direct ELISA	Corn, wheat, peanut butter	5-1000 pg/ 5 mg	El Nakib et al., 1981
AFM_1	Direct ELISA	Milk	25-1000 pg/ 0.01 mg	Hu et al., 1983
AFB_1	Direct competitive ELISA	—	12.5 pg/ 0.25-5.0 ng	Chen and Chen 1998a
AFB_1 AFB_2 AFG_1 AFG_2	Indirect competitive ELISA	Corn	12.5-250 pg/g	Chen and Chen 1998b
AFB_1	CARD-ELISA	Corn	0.1 pg/well	Bhattacharya et al., 1999
AF_s	ELISA	Corn and peanut	<1 ng/g	Yong and Cousin, 2001
AFB_1	Radioimmuno-assay (RIA)	Corn, wheat	250-5000 pg/ 5μg	El-Nakib et al., 1981
AFM_1	RIA	Milk	0.1-1.0 ng/ml/ 1.0μg	Rauch et al., 1987

methods (El-Nakib et al., 1981). For screening AFB_1, B_2, and G_2 contamination in cotton seeds and peanut butter (≥ 20 µg/kg) and in maize and raw peanuts (≥ 30 µg/kg) ELISA technique was suitable (Trucksess et al., 1989). Application of direct ELISA format for detection of AFM_1 in dairy products has received wide acceptance (Fremy and Chu, 1989), since milk samples can be used directly in the assay thus enabling the completion of the test within an hour. The sensitivity of detection by ELISA can be markedly enhanced (detection limit from 0.05 to 0.1 ppb to 10 to 25 ppb) by introducing a cleanup treatment with a C-18 reversed phase Sep-Pak cartridge (Hu et al., 1983).

A direct competitive ELISA (DC-ELISA) using high affinity PABs raised against AFB_1 was developed for rapid screening. The detection limit and the sensitivity of detection were 0.25 to 5.0 ng/ml and 12.5 pg/assay, respectively (Chen and Chen, 1998a). Quantitative analysis of AFB_1 using the indirect competitive ELISA (IDC-ELISA) test showed that cross-reactivities of aflatoxin analogues with AFB_1 antibody were 100, 34, 9.1, and 1.3 percent, respectively for AFB_1, B_2, G_1, and G_2. Analytical recoveries of AFB_1 added to rice and mixed feeds were 79 and 81.8 percent, respectively (Chen and Chen, 1998b). Aflatoxin produced by *A. parasiticus* in maize, rice, wheat, and peanuts was detected by ELISA. High correlation was found between the presence of *A. parasiticus* and aflatoxin production in these commodities (Tsai and Yu, 1999).

A novel signal amplification technology based on catalyzed reporter deposition (CARD) known as super-CARD amplification in ELISA was developed by Bhattacharya et al. (1999). The presence of AFB_1 in seed could be more rapidly detected as the total incubation time was reduced to 16 minutes compared with 50 minutes required for the CARD method. The increase in absorbance over the CARD method was approximately 400 percent and the limit of detection of AFB_1 by the super-CARD amplification method was 0.1 pg/well and the sensitivity enhancement was fivefold (Figure 13.1) (see Appendix section on detection of AFB_1 by super-CARD). The sensitivity of the super-CARD method was compared with competitive-ELISA format. The amounts of AFB_1 detected by these two methods were in good agreement. However, the super-CARD method could detect similar amounts of AFB_1 even at dilution of 1/25,000, whereas competitive ELISA could do so only at 1/1000 dilution (Table 13.3).

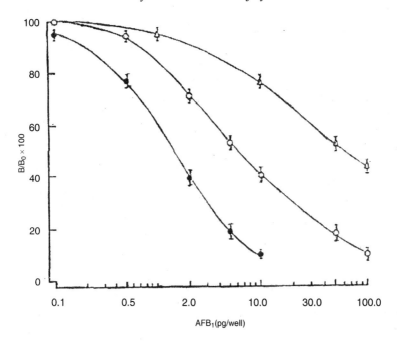

FIGURE 13.1. Dose-response curve of aflatoxin B_1 (AFB$_1$) under identical assay conditions: (●) super-CARD method; (O) CARD method; (△) nonamplified method (*Source:* Reprinted from Bhattacharya et al., 1999; with permission from Elsevier.)

Preharvest infection by *A. flavus* and contamination of maize with aflatoxin were detected by ELISA. The maize samples at the time of harvest had 12.35 to 20 ppb of aflatoxins. The concentrations of aflatoxins increased with increase in storage time, exceeding the limit accepted for consumption (Côrtes et al., 2000). AFB$_1$ was the predominant mycotoxin detected in the samples from the high-risk area in China where human primary hepatocellular carcinoma (PHC) was prevalent. The average daily intake of AFB$_1$ from corn in this area was estimated to be several times higher than the permissible level. Corn samples were simultaneously contaminated with fumonisins B$_1$, B$_2$, and B$_3$ (Li et al., 2001).

Aflatoxin contamination has been noted in spices also. By using indirect competitive ELISA technique, the presence of AFB$_1$ in 59 percent of 182 samples of pepper (chilli) was detected. The toxin was

TABLE 13.3. Determination of AFB_1 concentrations in extracts of seeds artificially inoculated with *Aspergillus parasiticus*

| Crop | Days after inoculation | Concentration of AFB_1 (μg/g of seeds) | |
		Competitive ELISA (1/1000 dilution)	Super-CARD (1/25,000 dilution)
Groundnut	8	16.8	16.4
	12	38.6	38.4
	16	58.0	58.3
Corn (Maize)	8	14.6	14.0
	12	33.0	32.6
	16	44.2	44.6

Source: Bhattacharya et al., 1999.

present at nonpermissible levels in 18 percent of the samples (Reddy et al., 2001).

Co-occurrence of aflatoxin and fumonisin has been observed in Brazilian corn hybrids. The corn samples from Central Western regions showed the presence of both mycotoxins (Ono et al., 2001).

A novel approach for the detection of fungi that produce aflatoxins in maize and peanuts was applied. *Aspergillus parasiticus* produces an extracellular antigen against which a specific antiserum was developed. This antiserum recognized not only *A. parasiticus* but also other species such as *A. flavus, A. sojae,* and *A. oryzae.* The amounts of aflatoxin present in the maize samples positively correlated with ELISA readings for the antigen of *A. parasiticus.* The antigen could be detected even before the production of aflatoxin by *A. parasiticus* in inoculated maize and peanut samples. The detection of the antigen of aflatoxin-producing pathogen by ELISA provides a potential monitoring system for the management of mycotoxins in agricultural commodities (Yong and Cousin, 2001).

Radioimmunoassay (RIA). RIA requires a radioactive mycotoxin marker to allow competition between the mycotoxin in the sample and the marker for the binding of the antibody. The mycotoxin-specific antibody is incubated with a solution of test sample or known standard and a constant amount of labeled toxin. After the separation of free and bound toxin, radioactivity in the fractions is determined by comparing the radioactivity levels with a standard graph that is plotted based on the ratio of radioactivities in the bound fraction and the free fraction versus the logarithm of the concentration of unlabeled standard toxin (Chu, 1988). The amount of aflatoxins in a sample is inversely proportional to the amount of radiolabeled aflatoxin in the solution. High specific activity tritiated AFB_1 is commercially available for use in RIA. RIA can detect as little as 0.05 to 0.5 ng of purified mycotoxin in a standard preparation and the lower limits of detection in food and feed samples by RIA is about 1 to 5 ppb (Chu 1988). The need for a marker toxin with high specific radioactivity, equipment for assessing radioactivity, and expertise for the handling and disposal of radioactive materials are limiting factors for large-scale adoption of RIA. Hence, it has been largely replaced by ELISA formats. RIA technique was employed for the assay of both AFB_1 and AFM_1 (Fukal et al., 1987; Rauch et al., 1987).

Immunoaffinity assay. In a study using this method, an antibody column was used to trap mycotoxins and aflatoxins AFB_1 and AFM_1 (Chu 1991). Affinity columns are prepared by adsorption of specific antibodies raised against the mycotoxin to be detected in the test sample on to a gel material contained in a small plastic cartridge. Aflatoxins are captured from the test solutions by immunospecific antibodies. The nonbinding impurities are washed from the cartridge and the aflatoxins are eluted with methanol. The aflatoxins are derivatized and quantified by solution fluorometry or transferred to the liquid chromatographic reversed-phase column for further separation and fluorescence assessment (Trucksess et al., 1991; Chu 1992). The immunoaffinity assay provides some advantages such as increased selectivity, the ability to trap the aflatoxins in large volumes of test samples (biological fluids), and integration with other analytical methods. However, the large amounts of antibodies required for the preparation of columns and the higher cost compared with ELISA and RIA techniques are limitations (Wilson et al., 1998).

Immunoaffinity column chromatography with fluorescence detection was employed for the detection of aflatoxins in maize-based products to be used for human consumption. Of the 27 samples tested, eight samples had trace levels (below 4 mg/kg) of aflatoxins, while one sweet corn sample had high levels (5 to 10 mg/kg) of aflatoxin (Candlish et al., 2000).

Commercial immunoassay kits. Several commercial immunoassay kits are available. Many of these kits are based on ELISA formats and the reactions are determined visually or by ELISA reader. The presence of aflatoxins B_1, B_2, G_1, and G_2 was detected by Aflatest in affinity column format (developed by Vicam, Watertown, Massachusetts), the detection limit being 10 ng/g. Agriscreen (Neogen Corp., Lansing, Michigan) in microwell format could detect 5 ng/g of aflatoxin in corn, peanuts, and cottonseed feed. One-Step ELISA (International Diagnostics, St. Joseph, Michigan) and Veratox (Neogen Corp., Lansing Michigan) was equally efficient as AGRISCREEN in detecting aflatoxin in food, feeds, and milk samples (Wilson et al., 1998).

Fiber-optic immunosensor assay. Fiber-optic immunosensor assay in noncompetitive format using the native fluorescence of aflatoxin was employed for the detection of AFB_1 in both spiked and naturally contaminated maize samples, while a competitive format was used for the assay of fumunosin B_1 (FB_1) (see the related discussion in the following section on fumonisins). The response of the sensor was directly proportional to the AFB_1 concentration, since the fluorescence of AFB_1 itself was measured. The sensor could detect as little as 2 ng/ml of AFB_1 in solution although technically, it is not an immunosensor, since the attachment of aflatoxin-specific antibodies was not required (Maragos and Thompson, 1999).

Fumonisins

Enzyme-linked immunosorbent assay (ELISA). Using cholera toxin as both hapten carrier protein and adjuvant, murine PABs to fumonisin B_1 (FB_1) were generated. In an indirect ELISA format, these PABs detected fumonisins with a sensitivity of 0.1µg/ml. The PABs cross-reacted with FB_2 (87 percent) and FB_3 (40 percent) indicating that the antibody recognized a section of the basic fumonisin molecule near the union of the two moieties (Azcona-Olivera, Abouzeid, Plattner, Norred, et al., 1992). The use of MABs in a direct ELISA format enhanced the sensitivity of the assay to 0.05 µg/ml (Azcona-Olivera, Abouzeid, Plattner, and Pestka, 1992). The results of ELISA tests performed on naturally contaminated and spiked corn samples were compared with those obtained with chemical methods (GC-MS, HPLC, and TLC). The results were generally well correlated in the case of artificially spiked corn samples. Higher estimations by ELISA tests for naturally contaminated corn samples were attributed to the presence of fumonisin precursors or unidentified metabolites (Pestka et al., 1994; Shelby et al, 1994; Sydenham, Shephard, et al., 1996). Further improvement in the sensitivity of ELISA could be achieved by employing highly specific polyclonal antibodies (Sydenham, Stockenström, et al., 1996).

An MAB-based direct competitive ELISA test for screening FB_1 in different cereals was developed. The selected MAB had mean cross-reactivities of 100 percent, 91.8 percent, and 209 percent, respectively, for FB_1, FB_2, and FB_3 and a detection limit of 7.6 ng/g was achieved by using this MAB (Barna-Vetró et al., 1999, 2000) (Figure 13.2) (see Ap-

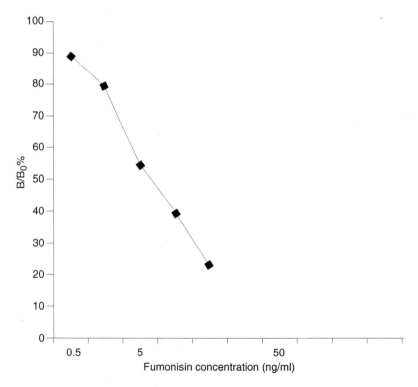

FIGURE 13.2. Dose-response curve of fumonisin B_1 (FB_1). The 50 percent inhibition value FB_1 is 5.4 ng/ml; the detection limit is 0.38 ng/ml. (*Source:* Reprinted with permission from Barna-Vetró et al., 2000; American Chemical Society, Journal of Agricultural and Food Chemistry.)

pendix section on determination of FB_1 with ELISA). PABs against fumonisin B_4 (FB_4) that showed good cross-reactivity with four major fumonisins were generated by immunizing rabbit with FB_4-keyhole limpet hemocyanin conjugate.

A sensitive direct competitive (CD)-ELISA for fumonisins was developed. This method can be used as a supplement to FB_1 antibody-based ELISA for screening *Fusarium* strains for the production of fumonisins (Christensen et al., 2000). A study to compare the extraction efficiency of solvents showed that an aqueous PO_4 buffer was quite suitable for extracting FB_1 from 16 field corn samples, thus eliminating the requirement of organic solvents for fumonisin extraction. Determinations of fumonisin B_1 concentrations by HPLC and ELISA were well correlated in fortified samples at known concentrations between 1 and 50 μg/ml (Kulisek and Hazebroek, 2000).

In a survey for detection of mycotoxins in Glasgow, Scotland, a total of 27 maize-based products meant for human consumption were tested by competitive ELISA. Fumonisins were detected in 30 percent of the samples at levels from 1 to 8 mg/kg and at levels below 1 mg/kg in another 30 percent. A sample of fine cornmeal contained the highest concentration (8 to 12 mg/kg) of fumonisins (Candlish et al., 2000). Fumonisin B_1 contamination due to incidence of *F. verticilioides* on maize was recorded in Cameroon. By employing (CD)-ELISA, the presence of FB_1 in maize samples was detected and concentrations varied from 300 to 26,000 ng/g (Ngoko et al., 2001).

Fiber-optic immunosensor assay. Evanescent wave-based fiber-optic immunosensors were employed in a competitive format to quantify FB_1 in both spiked and naturally contaminated maize samples. fumonisin MABs were covalently coupled to an optical fiber and the competition between FB_1 and FB_1-labeled fluorescein (FB_1-FITC) for the limited number of available binding sites on the fiber was determined. The signal generated in this assay was inversely proportional to the FB_1 levels. The results of fiber-optic immunosensor assays were well correlated with those using HPLC method for naturally contaminated maize samples, except when large amounts of other fumonisins that could cross-react with the immunosensor were present (Maragos and Thompson, 1999). The comparative efficacy of different immunoassays in detecting fumonisins are presented in Table 13.4.

TABLE 13.4. Sensitivity of immunoassays for detection of fumonisins

Mycotoxin	Immunoassay	Commodity	Applicable range (pg/assay)/ Detection limits (μg/kg)	References
Fumonisin B_1	Direct ELISA	Corn	10 ng/g	Usleber et al., 1994
Fumonisins	Competitive ELISA	Corn	10 mg /g	Candlish et al., 2000
	Indirect competitive ELISA	Corn	1-5 mg/g	Ono et al., 2000
Fumonisin B_1	Direct competitive ELISA	Corn	300-26,000 ng/g	Ngoko et al., 2001
Fumonisin B_1	Immunoaffinity column assay	Canned and frozen sweet-corn products	0.05-0.2 μg/g	Trucksess et al., 1995
Fumonisin B_1, B_2, and B_3	ELISA	Animal feed	50 ng/ml	Azcona-Olivera, Abouzeid, Plattner, and Pestka, 1992
Fumonisin B_1	Fiber-optic immunosensor assay	Corn	3.2-5.0 μg/g	Maragos and Thompson, 1999

Trichothecenes

Because of the difficulties in analyzing trichothecenes using conventional chemicals, development of immunoassays for these toxins have received greater attention. Production of highly specific antisera against T-2 toxin was achieved by obtaining T-2 protein conjugated and synthesized after T-2 hemisuccinate formation (Lee and Chu, 1981a). Immunoassays have been employed for the detection and quantification of trichothecenes in rice (Kemp et al., 1986), corn (Casale et al., 1988; Lee and Chu, 1981a), wheat (Xu et al., 1988; Lee and Chu, 1981b; Schubring and Chu, 1987), and barley (Teshima et al., 1990). The sensitivities of various immunoassays are presented in Table 13.5.

TABLE 13.5. Sensitivity of immunoassays for detection of trichothecenes

Mycotoxin	Immunoassay	Commodity	Applicable range/ Detection limits (μg/kg)	References
1. Deoxynival-enol (DON) vomitoxin	Competitive direct ELISA	Corn and wheat	12-1250 pg/ assay/10 μg/kg	Xu et al., 1988
	Competitive in-direct ELISA	Corn	100-1300 ng/g	Ngoko et al., 2001
2. Zearalenone	Competitive di-rect ELISA	Corn and wheat	25-2500 pg/ assay/ 100 μg/kg	Warner et al., 1986
	ELISA	Cereal grains and milk	15-500 pg/ assay	Azcona et al., 1990
3. T-2 toxins	Direct ELISA	Corn	2.5-200 pg/ assay/ 1.0 μg/kg	Pestka et al., 1981
	ELISA	Barley, wheat	6-5000 μg/pg	Lee et al., 1999
	Hit-and-run fluoro immunoassay	Cereals	25-50 ng/assay	Warden et al., 1987
	Flow-through enzyme immunoassay	Cereals	50 ng/g	Sibanda et al., 2000

Enzyme-linked immunosorbent assays (ELISA). Among immuno-assays, ELISA formats have been more widely used for the detection of trichothecenes. An MAB-based ELISA for the detection and quantification of DON in corn was effective. The sensitivity of the assay was in the range of 10 to 250 ng per assay (0.2 to 5.0 μg/ml) (Casale et al., 1988). Direct ELISA was shown to be more specific and sensitive than indirect ELISA in detecting DON in corn and wheat samples with detection limits of 10 ng/g (Xu et al., 1988). A commercial ELISA kit to detect and differentiate 3-acetyl-DON and DON was developed to test feeds, grains, and cereal products (RIDASCREEN DON, Rbiopharm GmbH) (Usleber et al., 1991). Another ELISA kit, AGRISCREEN, was useful for detection of DON in wheat samples (Putnam and Binkerd, 1992). A competitive direct ELISA was employed to detect DON (vomitoxin) and also the fungal pathogen

F. graminearum infecting maize. ELISA format was compared with gas chromatography-electron capture assay for determination of levels of vomitoxin (DON) in milled fraction of wheat. There was no difference between the levels of vomtoxin in wheat samples quantitated by ELISA or GC and hence ELISA was recommended for obtaining reliable results rapidly and economically in a commercial setting (Hart et al., 1998). A significant positive correlation was detected between the incidence of *F. graminearum* and contamination with DON (Ngoko et al., 2001).

Following inoculation of one control spikelet of spring wheat *(T. aestivum)* with *F. graminearum,* the amounts of DON in each spikelet and each internode of the rachis were determined by using specific MABs in ELISA. Whereas the spikelets below the point of inoculation accumulated high concentrations of DON, the spikelets above the point of inoculation contained much lower amounts of DON, indicating a faster downward movement of DON in the infected wheat spikes (Savard et al., 2000). Immunogold-labeling technique using specific antiserum revealed the localization of DON in the cell walls, cytoplasm, mitochondria, and vacuoles of the hyphae of *F. culmorum* (Kang et al., 2001).

Zearalenone is another trichothecene present in foods and feeds. ELISA formats have been employed to screen zearalenone and other important mycotoxins simultaneously. Specific MABs were produced to detect zearalenone contamination in cereal grains and milk (Azcona et al., 1990). In a collaborative study, an ELISA procedure available commercially for screening for zearalenone was employed. The protocol was used to assess zearalenone in corn, wheat, and feed by both visual and spectrophotometric methods, the detection limit being 800 ng/g (Bennett et al., 1994). The comparative efficacy of ELISA and HPLC was evaluated for the analysis of zearalenone in ground wheat grains. Application of ELISA permitted direct, highly sensitive measurements of zearalenone levels in crude extracts with methanol-water (70:30 v/v) without any cleanup. On the other hand, HPLC required a prior cleanup using GPC Biobeads S X 3 Soft gel, but the sensitivity of HPLC was higher than ELISA test (Radová et al., 2001).

A protocol for the production of high titer antiserum specific to zearalenone in chickens was developed by Pichler et al. (1998). The chickens were immunized by administering three injections of purified zearalenone. Five weeks after the start of immunization the antiserum with a titer of 1:76,000 was isolated from egg yolks without any adverse effects on the chickens. Using this antiserum in an indirect competitive ELISA format, zearalenone could be detected at a concentration range of 10 to 200 μg/ml.

Immunogold-labeling technique. Fusarium culmorum and *F. graminearum* are primarily involved in the destructive diseases of fusarium head blight (FHB) of wheat. Subcellular localization of fusarium toxins, deoxynivalenol (DON), and 3-acetyl deoxynivalenol (3-DON) elaborated by *F. culmorum* was studied by employing immunogold-labeling technique with antisera raised against these toxins. As the pathogen developed, higher concentrations of toxins were detected in the host cells, especially in the cells in close contact with the hyphae. Dense accumulation of gold particles was observed over the host cell wall in advance of the penetration peg (Figure 13.3A and B). In hyphal cells, the toxins were localized in the cytoplasm, mitochon-

(A)

(B)

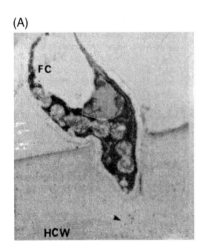
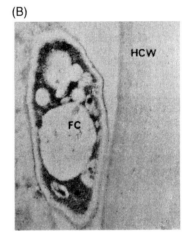

FIGURE 13.3. Immunolocalization of fusarium toxins (DON and 3-ADON) in wheat spike tissues infected by *Fusarium culmorum* 36 hours after inolculation: (A) gold particles are localized over the host cell wall (HCW) and in the hyphal cell wall (FC): dense accumulation of gold particles can be seen over the host cell wall in advance of the penetration peg, when anti-DON-serum-labeled gold particles are used. (B) With anti-3DON serum, gold particles can be detected over host cell wall (HCW) and hyphal cell wall (FC). (*Source:* Kang and Buchanauer, 1999; Reprinted with permission from Elsevier.)

dria, vacuoles, and cell wall. The presence of toxins in the cytoplasm, chloroplasts, plasmalemma, cell wall, vacuoles, and sometimes in the endoplasmic reticulum and ribosomes of the host cells was detected (Figure 13.4). In the xylem vessels, phloem sieve tube, and para-tracheal parenchyma cells in the rachis, a few gold particles were present in the cytoplasm and cell walls at ten days after infection, indicating that the toxins might be translocated upward through xylem vessels (Figure 13.5). The toxins were present in the pericarp tissues, pigment strands, aleurone cells, and starchy endosperm in infected kernels (Figure 13.6) (Kang and Buchenauer, 1999).

When infected by *F. culmorum* the responses of susceptible and resistant wheat spikes, were assessed by immunogold labeling technique. No differences in the lignin contents of healthy wheat spikes susceptible and resistant to *F. culmorum* were evident. β-(1,3)-glucan

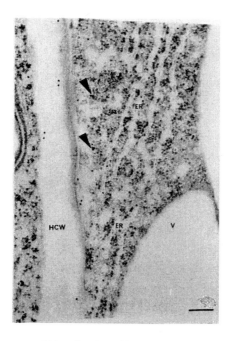

FIGURE 13.4. Immunogold localization of DON in the epidermal cell in the ovary of wheat infected by *Fusarium culmorum* at two days after inoculation: Gold particles can be observed over host cell wall (HCW), cytoplams, endoplasmic reticulum (ER), vacuole (V), and ribosomes (arrowheads). (*Source:* Kang and Buchanauer, 1999; Reprinted with permission from Elsevier.)

(A)   (B)

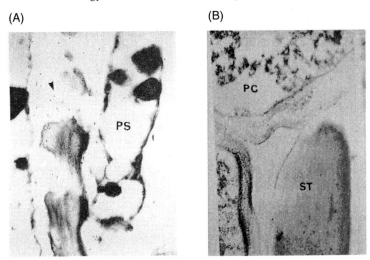

FIGURE 13.5. Immunogold localization of fusarium toxins in wheat spike tissues inoculated with *Fusarium culmorum:* (A) more intense labeling of gold particles on sieve pores than in the sieve plate and sieve wall (arrowheads); (B) labeling of the vessel wall and parenchyma cell wall with gold particles; (PC) paratracheal parenchyma; (ST) secondary thickening (*Source:* Kang and Buchanauer, 1999; Reprinted with permission from Elsevier.)

was detected in the appositions and papillae that were formed as structural barriers to the pathogen. The translucent areas of the sieve plate of uninoculated healthy lemma of a susceptible variety showed high density of gold particles, while a low density of gold particles was seen in the periplasmic space close to a hypha and on the fungal cell wall. On the other hand, in the resistant variety, the wall apposition was labeled densely with gold particles over translucent areas of the lemma. The lignin accumulated rapidly in the host cell walls of resistant spikes following infections while only a marginal increase in lignin content occurred in comparable susceptible spikes, indicating that the accumulation of lignin may have a role in the development of resistance to wheat FHB. Moreover, distinct differences in the toxin concentrations between susceptible and resistant tissues as revealed by density of gold labeling were observed (Figure 13.7). The presence of lower concentrations of DON in the infected spikes of resistant cultivars may result from activation of host defense mechanisms (Kang and Buchenauer, 2000a; Kang et al., 2001).

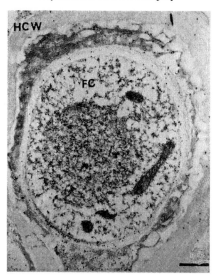

FIGURE 13.6. Detection of hyphal cell in a pigment strand of the infected kernel at six weeks after inoclulation by immunogold localization techinique HCW = host cell wall; FC = hyphal cell wall (*Source:* Kang and Buchanauer, 1999; Reprinted with permission from Elsevier.)

A significant advancement in the production of antibodies against antigens was achieved by developing transgenic plants. *Arabidopsis* plants were transformed with an antizearalenone single-chain Fv (scFv) DNA fragment. Plants transformed with the construct containing a PR-1b signal peptide sequence produced transgenic offspring. The antizearalenone scFv "plantibody" produced by the transgenic plants was able to bind zearalenone specifically with high affinity and was comparable with bacterially produced scFv antibody and the parent MAB. The presence of antizearalenone scFv was detected by electron microscopic immunogold-labeling technique, mainly in the cytoplasm and occasionally outside the cells. This expression of specific plantibodies in crops may lead to two possibilities: neutralization of mycotoxins in animal feeds and reduction in mycotoxin-associated crop diseases (Yuan et al., 2000).

The T-2 toxin produced by *F. sporotrichioides* in infected corn was detected by ELISA (Gendloff et al., 1984). The T-2 toxin, along with deoxynivalenol were detected in wheat and barley samples. These

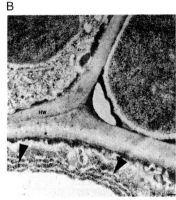

FIGURE 13.7. Immunogold localization of fusarium toxins in the infected lemma from susceptible and resistant wheat cultivars: (A) gold labeling over host cell wall (HW), plasmalemma (arrowheads), cytoplasm, endoplasmic reticulum (ER), and vacuole (V) of parenchyma cell in susceptible cultivar; (B) localization of gold particles over host cell wall (HW), cytoplasm, endoplasmic reticulum (arrowheads), and chloroplast of lemma of resistant cultivar; hyphal cell wall, cytoplasm, and mitochondria also labeled (*Source:* Kang and Buchanauer, 2000a; Reprinted with permission from Elsevier.)

mycotoxins present in cereals were associated with human red mold intoxications observed in China (Li et al., 1999).

The dietary intake of trichothecenes expressed as T-2 toxin equivalents by a Swedish population was studied. The intake of trichothecenes (T-2 toxin) in children with a high consumption of cereals was found to be close to the tolerable daily intake (TDI) values proposed by the international expert group, suggesting a critical monitoring of mycotoxins levels in foods (Thuvander et al., 2001).

In the "hit and run" fluoroimmunoassay developed by Warden et al. (1987) for the detection of T-2 toxin, an affinity column is prepared by conjugating T-2 toxin to Sepharose gel and is then equilibrated with fluorescein isothiocyanate (FITC)-labeled Fab fractions of the anti-T-2 toxin immunoglobulins. Test samples are injected into the column. The FITC-Fab, which elutes along with T-2 toxin in samples, is quantitated in a standard flow-through fluorometer. This assay requires specific monoclonal antibodies. However, the sensitivity is not very high. The sensitivity of detection of T-2 toxin in different cereals was enhanced by employing a membrane-based flow-through enzyme immunoassay with an internal reference (control). An Immunodyne ABC membrane was coated with goat anti-horseradish peroxidase (HRP) (2 µl) as internal control spot (1:1000) and rabbit antimouse as test spot (undiluted) immunoglobulin. The free binding sites in the membrane were blocked. In addition to the antibody-coated Immunodyne ABC membrane, the assay also comprises a plastic snap-fit device, absorbent cotton wool, mouse anti-T-2 MABs, and T-2-HRP conjugate. A significant difference was found between the color intensities of the internal control and samples contaminated with T-2 toxin (positive samples). The minimum detectable limit for visual assessment was 50 ng/g of cereal samples (Sibanda et al., 2000).

Ochratoxins

ELISA formats and radioimmunoassay (RIA) are useful for the detection of ochratoxin A (OA) in contaminated grains and feeds. The presence of OA in wheat (Lee and Chu, 1981a,b) and barley (Morgan et al., 1983) was detected by ELISA. Cereal samples from Bulgarian villages with a history of Balkan endemic nephropathy (BEN) were analyzed for ochratoxin A and citrinin. Highest levels of OA (maximum 140 ng/g) and citrinin (maximum 420 ng/g) were found in samples from the endemic villages. Cooccurrence of both mycotoxins was also noted (Vrabcheva et al., 2000). The contamination of black pepper, coriander seeds, powdered ginger, and turmeric powder with OA was established by ELISA. (Thirumala Devi et al., 2001) (Table 13.6) (see Appendix section on detection of OA).

Corn cultivars showed variations in OA contents which ranged from 0.06 to 0.07 µg/g in Diara composite and from 0.09 to 0.87 µg/g

TABLE 13.6. Detection of ochratoxin-A (OT-A) in spices by indirect competitive ELISA

Commoditiy	Number of samples tested	Range of OT-A (µg/kg)
Black pepper	26	15-69
Coriander seeds	50	10-51
Ginger powder	25	23-80
Turmeric powder	25	11-102

Source: Thirumala Devi et al., 2001.

in Suwan composite. The presence of ochratoxin B in these cultivars was also detected. Three isolates of *A. ochraceus* isolated from Diara composite were toxigenic and their contents of ochratoxin A and B were high (Shivendra Kumar and Roy, 2000).

A method to produce PABs in rabbits by applying a mixed anhydride reaction for immunogen synthesis and an activated ester method for production of the labeled antigen was developed. Ochratoxin A contamination in cereals and porcine blood was detected by using PABs in direct competitive ELISA format (Usleber et al., 2001). Radioimmunoassay (RIA) was effective in detecting ochratoxin A in barley (Rousseau et al., 1985), and cereals and feeds (Ruprich et al., 1988). A flow-through membrane-based enzyme immunoassay for the rapid detection of ochratoxin A in wheat samples was developed by Saeger and Van Peteghen (1999). In this method, a portable colorimeter was used to confirm visual observations and there was an inverse relationship between ochratoxin A levels and intensity of color developed.

Ergot alkaloids

Contamination of foods and feeds with ergot alkaloids may occur in two ways (1) the grain crop itself may be infected by *Claviceps* spp. (2) infected grasses growing as weeds in crops may be harvested with the grains. Ergot alkaloids may be detected by immunological techniques and the HPLC method. Among immunological techniques, direct/indirect competitive ELISA has been frequently used. The radioimmunoassay method has been largely replaced because of its need for radioisotopes. Immunoassays depend on the ability of antibody or immunoglobulin to bind with a specific target alkaloid. The specific antibodies can be produced by immunizing rabbits or mice with the alkaloid-antigenic carrier molecule (e.g., BSA) conjugate. By selecting different binding sites, antibodies with wide-spectrum or specific activity can be produced. ELISA formats are simple, sensitive, rapid, and less expensive compared to HPLC methods. Limitations of ELISA include quantitative inaccuracy due to cross-reactivity of the antibody with nontarget compounds in the sample. However, ELISA is best suited for rapid, large-scale screening of agricultural commodities (Shelby, 1999).

MANAGEMENT OF MYCOTOXIN HAZARDS

At least 25 percent of the world's food crops are estimated to be contaminated with mycotoxins, which poses a unique challenge to food safety since naturally occurring mycotoxins are unavoidable and unpredictable (Anonymous, 1996). It may not be possible to either destroy the entire stock of contaminated foods or divert it to non-human uses without seriously compromising the world's food supply. Hence, an imperative need exists for system(s) of risk management to avoid/lessen the potential risks associated with mycotoxins.

"Prevention is better than cure" is applicable to the management of mycotoxin hazards. Prevention through preharvest management seems to be the best method for controlling mycotoxin contamination. Among the preventive measures, development of resistant cultivars is the most important approach. Corn geneotypes resistant to aflatoxin production by *A. flavus* have been identified after extensive field testing (Widstrom et al., 1987). In another study, germplasm entries with greater resistance to aflatoxin contamination were selected and these lines exhibited resistance until the kernel pericarps were breached (Dunlap, 1995). Likewise, attempts have been made to select peanut germplasm showing resistance to aflatoxin. Improved methods of screening of peanut entries for resistance to aflatoxin have also been developed (Mehan et al., 1988; Holbrook et al., 1995). Other approaches include employing proper agronomic practices and harvesting at optimum stage of maturity followed by drying the produce to optimum moisture levels. The use of fungicides should be limited to minimum levels. Fungicides such as tebuconazole can reduce mycotoxin production if applied at the right time (Kang et al., 2001).

Application of biocontrol agents to reduce fungal populations producing mycotoxins and to detoxify mycotoxins already present in agricultural commodities is a preferable approach. Biological decontamination of aflatoxin by *F. aurantiacum* (NRRL B-184) was demonstrated. Aflatoxin-contaminated milk, oil, peanut butter, groundnuts, and corn were entirely freed of the toxin by use of *F. aurantiacum* (Ciegler et al., 1966). Degradation of aflatoxins in foods and other agricultural commodities by *A. parasiticus* (Doyle et al., 1982) and *Rhizopus* spp. (Cole et al., 1972) has been reported. The biocontrol agent *Trichoderma viride* effectively controlled several fungal pathogens, and suppressed fumonisin B$_1$ production by *F. moniliforme* affecting several cereals (Yates et al., 1999).

The methods of processing foods have essential roles in reducing the potential risks of mycotoxin-contaminated food commodities. Early detection and quantification of various mycotoxins in foods and feeds constitute the bases for decision-making processes by regulatory authorities. Elimination of all mycotoxin contamination is not practically possible. Hence, effective procedures such as physical removal of mold-damaged or incomplete kernels/seeds/nuts, chemical inactivation such as the ammoniation method, and use of additional chemical agents normally used in industrial processing may be adopted (López-García and Park, 1998).

Monitoring the levels of contamination in foods and feeds with control programs has been realized but not enforced in several countries, particularly developing countries. Since mycotoxin contamination may not be entirely prevented, small amounts of these compounds may be legally permitted, if they are unavoidable and are not injurious to health (Wood and Trucksess, 1998). In 1938, the U.S. Food and Drug Administration (FDA) passed the Federal Food, Drug, and Cosmetic Act. The regulations formulated by the FDA help both consumers and the food industry by providing guidelines to ensure safe and acceptable products. Dietary contamination levels are monitored and controlled by the FDA through field staff. Action levels for mycotoxins are determined based on scientific knowledge provided by analytical, toxicological, and other technological procedures. For example, the FDA action levels for total aflatoxins are 20 ng/g in all products except milk designated for humans, 300 ng/g in corn and peanut products for finishing beef cattle, and 0.5 ng/g in milk. Relatively simple, sensitive, and specific immunoassays such as ELISA have been developed, evaluated, and approved for qualitative screening and quantification of aflatoxins in foods and feeds. All countries should establish suitable agencies for ensuring the safety of food products available on the market.

SUMMARY

Several microbial pathogens infect seeds and propagative materials and adversely affect their ability to produce healthy and robust plants. In addition, the fungi belonging to the genera *Aspergillus* and

Fusarium produce different mycotoxins which pose hazards to humans and animals that consume contaminated grains and feeds. Among the techniques used to detect and quantify mycotoxins present in foodstuffs and feeds, immunological techniques are cost effective, rapid, and sensitive. ELISA formats and radio-immunoassay (RIA) have been used extensively for the detection of mycotoxins.

APPENDIX: PROTOCOLS FOR THE DETECTION OF MYCOTOXINS

Detection of Aflatoxin B_1 (AFB$_1$) Using Competitive ELISA (Makarananda and Neal, 1992)

1. Dilute the BSA-AFB$_1$ conjugate to have a dilution of 5 ng of protein/50 μl of PBS containing NaCl (8 g), KCl (0.2 g), Na$_2$HPO$_4$, 2 H$_2$O (1.15 g), and KH$_2$PO$_4$ (0.2 g) in one liter of distilled water; dispense aliquots of 50 μl to each well of the microtiter plates and incubate overnight at 37°C, ensuring the plates are completely dry, and store at –40°C, if required for use later.

2. Wash the plates four times with PBS (0.05 percent Tween-20 in PBS) by totally immersing the plates in PBS-T (800 ml for each plate) placed in a plastic box; dry the plates by vigorously banging inverted plates on several layers of (absorbent) filter paper.

3. Transfer 100 μl of PBS-gelatin (0.25 percent in PBS) to block nonspecific binding sites in the wells and incubate for 60 minutes at room temperature.

4. Prepare different dilutions of AFB$_1$ (standards) ranging from 0.01 to 100 ng/ml and pipette out 200 μl of each dilution into separate clean glass test tubes (2 ml); prepare the control (PBS alone without AFB$_1$).

5. Prepare dilutions of test samples appropriately and use 200 μl aliquots for each sample.

6. Prepare optimal dilution (predetermined range of serum dilutions) of rabbit anti-AFB$_1$ serum (1:10,000 using 2 μl in 20 ml of PBS-gelatin).

7. Add the diluted antibody to the AFB$_1$ dilutions (standards) and test samples at a ratio of 1:1 (antibody to sample); cover the tubes with contents and incubate at 37°C with constant shaking for 60 minutes.

8. Empty the microtiter plates (step 3); wash the plates twice with PBS-T as done earlier (step 2) and dry the plates for further use.

9. Transfer 50 μl of the mixture (step 7) to each well of the microtiter plates; cover the plates with plate-sealing tape and incubate at room

temperature for 90 minutes with constant shaking, using a microplate shaker.

10. Dilute rabbit anti-IgG-peroxidase conjugate (1:5000) using 4 µl of
    antibody in 20 ml of PBS-gelatin, approximately five minutes be-
    fore use and leave it in ice.

11. Empty the wells after removing sealing tape (step 9); wash the plates
    five times with PBS-T and dry the plates as done earlier (step 2).

12. Dispense 50 µl of diluted antirabbit IgG-peroxidase conjugate (as
    prepared in step 10) into each well; seal the plates with tape; incu-
    bate at room temperature for 90 minutes with constant shaking;
    wash the plates five times with PBS-T followed by washing with
    distilled water and dry the plates as done earlier.

13. Warm (approximately 40°C) 0.1 M sodium acetate buffer, pH 6.0
    (24.7 ml) for 30 minutes; add 250 ml of tetramethylbenzidine solu-
    tion (1 mg in 1 ml of DMSO); warm the mixture for another 30
    minutes and add 10 µl of 100 vol H_2O_2 to the mixture just before
    use and thoroughly mix by shaking.

14. Transfer substrate solution (50 µl) to each well; incubate at room
    temperature for 30 minutes and observe the development of blue
    color.

15. Stop the reaction by adding 50 µl of 2M H_2SO_4 to each well; color
    will change to yellow; allow the plates to remain undisturbed for 15
    minutes.

16. Record the absorbance at 450 nm using an ELISA reader and calcu-
    late the percentage inhibition using the control (without AFB_1)
    absorbance as the base.

Detection of Aflatoxin B_1 (AFB$_1$) in Seed Extracts by Super-CARD Method (Bhattacharya et al., 1999)

Preparation of Electron-Rich Proteins (EDC Method)

1. Dissolve casein (5 g) in 150 ml of sodium bicarbonate (150 mM/l);
   prepare five aliquots of 30 ml each; dilute each aliquot to 120 ml and
   add 25, 50, 100, 200, or 300 mg of tyramine hydrochloride separately
   to different aliquots.

2. Stir the mixtures for five minutes and adjust the pH to 6.0 with dil.
   HCl.

3. Add 1-ethyl-3 (3-dimethylaminopropyl) carbodiimide hydrochloride
   (EDC) 50, 100, 200, 300, or 300 mg, respectively, to the mixtures; stir
   the mixtures continuously for four hours at room temperature, while
   the pH is maintained at pH 6.0.

4. Dialyze the mixture against deionized water, followed by phosphate buffer (10 mM/l), pH 7.6.
5. Centrifuge the casein conjugates (I-V) to remove slight turbidity; lyophilize and characterize the conjugates spectroscopically at 275 nm or 293 nm (alkali). The conjugates (I-V) may contain 8, 18, 27, 39, and 54 mg of tyramine/g of casein, respectively.

Super-CARD (Catalyzed Reporter Deposition) Method

1. Dispense 100 μl of AFB_1-casein conjugate (0.02 μg/ml) in coating buffer consisting of sodium phosphate (50 mM/l, pH 7.6) to each well of microtiter plate; incubate at 4°C overnight and maintain the control without AFB_1-casein conjugate.
2. Wash the wells three times with washing buffer containing Tween-20 (0.5 ml/l).
3. Block the nonspecific sites with 100 μl of post-coating buffer (consisting of sodium carbonate/bicarbonate 50 mM/l, pH 9.6) containing electron-rich protein (0.2 percent); incubate the plates at 37°C for three hours and wash the wells three times with washing buffer.
4. Add to each well 25 μl of standard or sample in assay buffer consisting of sodium phosphate (50 mM/l), pH 7.6, containing NaCl (150 mM), bovine serum albumin (BSA) (1.0 g) and thimerosol (0.1 g per liter); add anti-AFB_1 antibody (50 μl of 1/50,000 dilution) in assay buffer to each well; cover the plate; mix the contents by gentle swirling and incubate for five minutes at room temperature.
5. Stop the reaction by emptying the wells using an aspirator (8-Channel) and wash three times with washing buffer.
6. Add goat antirabbit-HRP conjugate (50 μl of 1/6000 dilution) to each well; incubate for three minutes at room temperature and wash the plates with washing buffer.
7. For amplification, add biotinylated tyramine (B-T) (50 μl at 50 μM/l in Tris HCl buffer [50 mM/l], pH 8.0) containing H_2O_2 (0.01 percent) to each well; incubate for three minutes at room temperature and wash the plates with washing buffer.
8. Add avidin-HRP conjugate (50 μl at 1/5000 dilution in assay buffer) incubate for three minutes at room temperature and wash with washing buffer.
9. Add the substrate *O*-phenylene diamine hydrochloride (OPD) (50 μl) to each well and add 50 μl of sulfuric acid (4N) to each well.
10. Record the absorbance at 492 nm.

Competitive-ELISA

1. Coat the wells with 200 μl of AFB$_1$-casein conjugate (0.6 μg/ml); incubate for 16 hours at 4°C and wash the wells with washing buffer as in step 2 for Super-CARD.
2. Block nonspecific sites with casein (0.4 percent) in postcoating buffer.
3. Add 50 μl standard (1 pg to 1 ng/well) and anti-AFB$_1$ antibody (100 μl at 1/50,000 dilution); incubate for two hours at room temperature and wash the plates.
4. Add antirabbit IgG-HRP conjugate; incubate for 20 minutes at room temperature and wash the plates.
5. Dispense 150 μl of substrate solution-OPD to each well; stop reaction with sulfuric acid (4N) after 30 minutes and determine the absorbance at 492.

Preparation of Seed Extracts

1. Inoculate the seeds with *Aspergillus parasiticus* and incubate for 8, 12, or 16 days at 30°C.
2. Expose the inoculated seeds to a steam bath at 100°C for 30 minutes to kill the fungal pathogen; wash the seeds with sterile water and dry them on filter paper.
3. Homogenize the seed samples (1.0 g each) in methanol (10 ml/sample); centrifuge at 10,000 rpm for 15 minutes; collect the supernatant and store at −20°C.
4. Prepare dilutions with assay buffer before use (1/25,000 for Super-CARD and 1/2,000 for competitive ELISA).

Determination of Fumonisin B$_1$ in Cereals by ELISA (Barna-Vetró et al., 2000)

Preparation of Immunogen

1. Dissolve keyhole limpet hemocyanin (KLH) (10 mg) in 12 ml of NaCl (0.8 percent); add 1.3 ml of glutaraldehyde (GA) (12 percent); incubate the mixture for one day at room temperature; dialyze the excess GA for two days and centrifuge.
2. Dissolve fumonisin B$_1$ (FB$_1$) (1 mg) in 0.4 ml of methanol; add dropwise to the activated protein and incubate the mixture for 20 hours at room temperature.

3. Block the unreacted aldehyde groups by adding sodium borohydride (12 mg); stir the mixture for four hours; dialyze the conjugates against PBS for three days and store at −20°C.

Preparation of FB₁-BSA Conjugate

1. Dissolve *S*-acetylmercaptosuccinic anhydride (4.5 mg) in 0.1 ml of dimethyl formamide (DMF); dissolve BSA (15 mg) in 3 ml of phosphate buffer (0.2 M, pH 8.0) and add the BSA dropwise.
2. Stir the mixture for one hour at 4°C and dialyze for two days in PBS.
3. Dissolve E-maleimidocaproic acid *N*-hydroxysuccinimide ester (EMCS) (2.0 mg) in a mixture of methanol (0.2 ml) and phosphate buffer (0.2 ml, pH 8.0) and add FB₁ (2 mg in 0.2 ml of methanol).
4. Stir the mixture for 18 hours at 4°C; dilute the solution by adding distilled water (1.0 ml) and stir for another two hours at 4°C.
5. Remove the excess of EMCS by extraction with a mixture of hexane and ethylacetate (2:1) and separate the water fraction containing FB₁-EMCS complex.
6. Add *S*-methylmercaptosuccinylated BSA (10 mg) dissolved in PBS (2.5 ml) to 1.5 ml of solution containing hydroxylamine (0.33 M) in phosphate buffer (0.2 M, pH 7.3) and ethylenediaminotetraacetic acid (EDTA) (20 mM); stir the mixture for 30 minutes at room temperature and add FB₁-EMCS complex (2.0 mg) dropwise to the mixture.
7. Stir the mixture under nitrogen for 18 hours at room temperature; dialyze against PBS exhaustively and store at −20°C.

Preparation of FB₁-Peroxidase Conjugate

1. Dissolve HRP (8.0 mg) in distilled water (2.0 ml); add 0.2 ml sodium periodate (0.1 M) and stir the mixture for 20 minutes at room temperature.
2. Dialyze against sodium acetate buffer (1.0 mM, pH 4.4) overnight and adjust the pH to 9-10 with NaOH (0.1 M).
3. Mix this activated HRP with FB₁ (2.0 mg) dissolved in distilled water (0.5 ml) and stir the mixture for three hours at room temperature.
4. Stop the reaction with 0.2 ml of sodium borohydride solution (4.0 mg/ml); incubate the mixture overnight at 4°C; dialyze the conjugate against PBS (0.1 M) and store at −20°C.

Immunization and Hybridoma Production

1. Inject female Balb/c mice (two months old) subcutaneously with FB₁-GA-KLH conjugate (50 μg/50 μl) emulsified with equal vol-

umes of complete Freund's adjuvant (CFA); inject again after one month intraperitoneally with similar amounts of conjugates emulsified with complete and incomplete Freund's adjuvants mixture (CFA: IFA = 1:3 v/v).

2. Bleed the mice at 50 days after the first injection and assess the serum titers by indirect competitive (IC) ELISA.

3. Provide the third injection after seven days with same dose of conjugate and adjuvant, followed by a final intravenous injection at four days prior to fusion.

4. Fuse spleen cells (1×10^8) from a mouse producing antiserum for FB_1 with Sp2/0-Ag14 (from American Type Cell Collection CRL-1581) murine myeloma cell line using PEG 1600.

5. Fourteen days after fusion grow the selected hybridomas in hypoxanthine-aminopterine-thymidine (HAT) medium; identify the antibody-secreting hybridomas using both IC-ELISA and direct competitive (DC) ELISA and grow the positive hybridomas in hypoxanthine-thymidine medium (Chapter 4).

IC-ELISA

1. Coat the wells of the microplates with 100 μl of FB_1-EMCS-BSA (5 μg/ml) overnight at room temperature and wash three times with distilled water.

2. Add serially diluted murine serum (50 μl) to each well and coincubate with 50 μl of FB_1 (1 μg/ml) for one hour at room temperature; maintain controls with serially diluted serum with PBS (0 μg/ml of FB_1).

3. Wash the wells five times with distilled water; add 100 μl of goat anti-mouse Ig-HRP conjugate in PBS-Tween 20 (0.1 percent) to each well and incubate for one hour at room temperature.

4. Wash the wells five times with distilled water; add 150 μl of tetramethyl benzidine (TMB)/H_2O_2 substrate to each well and incubate for 15 minutes at room temperature.

5. Stop the reaction by adding 50 μl of sulfuric acid (6N) and record the color intensity/OD at 450 nm using an ELISA reader (Labsystems Multiscan PLUS, Helsinki, Finland).

DC-ELISA

1. Coat the microplate wells with 150 μl of rabbit Ig antimouse Ig (IgM + IgG + IgA, 15 μg/ml); add 100 μl of hybridoma supernatant or sample to wells in duplicate and incubate for one hour at room temperature and wash the plates.

2. Dispense 50 µl of PBS buffer (zero toxin combination Bo) in one well and 50 µl of FB_1 (500 ng/ml of PBS) in another well.
3. Immediately add 50 µl of FB_1-HRP conjugate (in working dilution) to each well; incubate for one hour at room temperature and wash four times.
4. Further steps as in IC-ELISA.

Detection of FB_1 in Spiked/Naturally Infected Cereals

1. Add 50 to 200 ng/g of pure FB_1 to finely ground cereal (wheat, maize, or rye) samples (5 g each) at one day prior to extraction.
2. Shake the samples with 20 ml of extraction solvent consisting of acetonitrile (50 parts), water (39 parts), 0.5 percent KCl (10 parts), and 6 percent H_2SO_4 (1 part) for two hours.
3. Centrifuge the extracts at 4,500 rpm for 30 minutes; dilute the supernatant to 1:5 with PBS-Tween 20 (0.1 percent); shake well and centrifuge again.
4. Test the supernatant by directly transferring to microplate well.

Detection of Ochratoxin-A (OA) by Indirect Competitive ELISA in Spices (Thirumala Devi et al., 2001)

Preparation of Samples

1. Grind the spices (black pepper, coriander, ginger, or turmeric) to a fine powder using a Waring blender; take three samples (each 15 g); extract with a mixture (75 ml) of KCl (0.5 percent) in methanol (70 percent); blend the extracts in a Waring blender; shake the extracts well for 30 minutes and filter through Whatman No. 41 filter paper.
2. Dilute the filtrate (by tenfold) using phosphate-buffered saline containing Tween-20 (0.05 percent) and BSA (0.2 percent) (PBST-BSA) just before ELISA test.

Indirect Competitive ELISA

1. Coat the wells of microtiter plates with purified OA at a concentration of 1 µg/ml of OA-BSA in 0.2 M sodium carbonate buffer, pH 9.6 (150 µl/well), and incubate overnight at 4°C.
2. Perform incubation at 37°C for one hour for subsequent steps.
3. Dilute OA-specific antiserum in PBST-BSA and hold the diluted antiserum for 45 minutes at 37°C; add aliquots of 50 µl to 100 µl of OA standards ranging from 100 ng/ml to 100 pg/ml.

4. Dispense 100 μl of samples (1:10 dilution) to each well containing 50 μl of OA antiserum (diluted to 1:10,000) and incubate.

5. Add goat antirabbit IgG conjugated to alkaline phosphatase (1:1000 dilution) and incubate.

6. Add the substrate *p*-nitrophenyl phosphate (1 mg/ml) in diethanolamine and incubate for one hour at room temperature.

7. Record the absorbance at 405 nm using an ELISA plate reader (Titertek Multiskan, Lab Systems, Finland).

8. Prepare standard curves by plotting $\log_{10}$ values of OA concentration against OD at 405 nm.

9. Express the concentrations of OA in μg/kg using the following formula: OA concentration (ng/ml) × dilution with buffer × extractant solvent volume used (ml) divided by the sample weight (g).

Chapter 14

Chemical Residues and Food Safety

The use of fungicides, pesticides, and herbicides for the management of crop diseases, pests, and weeds has been steadily increasing. Although effective against pathogens, they can be harmful to humans and animals through contaminated foods, feeds, and the environment. The presence of agrochemicals and industrial pollutants in almost all components of the environment is a reality. Fungicides, pesticides, and herbicides that are applied in different seasons make soil a reservoir of these agrochemicals. In turn, they affect soil microbes that are involved in the maintenance of soil fertility and crop development. Water sources may be contaminated by these chemicals and human foods and animals feeds can become contaminated through the food chain. Milk and other dairy products are especially vulnerable to contamination. Some fungicides are suspected to exert adverse effects on the endocrine system of vertebrates by mimicking or inhibiting endogenous hormones. A recombinant human androgen receptor (rhAR) assay in which the receptor was immobilized in microtiter plates via a specific antibody was employed for high throughput screening of pesticides in the environment. Among the 28 chemicals tested, fentin acetate exhibited a high binding affinity of 1.42 percent, indicating the hormonally active nature of this fungicide which may be present in vegetables, fish, and also on clothing (Bauer et al., 2002). Hence, appreciable efforts have been made to develop simple and rapid analytical methods to detect and quantify the residues of chemicals in foods and feeds. Immunological techniques are cost-effective and versatile and their potential for analysis of fungicide residues has been well-demonstrated.

DETECTION OF FUNGICIDES

As in mycotoxins (Chapter 13) fungicides or pesticides of small size (< 1000 Da) have to be covalently linked to a larger molecule or carrierlike protein to produce the immunogen that will induce the formation of specific antibodies in immunized animals such as rabbits (see Appendix section on the production of antiserum). The presence of the fungicide benomyl and its degradation product methyl-2-benzimidazole carbamate was detected by radioimmunoassay (RIA) in food crops (Newsome and Shields, 1981). An early attempt at analysis of a pesticide (parathion) by RIA was made by Ercegovich et al. (1981). The contents of metalaxyl in several vegetables were quantified by ELISA and it was possible to detect at levels as low as 0.1 mg/kg (Newson, 1985). Likewise, triadimefon at a concentration of 0.5 mg/kg could be detected in vegetables (Newson, 1986). The advantage of this immunoassay is that it does not require any cleanup procedure and the recoveries by ELISA tests were comparable with conventional gas chromatography (GC) analysis, after purification by adsorption chromatography.

Immunoassays have found increasing applications in the monitoring of chemical residues in agricultural commodities, foods, the environment, and in the body fluids of humans and animals. Furthermore, the portability, potential for parallel processing of samples, and amenability for automation have added to the wide applicability for mass screening studies either for monitoring regulatory compliance or for epidemiological studies. In addition, the sensitivity of conventional methods can be significantly increased by using immunoaffinity columns to concentrate and purify the analyte prior to the analysis by conventional techniques. The development of monoclonal antibodies has further enhanced the sensitivity and specificity of immunoassays.

Among the immunoassays employed for the analysis of the residues for the various fungicides, ELISA formats have been frequently applied (Table 14.1). Commercial ELISA kits have also been developed for rapid detection of fungicide residues. In dithiocarbamates, two types of haptens were synthesized. The symmetrical N-alkyl dithiocarbamate patterns of thiram with a spacer arm linked to one of the N-methyl terminal group represent one type. The second type shows one of the two symmetrical N-alkyl dithiocarbamate patterns

TABLE 14.1. Analysis of fungicide residues in foods and feeds by immunoassays

Fungicides	Assay format	Detection limit	Application	References
Benomyl	Radioimmunoassay (RIA)	0.5 mg/kg	Fruits	Newsome and Shields, 1981
	ELISA	0.35 ppm	Foods	Newsome and Collins, 1987
Metalaxyl	ELISA	0.1 mg/kg	Fruit	Newson, 1985
Benalaxyl	ELISA	0.5-24 ng/ml	Red wines	Rosso et al., 2000
Tradimefon	ELISA	0.1 mg/kg	Fruit	Newson, 1986
		0.5 ppm	Foods	
Chlorothalonil	ELISA	—	Cabbage and water	Takahashi et al., 1999
Triazole	ELISA	0.01-20.0 ppm	Wheat	Oxley, 1999
Thiabendazole	ELISA	0.05-0.2 ng/ml	Fruit juice	Abad et al., 2001
Tebuconazole	Competitive ELISA	0.02-20 mg/ml	Wheat	Danks et al., 2001
Thiram	Competitive ELISA	5 ng/ml	Lettuce	Queffelec et al., 2001

of thiram with a variable-length spacer arm linked to one sulfur atom. Polyclonal antibodies (PABs) that could recognize thiram were generated by immunization with a hapten of the first type. On the other hand, haptens of the second type were suitable for coating the plates as antigens to develop a competitive ELISA for the detection of thiram with a detection limit of 0.03 μg/ml. As no or little cross-reaction was observed with other dithiocarbamates, the PABs might be considered thiram specific (Gueguen et al., 2000). A competitive ELISA test was developed for the quantification of thiram in lettuce. The calibration curve for thiram had a linear range of 11 to 90 ng/ml and a detection limit of 5 ng/ml. The recovery of thiram from lettuce averaged 89 percent across the range of the ELISA test (Queffelec et al., 2001).

The competitive ELISA system developed for indirect monitoring of formaldehyde and the corresponding demethylated compounds was employed for the detection of the systemic fungicide myclobutanil (triazole). The antibodies recognizing myclobutanil exhibited only minor binding to corresponding N-alkylated derivatives of myclobutanil, suggesting the possibility of differential detection of N-methylated heterocycles (potentially including formaldehyde precursors) and their nonmethylated counterparts (Le et al., 1998). The performance of PAB-based ELISA tests employed for the detection of triazole fungicides such as tetraconazole, hexaconabole, penconazole, and propiconazole was frequently affected by two factors: (1) the use of coating conjugate characterized by a handle different in length and nature from that which is present in the immunizing conjugate, and (2) a low hapten/protein molar ratio (Forlani and Pagani, 1998).

Food products and alcoholic drinks produced from fruits treated with fungicides may have residues even after processing. The suitability of an ELISA format for detection and quantification of benalaxyl in red wine samples was evaluated. By using a rapid and simple cleanup step prior to ELISA, benalaxyl that spiked in red wine in the 0.5 to 24 ng/ml range could be assessed with good accuracy and precision up to 0.5 ng/ml. No false negatives or false positives were observed in this assay. In addition, a good correlation was found between the amounts of benalaxyl determined by ELISA and RP-HPLC methods (Rosso et al., 2000).

For the detection and quantification of thiabendazole (TBZ), a competitive indirect ELISA using a new MAB derived from a hapten functionalized at the nitrogen atom in the 1-position of the TBZ structure. The detection limit of the assay was 0.05 ng/ml. Samples of fruit juices were diluted with assay buffer and prepared even without extraction, cleanup, centrifugation, or filtering to remove fruit pulp. This ELISA format was effective for the detection of TBZ in orange, grapefruit, apple, and pear fruit juices (Abad et al., 2001) (see Appendix section on determination of TBZ).

Crop disease management may require the application of different fungicides at varying frequencies depending on the disease severity and prevailing weather conditions. Cost-effective methods are needed to detect fungicide levels in treated plants and to decide if there is sufficient concentration of fungicide to provide effective protection against fungal pathogens. The fungicide tebuconazole was applied to wheat for the control of *Septoria tritici*. Using ELISA, antibodies specific to tebuconazole were generated and employed to detect the fungicide in extracts of treated wheat leaves. The immunoassay detected teluconazole in the range of 0.01 to 20.0 ppm. This range of fungicide residue in plant material effectively restricted the development of the disease. Furthermore, the fungicide residue levels determined by ELISA were consistent with the doses of fungicide applied to the wheat crop. In addition, it was possible to accurately determine the decay of the active ingredient of the fungicide over time and also within the experiment. The residue levels in the wheat plants were reflected in the extent of disease control obtained. A prototype kit based on a laboratory test was developed for use under field conditions (Oxley, 1999). The possibility of employing immunodiagnosis as an aid for deciding the time of fungicide application for the control of wheat disease was indicated by Kendall, Hollomon, et al. (1998). The effectiveness of ELISA in optimal reduction of fungicide use was further demonstrated by Danks et al. (2001). A competitive ELISA developed for quantitative assessment of tebuconazole had a linear detection range between 0.02 to 20.0 μg/ml of tebuconazole in methanol (20 percent). The antibodies raised against two *p*-chloro substituted derivatives of tebuconazole exhibited little or no cross-reaction with seven other commercially important triazole compounds. The extracts of wheat leaf samples were tested at 48 hours after fungicide treatment by ELISA and GC-MS techniques. A highly significant correlation ($R^2 < 0.9821$) between the levels of fungicide residues determined by these methods was observed, thus validating the applicability of ELISA for residue determination (Danks, et al., 2001).

Several agrochemicals are applied to soil for the control of plant pathogens, pests, and weeds. These chemicals may persist in soil for varying periods and contaminate surface water. Immunological techniques are useful for measuring the levels of chemical residues. Commercially available ELISAs were evaluated and validated for application in water quality control in The Netherlands. The tests were conducted at three laboratories simultaneously. The carbendazim-specific ELISAs were found to be highly effective for discriminating the positive from negative samples and no false-negative values were indicated, although about 10 percent of the samples showed false-positive values. The carbendazim-specific ELISA kits have the potential for quantitative analyses in surface water (Meulenberg et al., 1999).

A commercially available immunoassay kit was used to determine the residues of the fungicide chlorothalonil and the pesticide diazinon in surface water in cabbage. Chlorothalonil was sprayed on cabbage at 400 ppm concentration followed by an artificial rainfall (60 mm/

hour). The levels of chlorothalonil deposited on cabbage and surface water were determined by immunoassay kits and conventional GC technique. The results obtained from both methods were nearly identical. Chlorothalonil deposited better on cabbages and soil surfaces than did diazinon (Takahashi et al., 1999).

Fungicides that are sprayed on leaves are deposited on plant cuticles and absorbed into plant tissues, if they have systemic activity. Plant cuticle-bound fungicide residues can be detected by immunoassays. Monoclonal antibodies (MABs) against the photo-addition products of chlorothalonil with olefinic compounds of plant cuticles were generated and used in an indirect competitive ELISA for the detection of free and bound chlorothalonil and its derivatives (Jahn et al., 1999). The photochemically cutin-bound residues of chlorothalonil in enzymatically isolated tomato and apple fruit cuticles were determined by an indirect competitive ELISA. The cuticles of tomatoes and apples that were irradiated for eight hours by simulated sunlight contained 0.03 and 0.068 mg/g, respectively, of photo-induced cutin-bound residues of wax-free cuticles (expressed as chlorothalonil). The cross-reactivities of the MAB with derivatives of chlorothalonil simulating different types of cuticle-bound residues were also determined (Jahn and Schwack, 2001).

Immunoassays provide the possibility of insight into the mechanisms of activation of defense reactions following the application of fungicides and other chemicals. The effect of application of the fungicide Mon 65500 (Latitude) as seed treatment on infection and colonization of wheat roots by *Gaeumannomyces graminis* var. *tritici (Ggt)* was studied. Immunogold-labeling technique revealed that β-(1,3)-glucanase and lignin were localized over cell walls of wheat roots. In the *Ggt*-infected wheat roots of the fungicide-treated plants, the enzyme level and lignin contents registered significant increases over the untreated control. The fungicide treatment appeared to indirectly induce the structural and biochemical defense reactions of wheat plants (Huang et al., 2001).

DETECTION OF PESTICIDES APPLIED AGAINST INSECT VECTORS

Several microbial plant pathogens, especially viral pathogens, depend on insects as vectors for their spread under natural conditions.

Immunoassays have been used for the rapid identification and differentiation of insects infesting plants. Two important tobacco pests, budworm *(Heliothis virescens)* and boll worm *(Helicoverpa zea),* could be distinguished using highly species-specific characteristic MABs for the eggs. This led to proper selection of insecticide and effective insecticide resistance management. No cross-reactivity could be observed between the eggs of these two pests thus indicating the specificity of immunological reaction (Zeng et al., 1998).

Monoclonal antibodies (MABs) have been employed to differentiate the whitefly *Trialeurodes vaporariorum* and *Bemisia tabaci* complex. An MAB capable of reacting strongly with both male and female of *T. vaporariorum* but not with either sex of *B. tabaci* in ELISA format was identified. Another MAB reacting strongly with the female of *B. tabaci* but not with *T. vaporariorum* of either sex was also detected. When these two MABs were used in combination, they provided a powerful tool for distinguishing the adults of the two whitefly species, especially at low levels of infestation of plant consignments in which pupae may not be easily located (Symondson et al., 1999).

Insecticides are applied to reduce populations of vector insects, although the possibility of reducing population levels that will effectively reduce disease spread is not feasible. Fungicide and pesticide use is on the increase at an annual rate of about 14 percent and the amount of agrochemicals to be used in the first decade of the new millennium is estimated to be more than 3 billion kg (Tseng, 1995). About 140 of over 200 fungicides introduced into agriculture are in current use. The global sales value of fungicides has been estimated to be 5.5 billion U.S. dollars (Brent, 2003). Immunological techniques have provided critical information on residues of pesticides in foods, feeds, and dairy products in addition to evaluation of food safety and cancer risk assessment. Both PABs and MABs specific to pesticides and their derivatives have been prepared. ELISA formats have been found to be as sensitive as conventional techniques for detection and quantification of pesticides (Table 14.2).

An analysis of chlorpyrifos in fruit and vegetable juices was performed by employing a commercial polyclonal plate ELISA protocol. A comparison of results obtained with ELISA tests and gas chromatography-atomic emission detector (GC-AED) technique revealed good agreement with a correlation coefficient of 0.967 (Bushway and Fan, 1998). The quantitative determination of carbaryl levels in fruit

TABLE 14.2. Analysis of pesticide residues in foods and feeds by immunoassays

Fungicides	Assay format	Detection limit	Application	References
Parathion	Radioimmunoassay (RIA)	10 μg/kg	Lettuce	Ercegovich et al., 1981
Endosulfan	ELISA	3 ng/ml	Water	Dreher and Podratzki, 1988
Carbaryl	ELISA	10-100 ng/ml	Fruit and vegetable juices	Nunes et al., 1999 Abad, Moreno, Pelegrí, et al ., 2001
Chlorpyrifos	ELISA	—	Fruit and vegetable juices	Bushway and Fan, 1998
Azinphos-methyl	ELISA	0.4 ng/ml	Fruit juices	Mercader and Montoya, 1997
Flycythrinate	Competitive ELISA	0.3 ng/ml	Apple and tea extracts	Nakata et al., 2001
Carbofuran	Indirect ELISA	0.2 ng/ml	Apple, grape, and pineapple fruit juices	Abad et al., 1997
	ELISA	10-200 ppb	Pepper, cucumber, strawberry, tomato, potato, orange, and apple	Moreno et al., 2001

and vegetable juices by ELISA could be carried out without any pretreatment. The accuracy of ELISA was comparable to liquid chromatography-post column fluorescence method (Nunes et al., 1999). Carbofuran residues in many fruits such as apple, orange, and strawberry and in vegetables such as pepper, cucumber, tomato, and potato were detected by direct and indirect ELISA formats using MABs characterized for affinity and specificity. The immunoassay was able to detect carbofuran at levels below the maximum residue limits (MRL) by simply diluting the sample without any cleanup. For assessment of carbofuran in nonpurified samples, ELISA results were closer to the true values of carbofuran than those detected by HPLC; the mean recovery values were 99 percent for ELISA (Abad et al., 1997; Abad et al., 1999; Moreno et al., 2001).

DETECTION OF PREDATION OF CROP PESTS

The urgent need for minimizing the use of pesticides for the management of insect pests is being realized and alternative approaches to limit these populations must be identified and implemented effectively. In this context, various predators of insect pests were evaluated and some potential effective species have been identified. Immuno-assays have been shown to be rapid, sensitive, and specific in determining levels of predation and have the potential for large-scale use under field conditions (Table 14.3). Some insects not only reduce crop yield levels directly, but also affect crop yield indirectly by acting as vectors of viruses and other microbial pathogens.

The rice brown planthopper *Nilaparvata lugens* causes severe hopper burn resulting in the rapid death of rice seedlings and mature plants. *Nilaparvata lugens* also transmits grassy stunt, ragged stunt, and wilted stunt diseases caused by viruses. The predatory activities of *Ummeliata insecticeps* and *Pirata subpiraticus* were determined by ELISA. The detection period for antigens of *N. lugens,* after the adult *U. insecticeps* had fed on three third to fifth instar nymphs, was 120 hours. The proportions of field-collected *U. insecticeps* and *P. subpiraticus* that were ELISA positive were 47.3 to 49.1 percent and 45.8 to 66.7 percent, respectively (Zhang et al., 1997, 1999). The antiserum raised against antigen from the female *N. lugens* distinguished the indi-

TABLE 14.3. Detection and quantification of predation on crop pests by immuno-assays

Crop/pests	Predators	Immunoassay format	References
A. Rice			
Nilaparvata lugens *Sogatella furcifera*	*Pirata subpiraticus* *Ummeliata insecticeps*	ELISA	Zhang et al., 1997, 1999
N. lugens	*P. subpiraticus*	ELISA	Lim and Lee, 1999
B. Cabbage and Cauliflower			
Pieris rapae	*Coleomegilla maculata lengi* *Nabis americoferus* *Orius insidiosus* *Paradosa milvina* *Phalangium opilio* *Stenolophus comma* *Trechus quadristriatus*	ELISA	Schmaedick et al., 2001
C. Cotton			
Pectinophora gossypiella	*Hippodamia convergens* *Geocoris punctipes* *Orius insidiosus*	ELISA (direct, indirect and sandwich) Dot-blot Western blot	Hagler, et al., 1997 Hagler, 1998 Naranjo and Hagler, 2001
Helicoverpa armigera	*Dicyphus tamaninii* *Macrolophus caliginosus* *Orius majusculus*	Indirect ELISA Dot-blot Squash blot	Agusti et al., 1999

viduals of *P. subpiraticus* that fed on female *N. lugens* from those that fed on other prey species, indicating the specificity of ELISA. *Pirata subpiracticus* collected from fields where *N. lugens* also occurred showed a higher frequency of positive reactions in ELISA than those collected in fields where *N. lugens* was not present, indicating the potential of ELISA for practical application (Lim and Lee, 1999).

Cotton is one among several high-value crops that receive heavy doses of pesticides. The contents of guts of predators were assayed for the presence of antigens of eggs of prey insects *(Pectinophora gossypiella)* by employing indirect ELISA and dot-blot immunoassays. Dot-blots were more reliable than indirect ELISA for detecting low concentrations of egg antigens in samples of the predator (lady beetles) *Hippodamia convergens* (Hagler et al., 1997). The amounts of pink bollworm *(P. gossypiella)* prey remains were evaluated by direct and indirect sandwich ELISA, dot-blot, and Western blot immunoassays, after the lady beetles had consumed one or five pink bollworms. DAS-ELISA was the most sensitive test for the detection of the prey remains (Hagler, 1998).

A model for predicting rates of predation of the pink bollworm was developed using ELISA format. This model was developed for *Geocoris punctipes* and *Orius insidiosus* attacking *P. gossypiella* (Naranjo and Hagler, 2001). An egg-specific MAB was used to detect *Helicoverpa armigera* egg antigens by indirect ELISA, dot-blot, and squash-blot assays. The gut contents of the mirids *Dicyphus tamaninii* and *Macrolophus caliginosus* and the anthoconid *Orius majusculus* were assayed after the predators had fed on *H. armigera* eggs. The presence of the egg antigens was detected in all samples tested, the dot-blot assay provided the best results (Agusti et al., 1999).

Pieris rapae is an important pest of cabbage and cauliflower crops. Several arthropod predators attack *P. rapae*. An ELISA test was employed to detect the remains of immature stages of *P. rapae* in the guts of predators. The first and fifth instars could be more readily detected than eggs or pupae or third instars of *P. rapae*. The whole body homogenates of predators also showed ELISA-positive reactions which, however, generally decreased with increasing size of the predator. The predator species tested included *Coleomegilla maculata-lengi, Nabis americoferus, Orius insidiosus, Pardosa milvina, Phalangium opilio, Stenolophus comma,* and *Trechus quadristriatus.* This study clearly reveals the high level of sensitivity and specificity of ELISA that may be useful to elucidate the role of arthropod predators and reduce the population of *P. rapae* in cabbage (Schmaedick et al., 2001).

SUMMARY

Several agrochemicals—fungicides, pesticides, and herbicides—are applied to soil and plants to protect crops against the adverse effects of plant pathogens, pests, and weeds. Although the deleterious effects of these chemicals on the environment are known, complete avoidance of their use has not been possible. Assessment of the residues of various chemicals in foods, feeds, and water has been precisely carried out using different immunological techniques. The results of immunoassays have highlighted the need for monitoring residues in food and feed and to adopt alternative strategies to reduce the use of agrochemicals. Effective use of predators of insects that are vectors of crop pathogens is a desirable approach to reduce pesticide use.

APPENDIX: IMMUNOASSAY PROTOCOLS

Production of Antiserum Against the Hapten of Carbendazim (Karanth et al., 1998)

Synthesis of Immunogen (Hapten)

1. Dissolve 2-aminobenzimidazole (2AB) (1.33 g, 0.01 M) in acetonitrile (40 ml) and then add succinic anhydride (1.0 g, 0.01 M) to the solution.
2. Stir the mixture for three hours at 40°C; separate the heavy precipitate and wash well with distilled water.
3. Suspend the residue in boiling methanol (50 ml), filter, and air dry.
4. Use the fine crystalline preparation, 2-succini-midobenzimidazole (2-AB hapten) for immunization.

Preparation of Hapten Conjugates with Carrier Proteins

1. Dissolve the 2AB in alkali and add BSA or ovalbumin in aqueous medium followed by the addition of 1-ethyl -3-(3-dimethylaminopropyl-carbodiimide.
2. Dialyze the conjugates against three changes of PBS.

Immunization

1. Inject a female rabbit epidermally at multiple sites with hapten-protein conjugate (1 mg) emulsified with Freund's complete adjuvant (1:1 ratio) and provide a booster dose with hapten-protein conjugate (500 μg equivalent of hapten protein) emulsified with incomplete Freund's adjuvant through intramuscular route.
2. After eight days, collect the blood (3.5 ml) from the marginal vein and check antibody titer.
3. Give a second booster dose by intramuscular injection.
4. After eight days, collect the blood (15-20 ml); separate the antiserum and purify the IgG, using Gamma bind sepharose column (Pharmacia).

Determination of Thiabendazole in Fruit Juices by Competitive Indirect ELISA (Abad et al., 2001)

1. Coat the wells in the microtiter plates with 100 μl of the hapten conjugate solution (0.1 μ/ml) in coating buffer containing carbonate-biocarbonate (50 mM), pH 9.6; incubate the plates overnight at room temperature and wash the wells with washing solution consisting of NaCl (0.15 M) containing Tween-20 (0.05 percent) four times.
2. Prepare the stock solution of thiabendazole (TBZ) (analytical standard from Riedel-de Han, Seelze, Germany) in dimethylsulfoxide; store at 4°C; prepare intermediate stock solutions (200 and 20 μg/ml) in dry *N,N*-dimethylformamide for fortifying fruit juice samples; prepare working standards in the range of 0.00051 to 40 ng/ml from the 20 μg/ml intermediate stock solution in assay buffer containing sodium phosphate (100 mM) and NaCl (137 mM), pH 7.2.
3. Fortify the fruit juice with intermediate stock solution of thiabendazole and dilute the fortified fruit juice to reach a concentration in the working range with assay buffer.
4. Dispense the samples in triplicates into the wells at 50 μl/well; add 50 μl of the monoclonal antibody specific to TBZ (0.6 μg/ml); incubate for one hour at room temperature and wash the wells with washing solution as done earlier.
5. Add 100 μl of peroxidase-labeled rabbit anti-mouse IgG to each well at a dilution of 1/2000 in PBST consisting of sodium phosphate (10 mM), NaCl (137 mM), KCl (2.7 mM), and Tween-20 (0.05 percent), pH 7.4; incubate for one hour at room temperature and wash the wells as done earlier.
6. Transfer to each well 100 μl of color developing solution containing *O*-phenylenediamine (2 mg/ml) and 0.012 percent H_2O_2 in citrate

buffer consisting of sodium citrate (25 mM) and sodium phosphate (62 mM), pH 5.4, and incubate for ten minutes at room temperature.

7. Stop the reaction with sulfuric acid (2.5 M) added at 100 μl/well and record the color intensity at 490 nm.

References

Chapter 1

Anonymous. 1994. *Microbial Diversity Research Priorities.* American Society for Microbiology, Washington, DC.

Cook, R.J. 2000. Advances in plant health management in the twentieth century. *Annual Review of Phytopathology,* 38:95-116.

Hawksworth, D.L. 1991. The fungal dimension of biodiversity: Magnitude, significance and conservation. *Mycological Research,* 95:641-655.

Narayanasamy, P. 2001. *Plant Pathogen Detection and Disease Diagnosis,* Second Edition. Marcel Dekker, Inc., New York.

Narayanasamy, P. 2002. *Microbial Plant Pathogens and Crop Disease Management.* Science Publishers, Enfield, New Hampshire.

Chapter 2

Atassi, M.Z. and Lee, C. 1978. The precise and entire antigenic structure of native lysozyme. *Biochemical Journal,* 171:429-434.

Hill, J.H., Benner, H.I., and Duesen, R.A. 1994. Rapid differentiation of soybean mosaic virus isolates by antigenic signature analysis. *Journal of Phytopathology,* 142:152-162.

Jayaram, C., Deusen, R.A. van, Eggenberger, A.L., Schwabacher, A.W., and Hill, J.H. 1998. Characteristic of a monoclonal antibody recognizing a DAG-containing epitope conserved in aphid transmissible potyvirus evidence that the DAG motif is in a defined conformation. *Virus Research,* 58:1-11.

Jerne, N.K. 1960. Immunological speculations. *Annual Review of Microbiology,* 14:341-358.

Sissons, J.G.P. and Oldstone, M.B.A. 1980. Antibody-mediated destruction of virus-infected cells. *Advances in Immunology,* 29:209-260.

Van Regenmortel, M.H.V. 1966. Plant virus serology. *Advances in Virus Research,* 12:207-271.

Van Regenmortel, M.H.V. 1982. *Serology and Immunochemistry of Plant Viruses.* Academic Press, New York.

Van Regenmortel, M.H.V. and Dubs, M.C. 1993. Serological procedures. In *Diagnosis of Plant Virus Diseases,* (ed.) R.E.F. Matthews, pp. 160-214. CRC Press, Boca Raton, Florida.

Chapter 3

Boonham, N. and Barker, I. 1998. Strain-specific recombinant antibodies to potato virus Y potyvirus. *Journal of Virological Methods,* 74:193-199.

Ey, P.L., Prowse, S.J., and Jenkin, C.R. 1978. Isolation of pure IgG 1, IgG 2a and IgG 2b immunoglobulins from mouse serum protein A-sepharose. *Immunochemistry,* 15:429.

Fischer, R., Schumann, D., Zimmermann, S., Drossard, J., Sack, M., and Schillberg, S. 1999. Expression and characterization of bispecific single-chain Fv-fragments produced in transgenic plants. *European Journal of Biochemistry,* 262: 810-816.

Franconi, R., Roggero, P., Pirazzi, P., Arias, F.J., Desiderio, A., Bitti, O., Pashkonlo, V.D., Mattei, B., Bracci, L., Masenga, V., et al. 1999. Functional expression in bacteria and plants of scFv antibody fragment against tospoviruses. *Immunotechnology,* 4:180-201.

Gough, K.C., Cockburn, W., and Whitelam, G.C. 1999. Selection of phage-displayed peptides that bind to cucumber mosaic virus coat protein. *Journal of Virological Methods,* 79:169-180.

Griep, R.A., van Twisk, C., van Beckhovan, J.R.C.M., van der Wolf, J.M., and Schotts, A. 1998. Development of specific recombinant monoclonal antibodies against lipopolysaccharide of *Ralstonia solanacearum* race 3. *Phytopathology,* 88:795-803.

Hardie, G. and Van Regenmortel, M.H.V. 1977. Isolation of specific antibody under conditions of low ionic strength. *Journal of Immunological Methods,* 15:305.

Harlow, E. and Lane, D. 1988. *Antibodies-A Laboratory Manual.* Cold Spring Harbor Laboratory, Cold Spring Harbor, New York.

Harper, K., Kerschbaumer, R.J., Ziegler, A., Macintosh, S.M., Cowan, G.H., Himmler, G., Mayo, M.A., and Torrance, L. 1997. A scFv-alkaline phosphatase fusion protein which detects potato leafroll luteovirus in plant extracts by ELISA. *Journal of Virological Methods,* 63:237-242.

He, X.H., Liu, S.J., and Perry, K.C. 1998. Identification of epitopes in cucumber mosaic virus using a phage-displayed random peptide library. *Journal of General Virology,* 79:3145-3153.

Khoudi, H., Laberge, S., Ferrillo, J.M., Bazin, R., Darveau, A., Castonguay, Y., Allard, G., Lemieux, R., and Vezina, 1999. Production of a diagnostic monoclonal antibody in perennial alfalfa plants. *Biotechnology and Bioengineering,* 64:135-143.

Köhler, G. and Milstein, C. 1975. Continuous cultures of fused cells secreting antibody of predefined specificity. *Nature (London),* 256:495-497.

Matthews, R.E.F. 1957. *Plant Virus Serology.* University Press, Cambridge, United Kingdom.

McKinney, M.M. and Parkinson, A. 1987. A simple, non-chromatographic procedure to purify immunoglobulins from serum and ascites fluid. *Journal of Immunological Methods,* 96:271.

Mernaugh, R. and Mernaugh, G. 1995. Methods for the production of mouse monoclonal antibodies. In *Molecular Methods in Plant Pathology,* (eds.) R.P. Singh and U.S. Singh, pp. 343-358. CRC Press, Boca Raton, Florida.

Narayanasamy, P. 2001. *Plant Pathogen Detection and Disease Diagnosis,* Second Edition. Marcel Dekker, Inc., New York.

Tavladoraki, P., Girotti, A., Donini, M., Arias, F.J., Mancini, C., Morea, V., Chiaraluce, R., Consalvi, V., and Benvenuto, E. 1999. A single-chain antibody fragment is functionally expressed in the cytoplasm of both *Escherichia coli* and transgenic plants. *European Journal of Biochemistry,* 262:617-624.

Toth, R.L., Harper, K., Mayo, M.A., and Torrance, L. 1999. Fusion proteins of single-chain variable fragments derived from phage display libraries are effective reagents for routine diagnosis of leafroll virus infection in potato. *Phytopathology,* 89:1015-1021.

Van Regenmortel, M.H.V. 1982. *Serology and Immunochemistry of Plant Viruses.* Academic Press, New York.

Van Regenmortel, M.H.V. and Dubs, M.C. 1993. Serological procedures. In *Diagnosis of Plant Virus Diseases,* (ed.) R.E.F. Matthews, pp. 160-214. CRC Press, Boca Raton, Florida.

Weiss, E. and Van Regenmortel, M.H.V. 1989. Use of rabbit Fab'-peroxidase conjugates prepared by the maleimide methods for detecting plant viruses by ELISA. *Journal of Virological Methods,* 24:11.

Ziegler, A., Mayo, M.A., and Torrance, L. 1998. Synthetic antigen from a peptide library can be an effective positive control in immunoassays for the detection and identification of two geminiviruses. *Phytopathology,* 88:1302-1305.

Chapter 4

Almeida, J.D., Stannard, L.M., and Pennington, T.H. 1965. A simple dark ground method for the recording of gel diffusions. *Archives Gesamte Virusforsch,* 17:330-334.

Bailey, B.A., Taylor, R., Dean, J.F.D., and Anderson, J.D. 1991. Ethylene biosynthesis-inducing endoxylanase is translocated through the xylem of *Nicotiana tabacum* cv. Xanthi plants. *Plant Physiology,* 97:1181.

Beesley, J.E. 1992. Preparation of gold probes. In *Immunochemical Protocols* (ed.) M.M. Manson, pp. 163-168. Humana Press, Totowa, New Jersey.

Clark, M.F. and Adams, A.N. 1977. Characteristics of the microplate method of enzyme-linked immunosorbent assay for the detection of plant viruses. *Journal of General Virology,* 34:475-483.

Clark, M.F. and Bar-Joseph, M. 1984. Enzyme immunosorbent assays in plant virology. *Methods in Virology,* 7:51-85.

Commoner, B. and Rodenberg, S.D. 1955. Relationship between tobacco mosaic virus and the non-virus proteins. *Journal of General Physiology*, 38:475-492.

Crowle, A.J. 1973. *Immunodiffusion,* Second Edition. Academic Press, New York.

Derrick, K.S. 1973. Quantitative assay for plant viruses using serologically specific electron microscopy. *Virology,* 56:652-653.

Dolares-Talens, A.C., Hill, J.H., and Durand, D.P. 1989. Application of enzyme-linked fluorescent assay (ELFA) to detection of lettuce mosaic virus in seeds. *Journal of Phytopathology,* 124:149-154.

Eun, A.J.C. and Wong, S.M. 1999. Detection of cymbidium mosaic potexvirus and odontoglossum ringspot tobamovirus using immuno capillary zone electrophoresis. *Phytopathology,* 89:522-528.

Hall, J.L. and Hawes, C. (eds.). 1991. *Electron Microscopy of Plant Cells.* Academic Press, New York.

Harris, N. 1994. Immunocytochemistry for light and electron microscopy. In *Plant Cell Biology-A Practical Approach,* (eds.) N. Harris and K.J. Oparka, pp. 157-176. IRL Press, Oxford.

Harris, N. and Oparka, K.J. (eds.). 1994. *Plant cell Biology,-A Practical Approach.* IRL Press, Oxford.

Hsu, H.T., Lawson, R.H., Lin, N.S., and Hsu, Y.H. 1995. Direct tissue blot immunoassay for analysis of plant pathogens. In *Molecular Methods in Plant Pathology,* (eds.) R.P. Singh and U.S. Singh, pp. 367-376. CRC Lewis Publishers, Boca Raton, Florida.

Khan, M.A. and Slack, S.A. 1978. Studies on the sensitivity of a latex agglutination test for the serological detection of potato virus S and potato virus X in Wisconsin. *American Potato Journal,* 55:627-637.

Milne, R.G. 1993. Solid phase immunoelectron microscopy of virus preparations. In *Immune Electron Microscopy for Virus Diagnosis,* (eds.) A.D. Hyat and B.T. Eaton, pp. 27-70. CRC Press, Boca Raton, Florida.

Mumford, R.A. and Seal, S.E. 1997. Rapid single-tube immunocapture RT-PCR for the detection of two yam potyviruses. *Journal of Virological Methods,* 69:73-79.

Narayanasamy, P. 2001. *Plant Pathogen Detection and Disease Diagnosis,* Second Edition. Marcel Dekker, Inc., New York.

Nolasco, G., de Blas, C., Torres, V., and Ponz, F. 1993. A method of combining immunocapture and PCR amplification in a microtiter plate for the detection of plant viruses and subviral pathogens. *Journal of Virological Methods,* 45:201-218.

Ragetli, H.W.J. and Weintraub, M. 1964. Immuno-osmophoresis, a rapid and sensitive method for evaluating viruses. *Science,* 144:1023-1024.

Roberts, I.M. and Harrison, B.D. 1979. Detection of potato leafroll and potato moptop viruses by immunosorbent electron microscopy. *Annals of Applied Biology,* 93:289-297.

Roland, J.C. and Vian, B. 1991. General preparation and staining of thin sections. In *Electron Microscopy of Plant Cells,* (eds.) J.L. Hall and C. Hawas, pp. 2- 66. Academic Press, New York.

Simmonds, D.H. and Cumming, B. 1979. Detection of lily symptomless viruses by immunodiffusion. *Phytopathology,* 69:1212-1215.

Slack, S.A. and Shepherd, R.J. 1975. Serological detection of seed-borne barley stripe mosaic virus by a simplified radical-diffusion technique. *Phytopathology,* 65:948-955.

Svircev, A.M., Gardiner, R.B., McKeen, W.E. Day, A.W., and Smith, R.J. 1986. Detection by protein A gold of antigens to *Botrytis cincera* in cytoplasm of infected *Vicia faba. Phytopathology,* 76:622-626.

Van Loon, L.C. and Van Kammen, A. 1970. Polyacrylamide disc electrophoresis of the soluble leaf proteins from *Nicotiana tabacum* var. "Samsun" and "Samsun NN" II Changes in protein constitution after infection with tobacco mosaic virus. *Virology* 40:199-211.

Van Regenmortel, M.H.V. 1982. *Serology and Immunochemistry of Plant Viruses.* Academic Press, New York.

van Slogteren, D.H.M. 1955. Serological micro-reactions with plant viruses under paraffin oil. *Proceedings of the Second Conference on Potato Virus Diseases,* pp. 51-54.

Vandenbosch, K.A. 1991. Immunogold labelling. In *Electron Microscopy of Plant Cells,* (eds.) J.L. Hall and C. Hawes, pp. 181-217. Academic Press, New York.

van der Vlugt, R.A.A., van der, Berendsen, M. and Koenraadt, B. 1997. Immunocapture reverse transcriptase PCR for detection of lettuce mosaic virus. In *Seed Health Testing Progress Toward the Twenty-first Century,* (eds.) J.D. Hutchins and J.C. Reeves, pp. 185-198. CAB International, Oxfordshire, United Kingdom.

Walkey, D.G.A., Lyons, N.F., and Taylor, J.D. 1992. An evaluation of a virobacterial agglutination test for the detection of plant viruses. *Plant Pathology,* 41:462-471.

Chapter 5

Bold, H.C., Alexopoulos, C.J., and Delevoryas, T. 1980. *Morphology of Plants and Fungi,* Fourth Edition. Harper and Row Publishers, New York.

Cassab, G.I. 1998. Plant cell wall proteins. *Annual Review of Plant Physiology and Plant Molecular Biology,* 49:281-309.

Chrispeels, M.J. 2003.The genetic basis of growth and development. In *Plants, Genes and Crop Biotechnology,* Second Edition, (eds.) M.J. Chrispeels and D.E. Sadava, pp. 182-211. Jones and Bartlett, Sudbury, Massachusetts.

Scagel, R.F., Bandoni, R.J., Maze, J.R., Rouse, G.E., Schofield, W.B., and Stein, J.R. 1984. *Plants: An Evolutionary Survey.* Wadsworth Publishing Company, Belmont, CA.

Chapter 6

Albersheim, P. and Anderson, A.J. 1971. Proteins from plant cell walls inhibit polygalacturonases secreted by plant pathogens. *Proceedings of the National Academy of Sciences, USA,* 68:1815-1819.

Andersen, I., Becker, W., Schlüter, K., Burges, J., Parthier, B., and Apel, K. 1992. The identification of leaf thionin as one of the main jasmonate-induced proteins of barley *(Hordeum vulgare). Plant Molecular Biology,* 19:193.

Andreu, A. and Daleo, G.R. 1988. Properties of potato lectin fractions isolated from different parts of the tuber and their effect on growth of *Phytophthora infestans. Physiological and Molecular Plant Pathology,* 32:323-333.

Antoniw, J.F., Ritter, C.E., Pierpoint, W.S., and Van Loon, L.C. 1980. Comparison of three pathogenesis-related proteins from plants of two cultivars of tobacco infected with TMV. *Journal of General Virology,* 30:375.

Balls, A.K., Hale, W.S., and Harris, T.H. 1942. A crystalline protein obtained from a lipoprotein of wheat flour. *Cereal Chemistry,* 19:279-288.

Barbieri, L., Aron, G.M., Irvin, J.D., and Stirpe, F. 1982. Purification and partial characterization of another form of the antiviral protein from the seeds of *Phytolacca americana* L. (Pokeweed). *Biochemical Journal,* 203:55-59.

Bi, Y.M., Cammue, B.P.A., Goodwin, P.H. Krishnaraj, S., and Saxena, P.K. 1999. Resistance to *Botrytis cinerea* in scented geranium transformed with a gene encoding the antimicrobial protein Ace-AMP1. *Plant Cell Reports,* 18:835-840.

Bohlmann, H. 1994. The role of thionins in plant protection. *Critical Reviews in Plant Sciences,* 13:1-16.

Bohlmann, H. 1999. The role of thionins in the resistance of plants. In *Pathogenesis-related Proteins in Plants,* (eds.) S.K. Datta and S. Muthukrishnan, pp. 207-234. CRC Press, Boca Ratan, Florida.

Bohlmann, H., Clansen, S., Behnke, S., Giese, H., Hiller, C., Reimann-Philipp, U., Schrader, G., Barkholt, V., and Apel, K. 1988. Leaf-specific thionins of barley—A novel class of cell wall proteins toxic to plant pathogenic fungi and possibly involved in the defence mechanism of plants. *The EMBO Journal,* 7:1559-1565.

Bramble, R. and Gade, W. 1985. Plant seed lectins disrupt growth of germinating fungal spores. *Physiologia Plantarum,* 64:402-408.

Broekaert, W.F., van Parijs, J., Laynes, F., Joos, H., and Peumans, W.J. 1989. A chitin-binding lectin from stinging nettle rhizomes with antifungal properties. *Science,* 245:1100-1102.

Brümmer, J., Thole, H., and Kloppstech, K. 1994. Hordothionins inhibit protein synthesis at the level of initiation in the wheat germ system. *European Journal of Biochemistry,* 219:425.

Callow, J.A. 1977. Recognition, resistance and the role of plant lectins in host-parasite interactions. *Advances in Botanical Research,* 4:1-49.

Carmona, M.J., Molina, A., Fernández, J.A., López-Fando, J.J., and García-Olmedo, F. 1993. Expression of the α-thionin gene from barley in tobacco confers enhanced resistance to bacterial pathogens. *The Plant Journal,* 3:457-462.

Carvalho, A.O., Machado, O.L.T., da Cunha, M., Santos, I.S., and Gomes, V.M. 2001. Antimicrobial peptides and immunolocalization of a LTP in *Vigna unguiculata* seeds. *Plant Physiology and Biochemistry,* 39:137-146.

Cassab, G.I. 1998. Plant cell wall proteins. *Annual Review of Plant Physiology and Plant Molecular Biology,* 49:281-309.

Cervone, F., de Lorenzo, G., Degrá, L., Salvi, G., and Bergami, M. 1987. Purification and characterization of a polygalacturonase-inhibiting protein from *Phaseolus vulgaris* L. *Plant Physiology,* 85:631-637.

Chen, Z.C., White, R.F., Antoniw, J.F., and Lin, Q. 1991. Effect of pokeweed antiviral protein (PAP) on the infection of plant viruses. *Plant Pathology,* 40:612-620.

Chesworth, J.M., Stuchbury, T., and Scaife, J.R. 1998. *An Introduction to Agricultural Biochemistry.* Champman and Hall, London.

Cho, H.J., Lee, S.J., Kim, S.J., and Kim, B.D. 2000. Isolation and characterization of cDNAs encoding ribosome-inhibiting protein from *Dianthus sinensis* L. *Molecules and Cells,* 10:135-141.

Chrispeels, M.J. and Raikhel, N.V. 1991. Lectins, lectin genes and their role in plant defenses. *The Plant Cell,* 3:1-9.

da Silva Conceicão, A. and Broekaert, W.F. 1999. Plant defensins. In *Pathogenesis-related Proteins in Plants,* (eds.) S.K. Datta and S. Muthukrishnan, pp. 247-260. CRC Press, Boca Raton, Florida.

Datta, K., Mutukrishnan, S., and Datta, S.K. 1999. Expression and function of PR-protein genes in trangenic plants in transgenic plants. In *Pathogenesis-Related Proteins in Plants,* (eds.) S.K. Datta and S. Muthukrishnan, pp. 261-277. CRC Press, Boca Raton, Florida.

de Lorenzo, G., Ito, Y., D'Ovidio, R., Cervone, F., Albersheim, P., and Darvill, A.G. 1990. Host-pathogen interactions. XXXVII Abilities of the polygalacturonase-inhibiting proteins from four cultivars of *Phaseolus vulgaris* to inhibit the endopolygalacturonases from three races of *Colletotrichum lindemuthianum. Physiological and Molecular Plant Pathology,* 36:421-435.

Dohm, A., Ludwig, C., Schilling, D., and Debener, T. 2001. Transformation of roses with genes for antifungal proteins. *Acta Horticulturae,* No. 547:27-33.

Epple, P., Apel, K., and Bohlmann, H. 1997. Overexpression of an endogenous thionin enhances resistance of *Arabidopsis* against *Fusarium oxysporum. Plant Cell,* 9:509.

Esquerré-Tugayé, M.T., Mazau, D., Pelissier, B., Roby, D., Rumeau, D., and Toppan, A. 1985. Induction of elicitors and ethylene of proteins associated to the defense of plants. In *Cellular Biology of Plant Stress,* (eds.) J.L. Key and T. Kosuge, pp. 459-473. Liss, New York.

Fakhoury, A.M. and Woloshuk, C.P. 2001. Inhibition of growth of *Aspergillus flavus* and fungal α-amylases by a lectin-like *Lab/lab-purpureus*. *Molecular Plant-Microbe Interactions,* 14:955-961.

Favaron, F., D'Ovidio, R., Proceddu, E., and Alghisi, P. 1994. Purification and molecular characterization of a soybean polygalacturonase-inhibiting protein. *Planta,* 195:80-87.

Filippone, M.P., Diaz-Ricci, J.C., Castagnaro, A.P., and Farias, R.N. 2001. Effect of fragarin on the cytoplasmic membrane of the phytopathogen *Calvibacter michiganensis*. *Molecular Plant-Microbe Interactions,* 14:925-928.

Fincher, G.B., Stone, B.A., and Clarke, A.E. 1983. Arabinogalactan-proteins: Structure, biosynthesis and function. *Annual Review of Plant Physiology,* 34:47-70.

Florack, D.E.A., Dirkse, W.G., Visser, B., Heidekamp, F., and Stiekema, W.J. 1994. Expression of biologically active hordothionin in tobacco: Effects of pre-and prosequences at the amino and carboxyl termini of the hordothionin precursor on mature protein expression and sorting. *Plant Molecular Biology,* 26:25-37.

Florack, D.E.A. and Stiekema, W.J. 1994. Thionins: Properties, possible biological roles and mechanisms of action. *Plant Molecular Biology,* 26:25-37.

Gao, A.G., Hakimi, S.M., Mittanck, C.A., Wu, Y., Woerner, B.M., Stark, D.M., Shah, D.M., Liang, J.H., and Rommens, C.M.T. 2000. Fungal pathogen protection in potato by expression of a plant defensin peptide. *Nature Biotechnology,* 18:1307-1310.

Garas, N.A. and Kuć, J. 1981. Potato lectin lyses zoospores of *Phytophthora infestans* and precipitates elicitors of terpenoid accumulation produced by the fungus. *Physiological Plant Pathology,* 18:227-237.

García-Olmedo, F., Carmona, M.J., López-Fando, J.J., Fernández, J.A., Castagnaro, A., Molina, A., Hernández-Lucas, C., and Carbonero, P. 1992. Characterization and analysis of thionin genes. In *Plant Gene Research Series: Genes Involved in Plant Defense,* (eds.) T. Boller and F. Meins, pp. 283-302. Springer-Verlag, Wien.

García-Olmedo, F., Rodriguez-Palenzuela, P., Hernández-Lucas, C., Ponz, F., Marana, C., Carmona, M.J., López-Fando, J., Fernández, J.A., and Carbonero, P. 1989. The thionins: A protein family that includes purothionins, viscotoxins and crambins. *Oxford Surveys of Plant Molecular and Cell Biology,* 6:31-60.

García-Olmedo, F., Salcedo, G., Sánchez-Monge, R., Hernández-Lucas, C., Carmona, M.J., López-Fando, J.J. Fernández, J.A., Gomez, L., Royo, J., García-Maroto, F., Castagnaro, A., and Carbonero, P. 1992. Trypsin/α-amylase inhibitors and thinonins: Possible defense proteins from barley. In *Barley: Genetics, Biochemistry, Molecular Biology and Biotechnology,* (ed.) P.R. Shewry, pp. 335-350. CAB International, Wallingford.

Habuka, N., Murakami, Y., Noma, M., Kudo, T., and Horikoshi, K.1989. Amino acid sequence of *Mirabilis* antiviral protein, total synthesis of its gene and expression in *Escherichia coli*. *Journal of Biological Chemistry,* 264:6629-6637.

Hejgaard, J., Jacobsen, S., and Svendsen, J. 1991. Two antifungal thaumatin-like proteins from barley grains. *FEBS Letters,* 291:127-131.

Hernandez-Lucas, C., Carbonero, P., and García-Olmedo, F. 1978. Identification and purification of a purothionin homologue from rye (*Secale cereale* L.). *Journal of Agricultural and Food Chemistry,* 26:794.

Huynh, Q.K., Borgmeyer, J.R., and Zobel, J.F. 1992. Isolation and characterization of a 22 kDa protein with antifungal properties from maize seeds. *Biochemistry and Biophysics Research Communications,* 182:1-5.

Inoushe, M., Hayashi, K., and Nevins, D.J. 1999. Polypeptide characteristics and immunological properties of exo- and endoglucanases purified from maize coleoptile cell walls. *Journal of Plant Physiology,* 154:334-340.

Irvin, J.D., Kelly, T., and Robertus, J.D. 1980. Purification and properties of a second antiviral protein from *Phytolacca americana* which inactivates eukaryotic ribosomes. *Archives of Biochemistry and Biophysics,* 200:418-428.

James, J.T. and Dubery, I.A. 2001. Inhibition of polygalacturonase from *Verticillium dahliae* by a polygalacturonase inhibiting protein from cotton. *Phytochemistry,* 57:149-156.

Johnston, D.J., Ramanathan, V., and Williamson, B. 1993. A protein from immature raspberry fruits which inhibits endopolygalacturonases from *Botrytis cinerea* and other microorganisms. *Journal of Experimental Botany,* 44:971-976.

Kragh, K.M., Nielsen, J.E., Nielsen, K.K., Dreboldt, S., and Mikkelsen, J.D. 1995. Characterization and localization of new antifungal cystein-rich proteins from *Beta vulgaris. Molecular Plant-Microbe Interactions,* 8:424.

Krishnaveni, S., Liang, G.H., Muthukrishnan, S., and Manickam, A. 1999. Purification and partical characterization of chitinases from sorghum seeds. *Plant Science (Limerick),* 144:1-7.

Landon, C., Sodano, P., Hétru, C., Hoffman, C., and Ptak, M. 1997. Solution structure of drosomycin, the first inducible antifungal protein from insects. *Protein Science,* 6:1878.

Leah, R., Tommerup, H., Svendsen, I., and Mundy, J. 1991. Biochemical and molecular characterization of three barley seed proteins with antifungal properties. *Journal of Biological Chemistry,* 226:1564-1573.

Lodge, J.K., Kaniewski, W.K., and Tumer, N.E. 1993. Broad-spectrum virus resistance in transgenic plants expressing pokeweed antiviral protein. *Proceedings of the National Academy of Sciences, USA,* 90:7089-7093.

Lord, E.M. and Sanders, L.C. 1992. Roles for the extracellular matrix in plant development and pollination: A special case of cell movement in plants. *Developmental Biology,* 153:16-28.

Machinandiarena, M.F., Olivieri, F.P., Daleo, G.R., and Oliva, C.R. 2001. Isolation and characterization of a polygalacturonase-inhibiting protein from potato leaves. Accumultion in response to salicylic acid, wounding and infection. *Plant Physiology and Biochemistry,* 39:129-136.

Maddaloni, M., Forlani, F., Balmas, V., Donini, G., Corazza, L., Fang, H., Pincus, S., and Motto, M. 1999. The role of b-32 protein in protecting plants against pathogens. In *Genetics and Breeding for Crop Quality and Resistance* (eds.) G.T. Scarascia Mugnozza, E. Porceddu, and M. Pagnotta, pp. 77-82. Kluwer Academic Publishers, Dordrecht, Netherlands.

Matsui, M., Toyosawa, I., and Fukuda, M. 1995. Purification and characterization of a glycine-rich protein from the aleurone layer of soybean seeds. *Bioscience, Biotechnology and Biochemistry,* 59:2231-2234.

Mauch, F., Mauch-Mani, B., and Boller, T. 1988. Antifungal hydrolases in pea tissue. II. Inhibition of fungal growth by combinations of chitinase and α-1,3-glucanase. *Plant Physiology,* 88:936.

Mirelman, D., Galun, E., Sharon, N., and Lotan, R. 1975. Inhibition of fungal growth by wheat germ agglutinin. *Nature,* 256:414-416.

Moreno, M., Segura, A., and García-Olmdeo, F. 1994. Pseudothionin-St1. A potato peptide active against potato pathogens. *European Journal of Biochemistry,* 223:135-139.

Nielsen, K. and Boston, R.S. 2001. Ribosome-inhibiting proteins: A plant perspective. *Annual Review of Plant Physiology and Plant Molecular Biology,* 52:785-816.

Piatak, M., Jr., and Habuka, N. 1992. Expression of plant-derived ribosome inactivating proteins in heterologous systems. In *Genetically Engineered Toxins,* (ed.) A.E. Frankel, pp. 99-131. Marcel Dekker, New York.

Radhajeyalakshmi, R., Meena, B., Thangavelu, R., Deborah, S.D., Vidhyasekaran, P., and Velazhahan, R. 2000. A 45kDa chitinase purified from pearl millet [*Pennisetum glacum* (L.) R.Br] shows antifungal activity. *Journal of Plant Diseases and Protection,* 107:605-616.

Richardson, M. 1991. Seed storage proteins: The enzyme inhibitors. *Methods in Plant Biochemistry,* 5:259-305.

Roberts, W.K. and Selitrennikoff, C.P. 1986. Isolation and partial characterization of two antifungal proteins from barley. *Acta Biochimica Biophysica,* 880:161-170.

Sanjukta, P., Mondal, A.K., and Sudhendu, M. 1999. Identification and characterization of the allergenic proteins of *Ricinus communis* L. pollen—A new approach. *Grana,* 38:311-315.

Sharrock, K.R. and Labavitch, J.M. 1994. Polygalacturonase inhibitors of Bartlett pear fruits: Differential effects on *Botrytis cinerea* polygalacturonase isozymes and influence on products of fungal hydrolysis of pear cell walls and on ethylene induction in cell culture. *Physiological and Molecular Plant Pathology,* 45:305-319.

Shelly Parveen, Savarni Tripathi, and Varma, A. 2001. Isolation and characterization of an inducer protein (Crip-31) from *Clerodendrum inerme* leaves responsible for induction of systemic resistance against viruses. *Plant Science,* 161:453-459.

Shewry, P.R. 1993. Barley seed proteins. In *Barley: Chemistry and Technology* (eds.) J. MacGregor and R. Bhatty, pp. 131-197. AACC, St. Paul, Minnesota.

Shewry, P.R. 1995. Plant storage proteins. *Biological Reviews,* 70:375-426.

Shewry, P.R. and Lucas, J.A. 1997. Plant proteins that confer resistance to pests and pathogens. *Advances in Botanical Research,* 26:136-192.

Singh, N.K., Bracker, C.A., Hasegawa, P.M., Handa, A.K., Buckel, S., Hermodson, M.A., Pfankoch, E., Regnier, F.E., and Bressan, R.A. 1987. Characterization of osmotin, a thaumatin-like protein associated with osmotic adaptation in plant cells. *Plant Physiology,* 85:529-536.

Skadsen, R.W., Sathish, P., and Kaeppler, H.F. 2000. Expression of thaumatin-like permatin PR-5 genes switches from ovary wall to aleurone in developing barley and oat seeds. *Plant Science (Limerick),* 156:11-22.

Stacey, N.J., Roberts, K., and Knox, J.P. 1990. Patterns of expression of the JIM4 arabinogalactan-protein epitopes in cell cultures and during somatic embryo-genesis in *Daucus carota* L. *Planta,* 180:285-292,

Stotz, H.U., Bishop, J.G., Bergmann, C.W., Koch, M., Albersheim, P., Darvill, A.G., and Labavitch, J.M. 2000. Identification of target amino acids that affect interactions of fungal polygalacturonases and their plant inhibitors. *Physiological and Molecular Plant Pathology,* 56:117-130.

Stotz, H.U., Contos, J.J.A., Powell, A.L.T., Bennett, A.B., and Labavitch, L.M. 1994. Structure and expression of an inhibitor of fungal polygalacturonases from tomato. *Plant Molecular Biology,* 25:607-617.

Sutherland, P., Hallett, I., Redgwell, Benhamou, N., and MacRae, E. 1999. Local-ization of cell wall polysaccharides during kiwi fruit *(Actinidia deliciosa)* ripen-ing. *International Journal of Plant Sciences,* 160:1099-1109.

Taira, T., Yamagami, T., Aso, Y., Ishiguro, M., and Ishihara, M. 2001. Localization, accumulation and antifungal activity of chitinases in rye *(Secale cereale)* seed. *Bioscience, Biotechnology and Biochemistry,* 65:2710-2718.

Terras, F.R.G., Eggermont, K., Kovaleva, V., Raikhel, N.V., Osborn, R.W., Kester, A., Rees, S.B., Vanderlayden, J., Cammue, B.P.A., and Broekaert, W.F. 1995. Small cysteine-rich antifungal proteins from radish: Their role in host defense. *The Plant Cell,* 7:573-588.

Toubart, P., Desiderio, A., Salvi, G., Cervone, F., Daroda, L., and de Lorenzo, O. 1992. Cloning and characterization of the gene encoding the endopolygalac-turonase-inhibiting protein (PGIP) of *Phaseolus vulgaris* L. *The Plant Journal,* 2:367-373.

Urdangarín, M.C., Norero, N.S., Broekaert, W.F., and de la Canal, L. 2000. A defensin gene expressed in sunflower inflorescence. *Plant Physiology and Bio-chemistry* (Paris), 38:253-258.

Van Parijs, J., Broekaert, W.F., Goldstein, I.J., and Peumans, W.J. 1990. Hevein: An antifungal protein from rubber tree *(Hevea brasiliensis)* latex. As cited in Chrispeels and Raikhel, 1991.

Vigers, A.J., Roberts, W.K., and Selitrennikoff, C.P. 1991. A new family of plant antifungal proteins. *Molecular Plant-Microbe Interactions,* 4:315.

Vivanco, J.M., Savary, B.J., and Flores, H.E. 1999. Characterization of two novel type I ribosome-inactivating proteins from the storage roots of Andean crop *Mirabilis expansa. Plant Physiology,* 119:1447-1456.

Wada, K. 1982. Localization of purothionins and genome expression in wheat plants. *Plant Cell Physiology,* 23:1357.

Wang, C.X., Croft, K.P.C., Jarlfors, U., and Hildebrand, D.F. 1999. Subcellular localization studies indicate that lipoxygenases 1 to 6 are not involved in lipid mobilization during soybean germination. *Plant Physiology,* 120:227-235.

Wang, X., Thoma, R.S., Carroll, S.A., and Duffin, K.L. 2002. Temporal generation of multiple antifungal proteins in primed seeds. *Biochemical and Biophysical Research Communications,* 292:236-242.

Yao, C., Conway, W.S., and Sams, C.E. 1995. Purification and characterization of a polygalacturonase-inhibiting protein from apple fruit. *Biochemistry and Cell Biology,* 85:1373-1377.

Ye, X.Y., Wang, H.X., and Ng, T.B. 2000. Sativin—A novel antifungal mirculin-like protein isolated from legumes of the sugar snap (*Pisum sativum* var *macrocarpon*). *Life Sciences,* 67:775-781.

Ye, Z.H., Song, Y.R., Marcus, A., and Varner, J.E. 1991. Comparative localization of three classes of cell wall proteins. *Plant Journal,* 1:175-183.

Ye, Z.H. and Varner, J.E. 1991. Tissue specific expression of cell wall proteins in developing soybean tissues. *Plant Cell,* 3:23-37.

Zhu, B., Chen, T.H.H., and Li, P.H. 1996. Analysis of late blight disease resistance and freezing tolerance in transgenic potato plants expressing sense and antisense genes for osmotin-like protein. *Planta,* 198:70-77.

Zoubenko, O., Hudak, K., and Tumer, N.E. 2000. A nontoxic pokeweed antiviral protein mutant inhibits pathogen infection via a novel salicylic acid-independent pathway. *Plant Molecular Biology,* 44:219-229.

Zoubenko, O., Uckun, F., Hur, Y., Chet, I., and Tumer, N. 1997. Plant resistance to fungal infection induced by nontoxic antiviral protein mutants. *Nature Biotechnology,* 15:992-996.

Chapter 7

Agastian, P., Kingsley, S.J., and Vivekanandan, M. 2000. Effect of salinity on photosynthesis and biochemical characteristics in mulberry genotypes. *Photosynthetica,* 38:287-290.

Antikainen, M. and Griffith, M. 1997. Antifreeze protein accumulation in freezing-tolerant cereals. *Physiologia Plantarum,* 99:423-432.

Artlip, T.S. and Funkhouser, E.A. 1995. Protein synthetic responses to environmental stresses. In *Handbook of Plant and Crop Physiology,* (ed.) M. Pessarakli, pp. 627-644. Marcel Dekker, New York.

Baker, E.H., Bradford, K.J., Bryant, J.A., and Rost, T.L. 1995. A comparison of desiccation-related proteins (Dehydrin and QP 47) in peas *(Pisum sativum). Seed Science Research,* 5:185-193.

Ben-Hayyim, G., Gueta-Dahan, Y., Avsian-Kretchmer, O., Weichert, H., and Fuessner, I. 2001. Preferential induction of a 9-lipoxygenese by salt-tolerant cells of *Citrus sinensis* L. *Planta,* 212:367-375.

Ben-Hayyim, G., Vaadia, Y., and Williams, B.G. 1989. Proteins associated with salt adoptation in citrus and tomato cells: Involvement of 26 kDa polypeptides. *Plant Physiology,* 77:332-340.

Bewley, J.D. and Larsen, K.M. 1982. Differences in the responses to water stress of growing and nongrowing regions of maize mesocotyls: Protein synthesis on total, free and membrane bound polyribosome fractions. *Journal of Experimental Botany,* 33:406-415.

Bhadula, S.K., Elthon, T.E., Habten, J.E., Helentjaris, T.G., Jiao, S., and Ristic, Z. 2001. Heat-stress induced synthesis of chloroplast protein synthesis elongation factor (EF-Tu) in heat tolerant maize line. *Planta,* 212:359-366.

Bray, E.A. 1995. Regulation of gene expression during abiotic stresses and the role of the plant hormone abscisic acid. In *Handbook of Plant and Crop Physiology,* (ed.) M. Pessarakli, pp. 733-752. Marcel Dekker, Inc., New York.

Breton, G., Vazquez-Tello, A., Danyluk, J., and Sarhan, F. 2000. Two novel intrinsic annexins accumulate in wheat membranes in response to low temperatures. *Plant and Cell Physiology,* 41:172-184.

Bruggemann, W., Klaucke, S., and Massakentel, K. 1994. Long-term chilling of young tomato plants under low light. 5. Kinetics and molecular properties of key enzyme for the Calvin cycle in *Lycopersicon esculentum* Mill. and *L. peruvianum* Mill. *Planta,* 194:160-168.

Burke, J.J., Hatfield, J.L., Klein, R.R., and Mullet, J.R. 1985. Accumulation of heat shock proteins in field grown cotton. *Plant Physiology,* 78:394-398.

Cassab, G.I. 1998. Plant cell wall proteins. *Annual Review of Plant Physiology and Plant Molecular Biology,* 49:281-309.

Claes, B., Dekeyser, R., Villarroel, R., Bulcke, M.V.D., Bauw, G., Montagu, M.V., and Caplan, A. 1990. Characterization of a rice gene showing organ specific expression in response to salt stress and drought. *Plant Cell,* 2:19-37.

Close, T.J. 1997. Dehydrins—A commonality in the response of plants to dehydration and low temperature. *Physiologia Plantarum,* 100:291-296.

Cordewener, J.H.G., Hause, G., Gorgen, E., Busink, R., Hause, B., Dons, H.J.M., Vanlammeran, A.A.M., Campagne, M.M.V., and Pechan, P. 1995. Changes in synthesis and localization of members of the 70 kDa class of heat-shock proteins accompany the induction of embryogenesis in *Brassica napus* L. microspores. *Planta,* 196:747-755.

Crafts-Brandner, S.J. and Law, R.D. 2000. Effect of heat stress on the inhibition and recovery of the ribulose-1,5- bisphate carboxylase/oxygenase activation state. *Planta,* 212:67-74.

Cramer, G.R., Schnudt, C.L., and Bidart, C. 2001. Analysis of cell wall hardening and cell wall enzymes of salt stressed maize *(Zea mays)* leaves. *Australian Journal of Plant Physiology,* 28:101-109.

de Ronde, J.A., Spreeth, M.H., and Cress, W.A. 2000. Effect of antisense L-Δ^1- pyrroline-5-carboxylate reductase transgenic soybean plants subjected to osmotic and drought stress. *Plant Growth Regulation,* 32:13-26.

Dreier, W., Schnarrenberger, C., and Borner, T. 1995. Light- and stress-dependent enhancement of amylolytic activities in white and green barley leaves-beta-amylases are stress-induced proteins. *Journal of Plant Physiology,* 145:342-348.

Du, Y.C., Kawamitsu, Y., Nose, A., Hiyane, S., Murayama, S., Wasano, K., and Uchida, Y. 1996. Effects of water stress on carbon exchange rate and activities of photosynthetic enzymes in leaves of sugarcane *(Saccharum* sp.). *Australian Journal of Plant Physioloy,* 23:719-726.

Dubey, R.S. 1999. Protein synthesis by plants under stressful conditions. In *Handbook of Plant and Crop Stress,* (ed.) M. Pessarakli, pp. 365-397. Marcel Dekker, Inc., New York.

Elenany, A.E. 1997. Shoot regeneration and protein synthesis in tomato tissue cultures. *Biologia Plantarum,* 39:303-308.

Elsamad, H.M.A. and Shaddad, M.A.K. 1997. Salt tolerance of soybean cultivars. *Biologia Plantarum,* 39:263-269.

García-Gómez, B.I., Campos, F., Hernandez, M., and Covarrubias, A.A. 2000. Two bean cell wall proteins more abundant during water deficit are high in proline and interact with a plasma membrane protein. *Plant Journal,* 22:277-288.

Gogorcena, Y., Iturbeormaetxe, I., Escuredo, P.R., and Becana, M. 1995. Antioxidant defenses against activated oxygen in pea nodules subjected to water stress. *Plant Physiology,* 108:753-759.

Griffith, M., Antikainen, M., Hon, W.C., Pihakaskimaunsbach, K., Yu, X.M., Chun, J.V., and Yang, D.S.C. 1997. Antifreeze proteins in winter rye. *Physiologia Plantarum,* 100:327-332.

Harrak, H., Chamberland, H., Plante, M., Bellemare, G., Lafontaine, J.G., and Tabaeizadeh, Z. 1999. A proline-, threonine-, and glycine-rich protein down-regulated by drought is localized in the cell wall of xylem elements. *Plant Physiology,* 121:557-564.

Harrington, H.M., Dash, S., Dharmasiri, N., and Dharmasiri, S. 1994. Heat-shock proteins—A search for functions. *Australian Journal of Plant Physiology,* 21: 843-855.

Heikkila, J.J., Papp, J.E.T., Schultz, G.A., and Bewley, J.D. 1984. Induction of heat shock protein messenger RNA in maize mesocotyls by water stress, abscisic acid and wounding. *Plant Physiology,* 76:270-274.

Hsiao, T.C. 1973. Plant responses to water stress. *Annual Review of Plant Physiology,* 24:519-570.

Hurkman, W.J. and Tanaka, C.K. 1987. The effects of salt on the pattern of protein synthesis in barley roots. *Plant Physiology,* 83:517-524.

Igarashi, Y., Yoshiba, Y., Sanada, Y., Yamaguchishinozaki, K., Wada, K., and Shinozaki, K. 1997. Characterization of the gene for delta (1)-pyrroline-5-carboxylate synthetase and correlation between the expression of the gene and salt tolerance in *Oryza sativa* L. *Plant Molecular Biology,* 33:857-865.

Kaiser, W.M. 1987. Effect of water deficit on photosynthetic capacity. *Physiologia Plantarum,* 71:142-149.

Kolesnichenko, A.V., Pobezhimova, T.P., and Voinikov, V.K. 2000. Cold-shock proteins in plants. *Fiziologiya Rastenii,* 47:624-630.

Kolesnichenko, A.V., Zykova, V.V., and Voinikov, V.K. 2000. A comparison of the immunochemical affinity of cytoplasmic, mitochondrial and nuclear proteins of winter rye (*Secale cereale* L.) to a 310 kDa stress protein in control plants and during exposure to cold stress. *Journal of Thermal Biology,* 25:203-209.

Korotaeva, N.E., Antipiria, A.I., Borovski, G.B. and Voinikov, V.K. 2001. Localization of low molecular weight heat shock proteins in cell compartments of maize, wheat and rye. *Maize Genetics Cooperation Newsletter,* No. 75:26-27.

Kumar, R.G., Kavita Shah, and Dubey, R.S. 2000. Salinity-induced behavioural changes in malate dehydrogenase and glutamate dehydrogenase activities in rice seedlings of differing salt tolerance. *Plant Science (Limerick),* 156:23-34.

Lásztity, D., Raćz, I., and Páldi, E. 1999. Effect of long periods of low temperature exposure on protein synthesis activity in wheat seedlings. *Plant Science (Limerick),* 149:59-62.

Levitt, J. 1980. *Responses of Plants to Environmental Stresses,* Volume I. Academic Press, New York.

Li, W.G. and Komatsu, S. 2000. Cold-stress-induced calcium-dependent protein kinase(s) in rice (*Oryza sativa* L.) seedling stem tissues. *Theoretical and Applied Genetics,* 101:355-363.

Mansfield, M.A. and Key, J.L. 1988. Cytoplasmic distribution of heat shock proteins in soybean. *Plant Physiology,* 86:1240-1246.

Martin, M.L. and Busconi, L. 2001. A rice membrane-bound calcium-dependent protein kinase is activated in response to low temperature. *Plant Physiology,* 125:1442-1449.

Mastrangelo, A.M., Baldi, P., Mare, C., Terzi, V., Galiba, G., Cattivelli, L., and Forzo, N.di. 2000. The cold-dependent accumulation of COR TMC-AP3 in cereals with contrasting frost tolerance is regulated by different mRNA expression and protein turnover. *Plant Science (Limerick),* 156:47-54.

Matsuba, K., Imaizumi, N., Kaneko, S., Samejima, M., and Ohsugi, R. 1997. Photosynthetic responses to temperature of phosphoenolpyruvate carboxykinase type C-4 species differing in cold sensitivity. *Plant Cell Environment,* 20:268-274.

Moons, A., Dekeyser, A., and Vanmontagu, M. 1997. A group 3 lea cDNA of rice, responsive to abscisic acid, but not to jasmonic acid, shows variety specific differences in salt response. *Gene,* 19:197-204.

Naot, D., Ben-Hayyim, G., Eshdat, Y., and Holland, D. 1995. Drought, heat and salt stress induce expression of a citrus homologue of an atypical late-embryogenesis *lea 5* gene. *Plant Molecular Biology,* 27:619-622.

Naqvi, S.M.S., Ozalp, V.C., Oktem, H.A., and Yucel, M. 1995. Salt induced synthesis of new proteins in the roots of rice varieties. *Journal of Plant Nutrition,* 18:1121-1137.

Örvar, B.L., Sangwan, V., Omann, F., and Dhindsa, R.S. 2000. Early steps in cold sensing by plant cells: The role of actin cytoskeleton and membrane fluidity. *Plant Journal,* 23:785-794.

Ouellet, F., Carpentier, É., Cope, M.J.T.V., Monroy, A.F., and Sarhan, F. 2001. Regulation of wheat actin-depolymerizing factor during cold acclimation. *Plant Physiology,* 125:360-368.

Pelah, D., Wang, W.X., Altman, A., Shoseyov, O., and Bartels, D. 1997. Differential accumulation of water stress-related proteins, sucrose synthase and soluble sugars in *Populus* species that differ in their water stress response. *Physiologia Plantarum,* 99:153-159.

Perezmolphebalch, E., Gidekel, M., Seguranieto, M., Herreraestrella, L., and Ochoaalejo, N. 1996. Effects of water stress on plant growth and root proteins in three cultivars of rice *(Oryza sativa)* with different levels of drought tolerance. *Physiologia Plantarum,* 92:284-290.

Popova, L.P., Stoinova, Z.G., and Maslenkova, L.T. 1995. Involvement of abscisic acid in photosynthetic process in *Hordeum vulgare* L. during salinity stress. *Plant Growth Regulation,* 14:211-218.

Preczewski, P.J., Heckathorn, S.A., Downs, C.A., and Coleman, J.S. 2000. Photosynthetic thermotolerance is quantitatively and positively correlated with production of specific heat-shock proteins among nine genotypes of *Lycopersicon* (tomato). *Photosynthetica,* 38:127-134.

Rajasekaran, L.R., Aspinall, D., and Paleg, L.G. 2000. Physiological mechanism of tolerance of *Lycopersicon* spp. exposed to salt stress. *Canadian Journal of Plant Science,* 80:151-159.

Ramagopal, S. 1993. Advances in understanding the molecular biology of drought and salinity tolerance in plants—The first decade. In *Advances in Plant Biotechnology and Biochemistry,* (eds.) M.L. Lodha, S.L. Mehta, S. Ramagopal, and G.P. Srivatsava, pp. 39-48. Indian Society of Biochemists, Kanpur, India.

Reddy, A.R. 1996. Fructose 2,6-biphosphate-modulated photosynthesis in sorghum leaves grown under low water regimes. *Phytochemistry,* 43:319-332.

Ritenour, M.A., Kochhar, S., Schrader, L.E., Hsu, T.P., and Ku, M.S.B. 2001. Characterization of heat shock protein expression in apple peel under field and laboratory conditions. *Journal of American Society for Horticultural Science,* 126: 564-570.

Saruyama, H. and Tanida, M. 1995. Effect of chilling on activated oxygen-scavenging enzymes in low-temperature-sensitive and tolerant cultivars of rice *(Oryza sativa* L.). *Plant Science,* 109:105-113.

Scott, N.S., Munns, R., and Barlow, E.W.R. 1979. Polyribosome content in young and aged wheat leaves subjected to drought. *Journal of Experimental Botany,* 30:905-911.

Singh, N.K., Bracker, C.A., Hasegawa, P.M., Handa, A.K., Buckel, S., Hermodson, M.A., Pfankoch, E., Regnier, F.E., and Bressan, R.A. 1987. Characterization of osmotin, a thaumatin-like protein associated with osmotic adaptation in plant cells. *Plant Physiology,* 85:529-536.

Singh, N.K., Handa, A.K., Hasegawa, P.M., and Bressan, R.N. 1985. Proteins associated with adaptation of cultured tobacco cells to NaCl. *Plant Physiology,* 79:126-137.

Singla, S.L. and Grover, A. 1994. Detection and quantitation of a rapidly accumulating and predominant 104 kDa heat-shock polypeptide in rice. *Plant Science,* 97:23-30.

Stewart, R.J., Sawyer, B.J.B., Bucheli, C.S., and Robinson, S.P. 2001. Polyphenoloxidase is induced by chilling and wounding in pineapple. *Australian Journal of Plant Physiology,* 28:181-191.

Tabaeizadeh, Z., Chamberland, H., Chen, R.D., Yu, L.X., Bellemare, G., and Lafontaine, J.G. 1995. Identification and immunolocalization of a 65 kDa drought-induced protein in cultivated tomato, *Lycopersicon esculentum. Protoplasma,* 186:208-219.

Uemura, M. and Yoshida, S. 1984. Involvement of plasma membrane alterations in cold acclimation of winter rye seedlings (*Secale cereale* L. cv. Puma). *Plant Physiology,* 75:818-821.

Vierling, E. 1991. The roles of heat shock proteins in plants. *Annual Review of Plant Physiology and Plant Molecular Biology,* 42:579-620.

Wasson, J.J., Reese, R.N., Schumacher, T.E., and Wicks, Z.W. 2000. Proline accumulation in response to dehydration and diurnal hydration cycles varies among maize genotypes. *Maydica,* 45:335-343.

Waters, E.R., Lee, G.J., and Vierling, E. 1998. Evolution, structure and function of small heat shock proteins in plants. *Journal of Experimental Botany,* 47:325-338.

Yu, L.X., Chamberland, H., Lafontaine, J.G., and Tabaeizadeh, Z. 1996. Negative regulation of gene expression of a novel proline-, threonine-, and glycine-rich protein by water stress in *Lycopersicon chiense. Genome,* 39:1185-1193.

Chapter 8

Albersheim, P., Jones, T.M., and English, P.D. 1969. Biochemistry of the cell wall in relation to infective processes. *Annual Review of Phytopathology,* 7:171-194.

Albersheim, P. and Valent, B.S. 1974. Host-pathogen interactions. VII. Plant pathogens secrete proteins which inhibit enzymes of the host capable of attacking the pathogen. *Plant Physiology,* 53:684.

Anil Kumar, Sing, A., and Garg, G.K. 1998. Development of seed immunoblot binding assay for the detection of Karnal bunt (*Tilletia indica*) of wheat. *Journal of Plant Biochemistry and Biotechnology*, 7:119-120.

Arie, T., Gouthu, S., Shimagaki, S., Kamakura, J., Kimura, M., Inoue, M., Takio, K., Ozaki, A., Yoneyama, K., and Yamaguchi, I. 1998. Immunological detection of endopolygalacturonase secretion by *Fusarium* in plant tissue and sequencing of its encoding gene. *Annals of Phytopathological Society, Japan*, 64:7-15.

Arie, T., Hayashi, Y., Yoneyma, K., Nagatani, A., Furuya, M., and Yamaguchi, I. 1995. Detection of *Fusarium* spp. in plants with monoclonal antibody. *Annals of Phytopathological Society, Japan*, 61:311-317.

Bateman, D.F., Van Etten, H.D., English, P.D., Nevins, P.J., and Albersheim, P. 1969. Susceptibility to enzymatic degradation of cell walls from bean plants resistant and susceptible to *Rhizoctonia solani*. *Plant Physiology, Lancaster*, 44:641-648.

Boller, T. 1995. Chemoperception of microbial signals in plant cells. *Annual Review of Plant Physiology and Plant Molecular Biology*, 46:189.

Bom, M. and Boland, G.J. 2000. Evaluation of polyclonal antibody-based immunoassays for detection of *Sclerotinia sclerotiorum* on canola petals and prediction of stem rot. *Canadian Journal of Microbiology*, 46:723-729.

Bonnet, P. 1988. Purifcation de divers filtrats de culture de *Phytophthora* et activités biologiques sur le tabac des differentes fractions. *Agronomie*, 8:347-350.

Bossi, R. and Dewey, F.M. 1992. Development of a monoclonal antibody-based immunodetection assay for *Botrytis cinerea*. *Plant Pathology*,41:472-482.

Boudart, G., Dechamp-Guillaume, G. and Laffitte, C. 1995. Elicitors and suppressors of hydroxyproline-rich glycoprotein accumulation are solubilized from plant cell walls by endopolygalacturonase. *European Journal of Biochemistry*, 232:449.

Bouterige, S., Robert, R., Bouchara, J.P., Marot-Leblond, A., Molinero, V., and Senet, J.M. 2000. Production and characterization of two monoclonal antibodies specific for *Plasmopara halstedii*. *Applied and Environmental Microbiology*, 66:3277-3282.

Bradford, M.M. 1976. A rapid and sensitive method for the quantification of microgram quantities of protein utilizing the principle of protein dye binding. *Annals of Biochemistry*, 72:248-257.

Cahill, D.M. and Hardham, A.R. 1994. A dipstick immunoassay for the specific detection of *Phytophthora cinnamomi* in soils. *Phytopathology*, 84:1284-1292.

Calvete, J.S. 1992. Function of cutinolytic enzymes in the infection of gerbera flowers by *Botrytis cinerea*. Thesis, University of Utrecht, The Netherlands.

Carter, J.P., Spink, J., Cannon, P.F., Daniels, M.J., and Osbourn, A.E. 1999. Isolation, characterization and avenacin sensitivity of a diverse collection of cereal root-colonizing fungi. *Applied and Environmental Microbiology*, 65:3364-3372.

Carzaniga, R., Bowyer, P., and O'Connell, R.J. 2001. Production of extracellular matrices during development of infection structures by the downy mildew *Peronospora parasitica. New Phytologist,* 149:82-93.

Carzaniga, R., Fiocco, D., Bowyer, P., and O'Connell, R.J. 2002. Localization of melanin in conidia of *Alternaria alternata* using phage display antibodies. *Molecular Plant-Microbe Interactions,* 15:216-224.

Centis, S., Guillas, I., Séjalon, N., Esequerre-Tugaye, M.T., and Dumas, B. 1997. Endopolygalacturonase genes from *Colletotrichum lindemuthianum:* Cloning of *CLPG2* and comparison of its expression to that of *CLPG1* during saprophytic and parasitic growth of the fungus. *Molecular Plant-Microbe Interactions,* 10:769-775.

Chen, L.M. and Chen, Y.X. 1998. Direct competitive ELISA screening method for aflatoxin B_1. *Journal of Nanjing Agricultural University,* 21:62-65.

Clement, J.A., Butt, I.M., and Beckett, A. 1993. Characterization of the extracellular matrix produced in vitro by urediniospores and sporelings of *Uromyces viciae-fabae. Mycological Research,* 97:594-603.

Clement, J.A., Martin, S.G. Porter, R., Butt, T.M., and Beckett, A. 1993. Germination and the role of extracellular matrix in adhesion of urediniospores of *Uromyces viciae-fabae* to synthetic surfaces. *Mycological Research,* 97:585- 593.

Comménil, P., Belingheri, L., and Dehorter, B. 1998. Antilipase antibodies prevent infection of tomato leaves by *Botrytis cinerea. Physiological and Molecular Plant Pathology,* 52:1-14.

Cooper, R.M., Longma, D., Campbell, A., Henry, M., and Lees, P.E. 1988. Enzymatic adaptation of cereal pathogens to the monocotyledonous primary wall. *Physiological and Molecular Plant Pathology,* 32:33-47.

Cooper, R.M. and Wood, R.K.S. 1975. Regulation of synthesis of cell wall-degrading enzymes by *Verticillium albo-atrum* and *Fusarium oxysporum* f. sp. *lycopersici. Physiological Plant Pathology,* 16:285-300.

Crowhurst, R.N. Binnie, S.J., Bowen, J.K., Hawthorne, B.T., Plummer, K.M., Rees-George, J., Rikkerink, E.H.A., and Templeton, M.D. 1997. Effect of disruption of a cutinase gene *(cut A)* on virulence and tissue specificity of *Fusarium solani* f. sp. *cucurbita* race 2 toward *Cucurbita maxima* and *C. moschata. Molecular Plant-Microbe Interactions,* 10:355-368.

Daroda, L., Hahn, K., Pashkoulov, D., and Benvenuto, E. 2001. Molecular characterization and *in planta* detection of *Fusarium moniliforme* endopolygalacturonase isoforms. *Physiological and Molecular Plant Pathology,* 59:317-325.

Darvill, A.G. and Albersheim, P. 1984. Phytoalexins and their elicitors: A defense against microbial infection in plants. *Annual Review of Plant Physiology,* 35: 243.

Debroah, S.D., Palaniswami, A., and Velazhahan, R. 2001. Differential induction of chitinase and β-1,3-glucanase in rice in response to inoculation with a pathogen *(Rhizoctonia solani)* and a non-pathogen (*Pestalotia palmarum*). *Acta Phytopathologica et Entomologica Hungarica,* 36:67-74.

Deising, H., Nicholson, R.L., Hang, M., Howard, R.J., and Mendgen, K. 1992. Adhesion pad formation and the involvement of cutinase and esterases in the attachment of uredospores to the host cuticle. *Plant Cell,* 4:1101-1111.

Delfosse, P., Reddy, A.S., Legréve, K., Thirumala Devi, K., Abdurahman, M.D., Maraite, H., and Reddy, D.V.R. 2000. Serological methods for the detection of *Polymyxa graminis,* an obligate root parasite and vector of plant viruses. *Phytopathology,* 90:537-545.

de Lorenzo, G., Ito, Y., D'Ovidio, R., Cervone, F., Albersheim, P. and Darvill, A.G. 1990. Host-pathogen interactions. XXXVII. Abilities of the polygalacturonase-inhibiting proteins from four cultivars of *Phaseolus vulgaris* to inhibit the endopolygalacturonases from three races of *Colletotrichum lindemuthianum.* *Physiological and Molecular Plant Pathology,* 36:421-435.

de Lorenzo, G., Salvi, G., Degra, L., D'Ovidio, R., and Cervone, F. 1987. Induction of extracellular polygalacturonase and its mRNA in the phytopathogenic fungus *Fusarium moniliforme. Journal of General Microbiology,* 133:3365-3373.

Devergne, J.C., Fort, M.A., Bonnet, P., Ricci, P., Vergnet, C., Delaunay, T., and Grosclaude, J. 1994. Immunodetection of elicitins from *Phytophthora* spp. using monoclonal antibodies. *Plant Pathology,* 43:885-886.

Dickman, M.B., Patil, S.S., and Kolattukudy, P.E. 1982. Purification, characterization and role in infection of an extracellular cutinolytic enzyme from *Colletotrichum gloeosporioides* Penz. on *Carica papaya* L. *Physiological Plant Pathology,* 20:333-347.

Dickman, M.B., Podila, G.K., and Kolattukudy, P.E. 1989. Insertion of cutinase gene into a wound pathogen enables it to infect intact host. *Nature,* 342: 446-448.

Durrands, P.K. and Cooper, R.M. 1988. The role of pectinases in vascular wilt disease as determined by defined mutants of *Verticillium albo-atrum. Physiological and Molecular Plant Pathology,* 32:363-371.

Duviau, M.P. and Kobrehel, K. 2000. Extracellular fungal proteinases target specific cereal proteins. In *Wheat Gluten,* (eds.) P.R. Shewry and A.S. Tatham, Special Publication No. 261, pp. 296-299. Royal Society of Chemists, Cambridge, United Kingdom.

El-Gendy, W., Brownleader, M.D., Ismail, H., Clarke, P.J., Gilbert, J., El-Bordiny, F., Trevan, M., Hopkins, J., Naldrett, M., and Jackson, P. 2001. Rapid deposition of wheat cell wall structural proteins in response to Fusarium-derived elicitors. *Journal of Experimental Botany,* 52:85-90.

Estrada-Garcia, M.T., Callow, J.A. and Green, J.R. 1990. Monoclonal antibodies to the adhesive cell coat secreted by *Pythium aphanidermatum* zoospores recognize 200×10^3 Mr. glycoproteins stored within large vesicles. *Journal of Cell Science,* 95:199-206.

Favaron, F., Destro, T., and D'Ovidio, R. 2000. Transcript accumulation of polygalacturonase-inhibiting protein (PGIP) following pathogen infections in soybean. *Journal of Plant Pathology,* 82:103-109.

Finnie, C., Andersen, C.H., Borch, J., Gjetting, S., Christensen, A.B., Boer, A.H. de. Thordal-Christensen, H., and Collinge, D.B. 2002. Do 14-3-3-proteins and plasma membrane H$^+$-ATPases interact in the barley epidermis in response to the barley powdery mildew fungus? *Plant Molecular Biology,* 49:137-147.

Fox, R.T.V. 1998. Plant disease diagnosis. In *The Epidemiology of Plant Diseases* (ed.) D. Gareth Jones, pp. 14- 41. Kluwer Academic Publishers, Dordrecht, The Netherlands.

Gabor, B.K., O'Gara, E.T., Philip, B.A., Horan, D.P., and Hardham, A.R. 1993. Specificities of monoclonal antibodies to *Phytophthora cinnamomi* in two rapid diagnostic assays. *Plant Disease,* 77:1189-1197.

Gergerich, R.C., Nannapaneni, R., and Lee, F.N. 1996. Accurate rice blast detection and identification using monoclonal antibodies. *Research Series of Arkansas Agricultural Experiment Station,* No. 453:126-132.

Giesbert, S., Lepping, H.B., Tenberge, K.B., and Tudzynski, P. 1998. The xylanolytic system of *Claviceps purpurea:* Cytological evidence for secretion of xylanases in infected rye tissue and molecular characterization of two xylanase genes. *Phytopathology,* 88:1020-1030.

Gleason, M.L., Ghabrial, S.A., and Ferriss, R.S. 1987. Serological detection of *Phomopsis longicolla* in soybean seeds. *Phytopathology,* 77:371-375.

Gough, K.C., Li, Y., Vaugan, T.J., Williams, A.J., Cockburn, W., and Whitelam, G.C. 1999. Selection of phage antibodies to surface epitopes of *Phytophthora infestans. Journal of Immunological Methods,* 228:97-108.

Graniti, A. 1989. Fusicoccin and stomatal transpiration. In *Host-specific Toxins* (eds.) K. Kohmoto and R. Durbin p. 143. Tottori University Press, Tottori, Japan.

Grote, D. and Claussen, W. 2001. Severity of root rot on tomato plants caused by *Phytophthora nicotianae* under nutrient-and light-stress conditions. *Plant Pathology,* 50:702-707.

Gubler, F., Hardham, A.R., and Duniec, J. 1989. Characterizing adhesiveness of *Phytophthora cinnamomi* zoospores during encystment. *Protoplasma,* 149:24-30.

Hahn, M., Jüngling, S., and Knogge, W. 1993. Cultivar-specific elicitation of barley defense reactions by the phytotoxic peptide NIP1 from *Rhynchosporium secalis. Molecular Plant-Microbe Interactions,* 6:745-754.

Hahn, M. and Mendgen, K. 1997. Characterization in *in planta*-induced rust genes isolated from a haustorium-specific cDNA library. *Molecular Plant-Microbe Interactions,* 10:427-434.

Hahn, M., Neef, U., Struck, C., Göttfert, M., and Mendgen, K. 1997. A putative amino acid transporter specifically expressed in haustoria of the rust fungus *Uromyces fabae. Molecular Plant-Microbe Interactions,* 10:438-445.

Ham, K.S., Albersheim, P., and Darvill, A.G. 1995. Generation of β-glucan elicitors by plant enzymes and inhibition of the enzymes by a fungal protein. *Canadian Journal of Botany,* 73:1100.

Ham, K.S., Wu, S.C., Darvill, A.G., and Albersheim, P. 1997. Fungal pathogens secrete an inhibitor protein that distinguishes isoforms of plant pathogenesis-related endo-β-1,3-glucanases. *Plant Journal,* 11:169.

Harrison, J.G., Rees, E.A., Barker, H., and Lowe, R. 1993. Detection of spore balls of *Spongospora subterranea* on potato tubers by enzyme-linked immunosorbent assay. *Plant Pathology,* 42:181-186.

Huang, L.L., Kang, Z., Heppner, C., and Buchenauer, H. 2001. Ultrastructural and immunocytochemical studies on effects of the fungicide Mon 65500 (Latitude ®) on colonization of roots of wheat seedlings by *Gaeumannomyces graminis* var. *tritici. Journal of Plant Diseases and Protection,* 108:188-203.

Hughes, H.B., Carzaniga, R., Rawlings, S.L., Green, J.R., and O' Connell, R.J. 1999. Spore surface glycoproteins of *Colletrotrichum lindemuthianum* are recognized by a monoclonal antibody which inhibits adhesion to polystyrene. *Microbiology (Reading),* 145:1927-1936.

Hutchison, K.A., Perfect, S.E., O'Connell, R.J., and Green, J.R. 2000. Immunomagnetic purification of *Colletotrichum lindemuthianum* appressoria. *Applied and Environmental Microbiology,* 66:3464-3467.

Isshiki, A., Akimitsu, K., Yamamoto, M., and Yamamoto, H. 2001. Endopolygalacturonase is essential for citrus black rot caused by *Alternaria citri,* but not for brown spot caused by *A. alternata. Molecular Plant-Microbe Interactions,* 14:749-757.

Jamaux, I.D. and Spire, D. 1999. Comparison of responses of ascospores and mycelium by ELISA with anti-mycelium and anti-ascospore antisera for the development of a method to detect *Sclerotinia sclerotiorum* on petals of oilseed rape. *Annals of Applied Biology,* 134:171-179.

Kageyama, K., Kobayashi, M., Tomita, M., Kubota, N., Suga, H., and Hyakumachi, M. 2002. Production and evaluation of monoclonal antibodies for the detection of *Pythium sulcatum* in soil. *Journal of Phytopathology,* 150:97-104.

Kang, Z. and Buchenauer, H. 1999. Immunocytochemical localization of fusarium toxins in infected wheat spikes by *Fusarium culmorum. Physiological and Molecular Plant Pathology,* 55:275-288.

Kang, Z. and Buchenauer, H. 2000a. Ultrastructural and cytochemical studies on cellulose-xylan and pectin degradation in wheat spikes infected by *Fusarium culmorum. Journal of Phytopathology,* 148:263-275.

Kang, Z. and Buchenauer, H. 2000b. Ultrastructural and immunocytochemical investigation of pathogen development and host responses in resistant and susceptible wheat spikes by *Fusarium culmorum. Physiological and Molecular Plant Pathology,* 57:255-268.

Kang, Z., Huang, L., and Buchenauer, H. 2000. Cytochemistry of cell wall component alterations in wheat roots infected by *Gaeumannomyces graminis* var. *tritici. Journal of Plant Diseases and Protection,* 107:337-351.

Karpovich-Tate, N., Spanu, P., and Dewey, F.M. 1998. Use of monoclonal antibodies to determine biomass of *Cladosporium fulvum* in infected tomato leaves. *Molecular Plant-Microbe Interactions,* 11:710-716.

Kendall, J.J., Hollomon, D.W., and Selley, A. 1998. Immunodiagnosis as an aid to timing of fungicide sprays for the control of *Mycosphaerella graminicola* on winter wheat in the U.K. *Brighton Crop Protection Conference: Pests and Diseases,* 2:701-706.

Kennedy, R., Wakeham, A.J., and Cullington, J.E. 1999. Production and immunodetection of ascospores of *Mycosphaerella brassicola:* Ringspot of vegetable crucifers. *Plant Pathology,* 48:297-307.

Keon, J.P.R., Byrde, R.J.W., and Cooper, R.M. 1987. Some aspects of fungal enzymes that degrade cell walls. In *Fungal Infection of Plants,* (eds.) G.F. Pegg and P.G. Ayres, pp. 133-157. Cambridge University Press, Cambridge, United Kingdom.

Kieffer, F., Lherminier, J., Simon-Plas, F., Nicole, M., Paynot, M., Elmayan, T., and Blein, J.P. 2000. The fungal elicitor cryptogein induces cell wall modifications on tobacco cell suspension. *Journal of Experimental Botany,* 51:1799-1811.

Kitagawa, T., Sakamoto, Y., Furumi, K., and Ogura, H. 1989. Novel enzyme immunoassays for specific detection of *Fusarium oxysporum* f. sp. *cucumerianum* and for general detection of various *Fusarium* species. *Phytopathology,* 79:165-168.

Kolattukudy, P.E. 1985. Enzymatic penetration of the plant cuticle by fungal pathogens. *Annual Review of Phytopathology,* 23:223-250.

Krishnaveni, S., Liang, G.H., Muthukrishnan, S., and Manickam, A. 1999. Purification and partical characterization of chitinases from sorghum seeds. *Plant Science (Limerick),* 144:1-7.

Kronstad, J.W. 1997. Virulence and cAMP in smuts, blasts and blights. *Trends in Plant Science,* 2:193-199.

Kumar, K.K. and Parameswaran, K.P. 1998. Characterization of storage protein from selected foxtail millet [*Setaria italica* (L.) Beauv]. *Journal of Science of Food and Agriculture,* 77:535-542.

Kutilek, V., Lee, R., and Kitto, G.B. 2001. Development of immunochemical techniques for detecting Karnal bunt in wheat. *Food and Agricultural Immunology,* 13:103-114.

Kwon, Y.H. and Epstein, L. 1993. A 90 kDa glycoprotein associated with adhesion of *Nectria haematococca* malcroconidia to substrate. *Molecular Plant-Microbe Interactions,* 6:481-487.

Lagerberg, C. 1996. Comparison of polyclonal ELISA with the seed-blotter, fluorescence and agar plate methods for detection and quantification of seed-borne *Septoria nodorum* in wheat. *Seed Science and Technology,* 24:585-588.

Lawton, M.A., Beck, J., Potter, S., Ward, E., and Ryals, J. 1994. Regulation of cucumber class III chitinase gene expression. *Molecular Plant-Microbe Interactions,* 7:48-57.

Lee, Y.K., Hippe-Sanwald, S., Jung, H.W. Hong, J.K., Hause, B., and Hwang, B.K. 2000. In situ localization of chitinase mRNA and protein in compatible and incompatible interactions of pepper stems with *Phytophthora capsici. Physiological and Molecular Plant Pathology,* 57:111-121.

Leubner-Metzger, G. and Meins, F., Jr. 1999. Functions and regulation of plant β-1,3-glucanases (PR-2). In *Pathogenesis-Related Proteins in Plants*, (eds.) S.K. Datta and S. Muthukrishnan, pp. 49-76. CRC Press, Boca Raton, Florida.

Li, S.X., Hartman, G.L., Lee, B.S., and Widholm, J.W. 2000. Identification of a stress-induced protein in stem exudates of soybean seedlings root infected with *Fusarium solani* f. sp. *glycines*. *Plant Physiology and Biochemistry*, 38:803-809.

Linfield, C.A., Kenny, S.R., and Lyons, N.F. 1995. A serological test for detecting *Botrytis allii*, the cause of neck rot of onion bulbs. *Annals of Applied Biology*, 126:259-268.

Lyons, N.F. and White, J.G. 1992. Detection of *Pythium violae* and *Pythium sulcatum* in carrots with cavity spot using competition ELISA. *Annals of Applied Biology*, 120:235-244.

Macko, V., Wolpert, T.J., Acklin, W., and Anigoni, D. 1989. Biological activities of structural variants of host-selective toxins of *Cochlioboluse victoriae*. In *Phytotoxins and Plant Pathogenesis*, (eds.) A. Graniti, R.D. Durbin, and A. Ballio, pp. 31-41. Springer-Verlag, Berlin.

Manandhar, H.K., Mathur S.B., Smedegaard-Petersen, V., and Thordal-Christensen, H. 1999. Accumulation of transcripts for pathogenesis-related proteins and peroxidase in rice plants triggered by *Pyricularia oryzae*, *Bipolaris sorokiniana* and UV light. *Physiological and Molecular Plant Pathology*, 55:289-295.

Martin, R.R., James, D., and André Lévesque, C. 2000. Impacts of molecular diagnostic technologies on plant disease management. *Annual Review of Phytopathology*, 38:207-239.

Mercure, E.W., Kunoh, H., and Nicholson, R.L. 1994. Adhesion of *Colletotrichum graminicola* to corn leaves: A requirement for disease development. *Physiological and Molecular Plant Pathology*, 45:407-420.

Mercure, E.W., Leite, B., and Nicholson, R.L. 1994. Adhesion of ungerminated conidia of *Colletotrichum graminocla* to artificial hydrophobic surfaces. *Physiological and Molecular Plant Pathology*, 45:421-440.

Milat, M.L., Ducruet, J.M., Ricci, P., Marty, F., and Blein, J.P. 1991. Physiological and structural changes in tobacco leaves treated with cryptogein, a proteinaeous elicitor from *Phytophthora cryptogea*. *Phytopathology*, 81:1364-1368.

Moreau, R.A. and Rawa, D. 1984. Phospholipase activity in cultures of *Phytophthora infestans* and in infected leaves. *Physiological Plant Pathology*, 24:187-199.

Murillo, I., Cavallarin, L., and San Segundo, B. 1999. Cytology of infection of maize seedlings by *Fusarium moniliforme* and immunolocalization of the pathogenesis-related PRms protein. *Phytopathology*, 89:737-747.

Murillo, I., Jaeck, E., Cordero, M.J., and San Segundo, B. 2001. Transcriptional activation of a maize calcium-dependent protein kinase gene in response to fungal elicitors and infection. *Plant Molecular Biology*, 45:145-148.

Nallathambi, P., Padmanaban, P., and Mohanraj, D. 2001. Standardization of an indirect ELISA technique for detection of *Ustilago scitaminea* Syd. Causal agent of sugarcane smut disease. *Journal of Mycology and Plant Pathology*, 31:76-78.

Nannapaneni, R., Gergerich, R.C., and Lee, F.N. 2000. Technology for rapid detection, identification and quantifiction of rice blast fungus *Pyricularia grisea*. *Arkansas Agricultural Experiment Station Research Series*, No. 476:480-485.

Narayanasamy, P. 2001. *Plant Pathogen Detection and Disease Diagnosis*, Second Edition. Marcel Dekker, Inc., New York.

Nicholson, R.I. and Epstein, L. 1991. Adhesion of fungi to the plant surface: Prerequisite for pathogenesis. In *The Fungal Spore and Disease Initiation in Plants and Animals*, (eds.) G.T. Cole and H.C. Hoch, pp. 3-23. Plenum Press, New York.

Olsson, C.H.B. and Heiberg, N. 1997. Sensitivity of the ELISA test to detect *Phytophthora fragariae* var. *rubi* in raspberry roots. *Journal of Phytopathology*, 145:285-288.

Orihara, S. and Yamamoto, T. 1998. Detection of resting spores of *Plasmodiophora brassicae* from soil and plant tissues by enzyme immunoassay. *Annals of Phytopathological Society, Japan*, 64:569-573.

Orsomando, G., Lorenzi, M., Raffaelli, N., Rizza, M.D., Mezzetti, B., and Ruggieri, S. 2001. Phytotoxic protein PcF-purification, characterization and cDNA sequencing of a novel hydroxyproline-containing factor secreted by the strawberry pathogen *Phytophthora cactorum*. *Journal of Biological Chemistry*, 276:21578-21584.

Pascholati, S.F., Deising, H., Leite, B., Anderson, D., and Nicholson, R.L. 1993. Cutinase and nonspecific esterase activities in the conidial mucilage of *Colletotrichum graminicola*. *Physiological and Molecular Plant Pathology*, 42:37-51.

Pekárová, B., Krátká, J., and Slováèek, J. 2001. Utilization of immunochemical methods to detect *Phytophthora fragariae* in strawberry plants. *Plant Protection Science*, 37:57-65.

Perfect, S.E., Green, J.R., and O'Connell, R.J. 2001. Surface characteristics of necrotrophic secondary hyphae produced by the bean anthracnose fungus *Colletotrichum lindemuthianum*. *European Journal of Plant Pathology*, 107:813-189.

Petersen, A.B., Olson, L.W., and Rosendahl, S. 1996. Use of polyclonal antibodies to detect oospores of *Aphanomyces*. *Mycological Research*, 100:495-499.

Pietro, A.D. and Roncero, M.I. 1998. Cloning, expression and role in pathogenicity of *Pg1* encoding the major extracellular endopolygalacturonase of the vascular wilt pathogens *Fusarium oxysporum*. *Molecular Plant-Microbe Interactions*, 11:91-98.

Plantiño Álvarez, B., Rodríguez Cámara, M.C., Rodríguez Fernández, T., González Jaen, M.T., and Vázquez Estévez, C. 1999. Immunodetection of an exopolygalacturonase in tomato plants infected with *Fusarium oxysporum* f. sp. *radicislycopersici*. *Boleín de Sanidad Vegetal, Plagas*, 25:529-536.

Plasencia, J., Jemmerson, R., and Banttari, E.E. 1996. Production and characterization of monoclonal antibodies to *Verticillium dahliae* and development of a quantitative immunoassay for fungal biomass. *Phytopathology,* 86:170-176.

Podila, G.K., Dickman, M.B., Rogers, L.M., and Kolattukudy, P.E. 1989. Regulation of expression of fungal genes by plant signals. In *Molecular Biology of Filamentous Fungi,* (eds.) H. Nevalainen and M. Penttila, pp. 217-226. Foundation for Biotechnical Industrial Fermentation Research.

Pryce-Jones, E., Carver, T., and Gurr, S.J. 1999. The roles of cellulose enzymes and mechanical force in host penetration by *Erysiphe graminis* f. sp. *hordei. Physiological and Molecular Plant Pathology,* 55:175-182.

Radhajeyalakshmi, R., Meena, B., Thangavelu, R., Deborah, S.D., Vidhyasekaran, P., and Velazhahan, R. 2000. A 45kDa chitinase purified from pearl millet [*Pennisetum glacum* (L.) R.Br] shows antifungal activity. *Journal of Plant Diseases and Protection,* 107:605-616.

Rha, E., Park, H.J., Kim, M.O., Chung, Y.R., Lee, C.W., and Kim, J.W. 2001. Expression of exo-polygalacturonases in *Botrytis cinerea. FEMS Microbiology Letters,* 201:105-109.

Ricci, P., Bonnet, P., Huet, J.C., Sallantin, M., Beauvais-Cante, F., Bruneteau, M., Billard, V., Michel, G., and Pernollet, J.C. 1989. Structure and activity of proteins from pathogenic fungi *Phytophthora* eliciting necrosis and acquired resistance in tobacco. *European Journal of Biochemistry,* 183:555-563.

Robold, A.V. and Hardham, A.R. 1998. Production of species specific antibodies that react with surface components on zoospores and cysts of *Phytophthora nicotianae. Canadian Journal of Microbiology,* 44:1161-1170.

Rogers, L.M., Kim, Y.K., Guo, W.J., González-Candelas, L., Li, D.X., and Kolattukudy, P.E. 2000. Requirement for either a host- or/pectin-induced pectate lyase for infection of *Pisum sativum* by *Nectria haematococca. Proceedings of the National Academy of Sciences, USA.,* 97:9813-9818.

Ruiz, E. and Ruffner, H.P. 2002. Immunodetection of *Botrytis* specific invertase in infected grapes. *Journal of Phytopathology,* 150:76-85.

Rumbolz, J., Kassemeyer, H.H., Steinmetez, V., Deising, H.B., Mendgen, K., Mathys, D. Wirtz, S., and Guggenheim, R. 2000. Differentiation of infection structures of the powdery mildew fungus *Uncinula necator* and adhesion to the host cuticle. *Canadian Journal of Botany,* 78:409-412.

Salinas, J. and Schots, A. 1994. Monoclonal antibodies based immunofluorescence test for detection of conidia of *Botrytis cinerea* on cut flowers. *Phytopathology,* 84:351-356.

Schäfer, W. 1994. Molecular mechanisms of fungal pathogenicity to plants. *Annual Review of Phytopathology,* 32:461-477.

Segarra, C.I., Casalongué, C.A., Pinedo, M.L., Cordo, C.A., and Conde, R.D. 2002. Changes in wheat leaf extracellular proteolytic activity after infection with *Septoria tritici. Journal of Phytopathology,* 150:105-111.

Sela-Burrlage, M.B., Epstein, L., and Rodrigez, R.J. 1991. Adhesion of ungerminated *Colletotrichum musae* conidia. *Physiological and Molecular Plant Pathology,* 39:345-352.

Shaykh, M., Soliday, C., and Kolattukdy, P.E. 1977. Proof for the production of cutinase by *Fusarium solani* f. sp. *pisi* during penetration into its host, *Pisum sativum. Plant Physiology,* 60:170-172.

Sheba, M.J.I., Huang, J.K., Chopra, R.K., and Muthukrishnan, S. 1994. Isolation and characterization of a barley chitinase genomic clone: Expression in powdery mildew infected barley. *Journal of Plant Biochemistry and Biotechnology,* 3:91-95.

Strobel, G.A. 1963. A xylanase system produced by *Diplodia viticola. Phytopathology,* 53:592-596.

Svircev, A.M., Gardiner, R.B., McKeen, W.E. Day, A.W., and Smith, R.J. 1986. Detection by protein A-gold of antigens to *Botrytis cincera* in cytoplasm of infected *Vicia faba. Phytopathology,* 76:622-626.

Svircev, A.M., Smith, R.J., Zhou, T., and Day, A.W. 2000. Localization of fungal fimbriae by immunocytochemistry in pathogenic and nonpathogenic isolates of *Venturia inaequalis. Canadian Journal of Microbiology,* 46:800-808.

Sweigard, J.A., Chumley, F.G., and Valent, B. 1992. Disruption of a *Magnaporthe grisea,* cutinase gene. *Molecular and General Genetics,* 232:181-190.

Tan, M.K., Timmer, L.W., Broadbent, P., Priest, M., and Cain, P. 1996. Differentiation by molecular analysis of *Elsinoe* spp. Causing scab diseases of citrus and its epidemiological implications. *Phytopathology,* 86:1039-1044.

Tani, T. 1965. Studies on the phytopathological physiology of kaki anthracnose with special reference to the role of pectic enzyme in the symptom development on kaki fruits. *Memoirs of Faculty of Agriculture, Kagawa University,* 18:1.

Tani, T. 1967. The relation of soft rot caused by pathogenic fungi to pectic enzyme production by the host. In *The Dynamic Role of Molecular Constituents in Plants-Parasite Interaction,* (eds.) C.J. Mirocha and I. Uritani, p. 40. The American Phytopathological Society, St. Paul, Minnesota.

ten Have, A., Mulder, W., Visser, J., and Van Kan, J.A.L. 1998. The endopolygalacturonase gene *Bcpg1* is required for full virulence of *Botrytis cinerea. Molecular Plant-Microbe Interactions,* 11:1009-1016.

Tenberge, K.B. 1999. Host wall alterations by *Claviceps purpurea* during infection of rye: Molecular cytology of a host-pathogen interaction. *Scanning,* 21:111-112.

Tenberge, K.B., Brockmann, B., and Tudzynski, P. 1999. Immunogold localization of an extracellular β-1,3-glucanase of the ergot fungus *Claviceps purpurea* during infection of rye. *Mycological Research,* 103:1103-1118.

Thornton, C.R., Dewey, F.M. and Gilligan, C.A. 1993. Development of monoclonal antibody based immunological assays for the detection of live propagules of *Rhizoctonia solani* in soil. *Plant Pathology,* 42:763-773.

Timmer, L.W., Menge, J.A., Zitko, S.E., Pond, E., Miller, S.A., and Johnson, E.L. 1993. Comparison of ELISA techniques and standard isolation methods for *Phy-*

tophthora detection in citrus orchards in Florida and California. *Plant Disease,* 77:791-796.

Umemoto, N., Kakitani, M., Iwamatsu, A., Yoshikawa, M., Yamaoka, N., and Ishida, I. 1997. The structure and function of a soybean β-glucan-elicitor-binding protein. *Proceedings of the National Academy of Sciences, USA,* 94:1029.

Utomo, C. and Niepold, F. 2000. Development of diagnostic methods for detecting *Ganoderma*-infected oil palms. *Journal of Phytopathology,* 148:507-514.

Van Kan, J.A.L., Van't Klooster, J.W., Wagemakers, A.M., Dees, D.C.T., van der Vlugt, R.A.A., and Bergmans, C.J.B. 1997. Cutinase A of *Botrytis cinerea* is expressed, but not essential during penetration of gerbera and tomato. *Molecular Plant-Microbe Interactions,* 10:30-38.

Van Loon, L.C. 1999. Occurrence and properties of plant pathogenesis-related proteins. In *Pathogenesis-Related Proteins in Plants,* (eds.) S.K. Datta and S. Muthukrishnan, pp. 1- 19. CRC Press, Boca Raton, Florida.

Velazhahan, R., Samiyappan, R., and Vidhyasekaran, P. 2000. Purification of an elicitor-inducible antifungal chitinase from suspension-cultured rice cells. *Phytoparasitica,* 28:131-139.

Vikrant Gupta, Anil Kumar, Lakhchaura, B.D., and Garg, G.K. 2001. Generation of anti-teliospores antibodies for immunolocalization and characterization of antigenic epitopes of teliospores of Karnal bunt *(Tilletia indica)* of wheat. *Indian Journal of Experimental Biology,* 39:686-690.

Viswanathan, R., Padmanaban, P., Mohanraj, D., and Jothi, R. 2000. Indirect-ELISA technique for the detection of the red rot pathogen in sugarcane (*Saccharum* spp. Hybrid) and resistance screening. *Indian Journal of Agricultural Sciences,* 70:308-311.

Vöhringer, G. and Sander, G. 2001. Comparison of antibodies in chicken egg yolk (IgY) and rabbit (IgG) for quantitative strain detection of *Colletotrichum falcatum* and *Fusarium subglutinans. Journal of Plant Diseases and Protection,* 108:39-48.

Wakeham, A.J. and White, J.G. 1996. Serological detection in soil of *Plasmodiophora brassicae* testing spores. *Physiological and Molecular Plant Pathology,* 48:289-303.

Wang, H., Li, J., Bostock, R.M., and Gilchrist, D.G. 1996. Apoptosis: A functional paradigm for programmed cell death induced by a host-selective phytotoxin and invoked during development. *Plant Cell,* 8:375-391.

Wang, P., Matthews, D.E., and Van Etten, H.D. 1992. Purification and charcterization of cyanide hydratase from the phytopathogenic fungus *Gloeocercospora sorghi. Archives of Biochemistry and Biophysics,* 298:569-575.

Wevelsiep, L., Rüpping, E., and Knogge, W. 1993. Stimulation of barley plasma lamma H+-ATPase by phytotoxic peptides from the fungal pathogen *Rhynchosporium secalis. Plant Physiology,* 101:297-301.

Wolpert, T.J. and Macko, V. 1989a. Specific binding of victorin to a 100 kDa protein from oats. *Proceedings of the National Academy Sciences, USA,* 86:4092-4096.

Wolpert, T.J. and Macko, V. 1989b. Victorin-binding to proteins in susceptible and resistant oat genotypes. In *Phytotoxins and Plant Pathogenesis*, (eds.) A. Graniti, R.D. Durbin, and A Ballio, p. 39. Springer-Verlag, Berlin.

Xu, H.X. and Mendgen, K. 1997. Targeted cell wall degradation at the penetration site of cowpea rust basidiosporelings. *Molecular Plant-Microbe Interactions*, 10:87-94.

Xu, J.R. and Hamer, J.E. 1996. MAP-kinase and cAMP signalling regulate infection structure formation and pathogenic growth in the rice blast fungus *Magnaporthe grisea*. *Gene and Development*, 10:2696-2706.

Xu, J.R., Urban, M., Sweigard, J.A., and Hamer, J.E. 1997. The *CPKA* gene for *Magnaporthe grisea* is essential for appressorial penetration. *Molecular Plant-Microbe Interactions*, 10:187-194.

Yabuta, T. and Hayashi, T. 1939. Biochemical studies on "bakanae" fungus of rice II. Isolation of gibberellin the active principle which produces slender rice seedlings. *Journal of Agricultural Chemistry Society, Japan*, 15:257.

Yakoby, N., Freeman, S., Dinoor, A., Keen, N.T., and Prusky, D. 2000. Expression of pectate lyase from *Colletotrichum gloeosporioides* in *C. magna* promotes pathogenicity. *Molecular Plant-Microbe Interactions*, 13:887-889.

Yamada, T., Tsukamoto, H., Shiraishi, T., Nomura, T., and Oku, H. 1990. Detection of indoleacetic acid biosynthesis in some species of *Taphrina* causing hyperplastic diseases in plants. *Annals of Phytopathological Society, Japan*, 56:532.

Chapter 9

Aljanabi, S.M., Parmessure, Y., Moutia, Y., Saumtally, A., and Dookun, A. 2001. Further evidence of the association of a phytoplasma and a virus with yellow leaf syndrome in sugarcane. *Plant Pathology*, 50:628-636.

Ariovich, D. and Garnett, H.M. 1989. The use of immunogold staining technique for detection of a bacterium associated with greening diseased citrus. *Phytopathology*, 79:382-384.

Berg, M., Davies, D.L., Clark, M.F., Vetten, H.J., Maier, G., Marcone, C., and Seemüller, E. 1999. Isolation of the gene encoding one immunodominent membrane protein of the apple proliferation phytoplasma and expression and characterization of the gene product. *Microbiology (Reading)*, 145:1937-1943.

Berisha, B., Chen, Y.D., Zhang, G.Y., Xu, B.Y., and Chen, T.A. 1998. Isolation of Pierce's disease bacteria from grapevines in Europe. *European Journal of Plant Pathology*, 104:427-433.

Bonas, V., van den Ackerveken, G., Büttner, D., Hahn, K., Marios, E., Nennstiel, D., Noel, L., Rossier, O., and Szurek, B. 2000. How the bacterial plant pathogen *Xanthomonas campestris* pv. *vesicatoria* conquers the host? *Molecular Plant Pathology*, 1:73-76.

Bragard, C. and Verhoyen, M. 1993. Monoclonal antibodies specific for *Xanthomonas campestris* bacteria pathogenic on wheat and other small grains in comparison with polyclonal antisera. *Journal of Phytopathology*, 139:217-228.

Brown, I.R., Mansfield, J.N., Taira, S., Roine, E., and Romantschuk, M. 2001. Immunocytochemical localization of Hrp A and Hrp Z supports a role for the Hrp pilus in the transfer of effector proteins from *Pseudomonas syringae* pv. *tomato* across the host plant cell wall. *Molecular Plant-Microbe Interactions,* 14:394-404.

Che, F.S., Nakajima, Y., Tanaka, N., Iwano, M., Yoshida, T., Takayama, S., Kadota, I., and Isogai, A. 2000. Flagellin from an incompatible strain of *Pseudomonas avenae* induces a resistance response in cultured rice cells. *Journal of Biological Chemistry,* 275:32347-32356.

Chen, T.A. and Jiang, X.F. 1988. Monoclonal antibodies against the maize bushy stunt agent. *Canadian Journal of Microbiology,* 34:6-11.

Chumakov, M.I., Dykman, L.A., Bogatyrev, V.A., and Kurbanova, I.V. 2001. Investigation of the cell surface structures of agrobacteria involved in bacterial and plant interaction. *Microbiology (Moscow),* 70:232-238.

Das, A.K. and Mitra, D.K. 1999. Detection of brinjal little leaf phytoplasma by monoclonal antibodies. *Journal of Mycology and Plant Pathology,* 29:48-52.

Davis, R.E. and Fletcher, J. 1983. *Spiroplasma citri* in Maryland: Isolation from field grown plants of horseradish with brittle root symptoms. *Plant Disease,* 67:900-903.

De Boar, S.H., Cupples, D.A., and Gitaitis, R.D. 1996. Detecting latent bacterial infections. *Advances in Botanical Research,* 23:27-57.

De Boer, S.H. and Hall, J.W. 2000. Proficiency testing in a laboratory accreditation program for the bacterial ring rot pathogen of potato. *Plant Disease,* 84:649-653.

de Lima, J.E.O., Miranda, V.S., Hartung, J.S., Brlansky, R.H., Coutinho, A., Roberto, S.R., and Carlos, E.F. 1998. Coffee leaf scorch bacterium: Axenic culture, pathogenicity and comparison with *Xylella fastidiosa* of citrus. *Plant Disease,* 82:94-97.

Doan, C.N., Caughron, M.K., Myers, J.C., Breakfield, N.W., Oliver, R.L., and Yoder, M.D. 2000. Purification, crystallization and X-ray analysis of crystals of pectate lyase A from *Erwinia chrysanthemi. Acta Crystallographica, Section D, Biological Crystallography,* 56:351-353.

Elphinstone, J.G. and Stanford, H. 1998. Sensitivity of methods for the detection of *Ralstonia solanacearum* in potato tubers. *Bulletin OEPP,* 28:69-70.

Eriksson, A.R., Anderson, R.A., Pirhonen, M., and Palva, E.T. 1998. Two component regulators involved in the global control of virulence of *Erwinia carotovora* subsp. *carotovora. Molecular Plant-Microbe Interactions,* 11:743-752.

Errampalli, D. and Fletcher, J. 1993. Production of monospecific polyclonal antibodies against aster yellows mycoplasma-like organism-associated antigen. *Phytopathology,* 83:1279-1282.

Fleischer, S.J., de Mackiewicz, D., Gildow, F.E., and Lukezic, F.L. 1999. Serological estimates of the seasonal dynamics of *Erwinia tracheiphila* in *Acalyamma vittata* (Coleoptera: Chrysomelidae). *Environmental Entomology,* 28: 470-476.

Fogliano, V., Gallo, M., Vinale, F., Ritieni, A., Randazzo, G., Greco, M.L., Lops, R., and Graniti, A. 1999. Immunological detection of syringopeptins produced by *Pseudomonas syringae* pv. *lachrymans*. *Physiological and Molecular Plant Pathology*, 55:255-261.

Franken, A.A.J.M., Zilverentant, J.F., Boonekamp, P.M., and Schots, A. 1992. Specificity of polyclonal and monoclonal antibodies for the identification of *Xanthomonas campestris* pv. *campestris*. *Netherlands Journal of Plant Pathology*, 98:81-94.

Frommel, M.I. and Pazos, G. 1994. Detection of *Xanthomonas campestris* pv. *undulosa* in infested wheat seeds by combined liquid medium enrichment and ELISA. *Plant Pathology*, 43:589-596.

Gallo, M., Fogliano, V., Ritieni, A., Peluso, L., Greco, M.L., Lops, R., and Graniti, A. 2000. Immunoassessment of *Pseudomonas syringae* lipodepsipeptides (syringomycins and syringopeptins). *Phytopathologia Mediterranea*, 39:410-416.

García-Salazar, C., Gildow, F.E., Fleischer, S.J., Cox-Foster, D., and Lukezic, F.L. 2000. ELISA versus immunolocalization to determine the association of *Erwinia tracheiphila* in *Acalyamna vittatum* (Coleoptera: Chrysomelidae). *Environmental Entomology*, 29:542-550.

Griep, R.A., van Twisk, C., van Beckhovan, J.R.C.M., van der Wolf, J.M., and Schotts, A. 1998. Development of specific recombinant monoclonal antibodies against lipopolysaccharide of *Ralstonia solanacearum* race 3. *Phytopathology*, 88:795-803.

Guo, Y.H., Cheng, Z.M., Walla, J.A., and Zhang, Z. 1998. Diagnosis of X-disease phytoplasma in stone fruits by a monoclonal antibody developed directly from a woody plant. *Journal of Environmental Horticulture*, 16:33-57.

Güven, K. and Mutlu, M.B. 2000. Development of immunomagnetic separation technique for isolation of *Pseudomonas syringae* pv. *phaseolicola*. *Folia Microbiologica*, 45:321-324.

He, S.Y. 1998. Type III protein secretion systems in plant and animal pathogenic bacteria. *Annual Review of Phytopathology*, 36:363-392.

Henriquez, N.P., Kenyone, L., and Quiroz, L. 1999. Corn stunt complex mollicutes in Belize. *Plant Disease*, 83:77.

Hoy, J.W., Grisham, M.P., and Damann, K.E. 1999. Spread and increase of ratoon stunting disease of sugarcane and comparison of disease detection methods. *Plant Disease*, 83:1170-1175.

Hseu, S.H. and Lin, C.Y. 2000. Serological detection of *Xanthomonas axonopodis* pv. *dieffenbachiae* in Taiwan. *Plant Protection Bulletin* (Taipei), 42:97-106.

Hsu, H.T., Lawson, R.H., Lin, N.S., and Hsu, Y.H. 1995. Direct tissue blot immunoassay for analysis of plant pathogens. In *Molecular Methods in Plant Pathology*, (eds.) R.P. Singh and U.S. Singh pp. 367-376. CRC Lewis Publishers, Boca Raton, Florida.

Hsu, H.T., Lee, I.M., Davis, R.E., and Wang, Y.C. 1990. Immunization for generation of hybridoma antibodies specifically reacting with plants infected with a mycoplasma-like organism (MLO) and their use in detection of MLO antigens. *Phytopathology,* 80:946-950.

Hu, W.Q., Yuan, J., Jin, Q.L., Hart, P., and He, S.Y. 2001. Immunogold labeling of Hrp pili of *Pseudomonas syringae* pv. *tomato* assembled in minimal medium and *in planta. Molecular Plant-Microbe Interactions,* 14:234-241.

Jim, K.S., Kang, I.B., Ko, K.I., Lee, E.S., Heo, J.Y., Kang, Y.K., and Kim, B.K. 2001. Detection of *Xanthomonas axonopodis* pv. *citri* on citrus fruits using enzyme-linked immunosorbent assay. *Plant Pathology Journal,* 17:62-66.

Jones, W.T., Harvey, D., Zhao, Y.F., Mitchell, R.E., Bender, C.L., and Reynolds, P.H.S. 2001. Monoclonal antibody-based immunoassays for the phytotoxin coronatine. *Food and Agricultural Immunology,* 13:19-32.

Kokošková, B., Pánková, I., and Krejzar, V. 2000. Characteristics of polyclonal antisera for detection and determination of *Clavibacter michiganensis* subsp. *insidiosus. Plant Protection Science,* 36:46-52.

Kumar, R.B., Xie., Y.H., and Anath Das, 2000. Subcellular localization of *Agrobacterium tumefaciens* T-DNA transport pore proteins: Vir B8 is essential for the assembly of the transport pore. *Molecular Microbiology,* 36:608-617.

Li, W.B., Zreik, L., Fernandes, N.G., Miranda, V.S., Teixeira, D.C., Ayres, A.J., Granier, M., and Bové, J.M. 1999. A triply cloned strain of *Xylella fastidiosa* multiplies and induces symptoms of citrus variegated chlorosis in sweet orange. *Current Microbiology,* 39:106-108.

Li, X., Wong, W.C., and Hayward, A.C. 1993. Production and use of monoclonal antibodies to *Pseudomonas andropogonis. Journal of Phytopathology,* 138: 21-30.

Lin, C.P. and Chen, T.A. 1985. Production of monoclonal antibodies against *Spiroplasma citri. Phytopathology,* 75:845-851.

Lin, C.P. and Chen, T.A. 1986. Comparison of monoclonal antibodies and polyclonal antibodies in detection of aster yellows mycoplasma-like organism. *Phytopathology,* 76:45-50.

Lindgren, P.B. 1997. The role of *hrp* genes during plant bacterial interactions. *Annual Review of Phytopathology,* 35:129-152.

Liu, Z.J., Luo, H.L., Ma, J.H., Wang, X.L., Zhang, J.N., and Huan, Z.R. 1999. The diagnosis of Huanglongbing disease of citrus. *Forest Studies in China,* 1:10-15.

Loi, N., Ermacora, P., Carraro, L., Osler, R., and Chen, T.A. 2002. Production of monolonal antibodies against apple proliferation phytoplasma and their use in serological detection. *European Journal of Plant Pathology,* 108:81-86.

Lyons, N.F. and Taylor, J.D. 1990. Serological detection and identification of bacteria from plants by the conjugated *Staphylococcus aureus* slide agglutination test. *Plant Pathology,* 39:584-590.

Malandrin, L. and Samson, R. 1999. Serological and molecular size characterization of flagellins of *Pseudomonas syringae* pathovars and related bacteria. *Systematic and Applied Microbiology,* 22:534-545.

Mazarei, M., Hajimorad, M.R., and Kerr, A. 1992. Specificity of polyclonal antibodies to different antigenic preparations of *Pseudomonas syringae* pv. *pisi* strain UQM 551 and *Pseudomonas syringae* pv. *syringae* strain L. *Plant Pathology,* 41:437-443.

McGarvey, J.A., Denny, T.P., and Schell, M.A. 1999. Spatial-temporal and quantitative analysis of growth and EPSI production by *Ralstonia solanacearum* in resistant and susceptible tomato cultivars. *Phytopathology,* 89:1233-1239.

Mguni, C.M., Mortensen, C.N., Keswani, C.L., and Hockenhull, J. 1999. Detection of the black rot pathogen *Xanthomonas campestris* pv. *campestris* and other xanthomonads in Zimbabwean and imported *Brassica* seed. *Seed Science and Technology,* 27:447-454.

Mitchell, R.E. and Young, H. 1978. Identification of chlorosis inducing toxin of *Pseudomonas glycinea* as coronatine. *Phytochemistry,* 17:2028.

Mráz, I., Pánková, I., Petrzik, K., and Šíp M. 1999. *Erwinia* and *Pseudomonas* bacteria can be reliably screened by an improved serological agglutination test. *Journal of Phytopathology,* 147:429-431.

Najar, A., Bouachem, S., Danet, J.L., Saillard, C., Garnier, M., and Bové, J.M. 1998. Presence of *Spiroplasma citri,* the pathogen responsible for citrus, stubborn disease and its vector leafhopper *Circulifer haematoceps* in Tunisia: Contamination of both *C. haematoceps* and *C. opacipennis. Fruits (Paris),* 53:391-396.

Narayanasamy, P. 2001. *Plant Pathogen Detection and Disease Diagnosis,* Second Edition. Marcel Dekker Inc., New York.

Nasser, W., Faelen, M., Hugouvieux-Cotte-Pattat, N., and Reverchon, S. 2001. Role of the nucleoid-associated protein H-NS in the synthesis of virulence factors in the phytopathogenic bacterium *Erwinia chrysanthemi. Molecular Plant-Microbe Interactions,* 14:10-20.

Nishiyama, K., Sakai, R., Ezuka, A., Ichihara, A., Shiraishi, K., and Sakamura, S. 1977. Phytotoxic effect of coronatine in haloblight lesions of Italian ryegrass. *Annals of Phytopathological Society, Japan,* 42:219.

Ovod, V.V., Knirel, Y.A., Samson, R., and Krohn, K.J. 1999. Immunochemical characterization and taxonomic evaluation of the O-polysaccharides of the lipopolysaccharides of *Pseudomonas syringae* serogroup 01 strain. *Journal of Bacteriology,* 181:6937-6947.

Pasichnyk, L.A. 2000. Antigenic properties of bacteria of pathovars of *Pseudomonas syringae* affecting cereals. *Mikrobiologichnii Zhurnal,* 62:18-22.

Patil, S.S., Kolattukudi, P.F., and Dimond, A.E. 1970. Inhibition of ornithine carbamoyl-transferase from bean plant by the toxin of *Pseudomonas phaseolicola. Plant Physiology,* 46:752.

Perino, C., Gaudriault, S., Vian, B., and Barny, M.A. 1999. Visualization of harpin secretion *in planta* during infection of apple seedlings by *Erwinia amylovora. Cellular Microbiology,* 1:131-141.

Perombelon, M.C.M., Bertheau, Y., Cambra, M. Frechon, D., Lopez, M.M., Niepold, F., Persson, P., Sletten, A., Toth, I.K., van Vuurde, J.W.L., et al. 1998.

Microbiological, immunological and molecular methods suitable for commercial detection and quantification of the black leg pathogen *Erwinia carotovora* subsp. *atroseptica*. *Bulletin, EPPO*, 28:141-155.

Perombelon, M.C.M. and Hynan, L.J. 1995. Serological methods to quantify potato seed contamination by *Erwinia carotovora* subsp. *atroseptica*. *Bulletin, EPPO*, 25:195-202

Pooler, M.R., Myung, I.S., Bentz., J., Sherald, J., and Hartung, J.S. 1997. Detection of *Xylella fastidiosa* in potential insect vectors by immunomagnetic separation and nested polymerase chain reaction. *Letters in Applied Microbiology*, 25:123-126.

Preston, J.F., Rice, J.D., Ingram, L.O., and Keen, N.T. 1992. Differential depolymerization mechanisms of pectate lyases secreted by *Erwinia chrysanthemi* EC 16. *Journal of Bacteriology*, 174:2039-2042.

Priou, S., Gutarra, L., and Aley, P. 1999. Highly sensitive detection of *Ralstonia solanacearum* in latently infected potato tubers by post enrichment enzyme-linked immunosorbent assay on nitrocellulose membrane. *Bulletin OEPP*, 29: 117-125.

Pugsley, A.P. 1993. The complete general protein secretory pathway in Gram-negative bacteria. *Microbiology Review*, 57:50-108.

Purcell, A.H., Saunders, S.R., Hendson, M., Grebus, M.E., and Henry, M.J. 1999. Causal role of *Xylella fastidiosa* in oleander leaf scorch disease. *Phytopathology*, 89:53-58.

Rabenstein, F., Nachtigall, M., and Boll, E. 1999. Production and characterization of monoclonal antibodies against *Xanthomonas campestris* pv. *campesris*. *Beiträge zur Züchtungforschung Bundesanstalt für. Züchtungs-forschung an Kulturpflanzen*, 5:4-6.

Rott, P., Davis, M.J., and Baudin, P. 1994. Serological variability in *Xanthomonas albilineans*, causal agent of leaf scald disease of sugarcane. *Plant Pathology*, 43:344-349.

Schrammeijer, B., Risseeuw, E., Pansegrau, W., Regensburg-Tuink, T.J.G., Crosby, W.L., and Hooykaas, P.J.J. 2001. Interaction of the virulence protein VirF of *Agrobacterium tumefaciens* with plant homologs of the yeast SkP[1] protein. *Current Biology*, 11:258-262.

Shan, Z.H., Liao, B.S., Tan, Y.J., Li, D., Lei, Y., and Shen, M.Z. 1997. ELISA technique used to detect latent infection of groundnut by bacterial wilt *(Pseudomonas solanacearum)*. *Oilcrops of China*, 19:45-47.

Shigaki, T., Nelson, S.C., and Alvarez, A.M. 2000. Symptomless spread of blight-inducing strains of *Xanthomonas campestris* pv. *campestris* on cabbage seedlings in misted seed beds. *European Journal of Plant Pathology*, 106:339-346.

Sinden, S.L. and Durbin, R.D. 1968. Glutamine synthetase inhibition: Possible mode of action of wild fire toxin from *Pseudomonas tabaci*. *Nature (London)*, 219:379.

Singh, U., Trevors, C.M., de Boer, S.H., and Janse, J.D. 2000. Fimbrial-specific monoclonal antibody based ELISA for European potato strains of *Erwinia chrysanthemi* and comparison to PCR. *Plant Disease,* 84:443-448.

Slack, S.A., Drennan, J.L., Westra, A.A.G., Gudmestad, N.C., and Oleson, A.E. 1996. Comparison of PCR, ELISA and DNA hybridization for the detection of *Clavibacter michiganensis* subsp. *sepedonicus* in field-grown potatoes. *Plant Disease,* 80:519-524.

Srinivasulu, B. and Narayanasamy, P. 1995. Serological detection of phyllody disease MLO in sesamum and leafhopper *Orosius albicinctus. Journal of Mycology and Plant Pathology,* 25:154-157.

Thomas, M.D., Langston-Unkefer, P.J., Uchytil, T.F., and Durbin, R.D. 1983. Inhibition of glutamine synthetase from pea by tabtoxinine β-lactam. *Plant Physiology, Lancaster,* 71:912-915.

van der Wolf, J.M., van Bekkum, P.J., van Elas, J.D., Nijhuis, E.H., Vriend, S.G.C., and Ruissen, M.A. 1998. Immunofluorescence colony staining and selective enrichment in liquid medium for studying the population dynamics of *Ralstonia solnacearum* (race 3) in soil. *Bulletin OEPP,* 28:71-79.

van der Wolf, J.M., Hyman, L.J., Jones, D.A.C., Grevesse, C., van Beckhoven, J.R.C.M., van Vuurde, J.W.L., and Pérombelon, M.C.M. 1996. Immunomagnetic separation of *Erwinia crotovora* subsp. *atroseptica* from potato peel extracts to improve detection sensitivity on a crystal violet pectate medium or by PCR. *Journal of Applied Bacteriology,* 80:487-495.

Veena, M.S. and van Vuurde, J.W.L. 2002. Indirect immunofluorescence colony staining method for detecting bacterial pathogens of tomato. *Journal of Microbiological Methods,* 49:11-17.

Vijayanand, D., Shylaja, M.D., Krishnappa, M., and Shetty, H.S. 1999. An approach to obtain specific polyclonal antisera to *Xanthomonas campestris* pv. *cyanopsidis* and its potential application in indexing of infected seeds of guar. *Journal of Applied Microbiology,* 87:711-717.

Viswanathan, R. 1997. Detection of phytoplasmas associated with grassy shoot disease of sugarcare by ELISA techniques. *Journal of Plant Diseases and Protection,* 104:9-16.

Viswanathan, R. and Alexander, K.C. 1995. Production of polyclonal antisera to grassy shoot disease (MLO) of sugarcane and their use in disease detection. In *Detection of Plant Pathogens and Their Management,* (eds.) J.P. Verma, A. Varma, and Dinesh Kumar, pp. 153-158. Angkor Publishers, New Delhi.

Walcott, R.R. and Gitaitis, R.D. 2000. Detection of *Acidovorax avenae* subsp. *citrulli* in watermelon seed using immunomagnetic separation and the polymerase chain reaction. *Plant Disease,* 84:470-474.

Wang, Z.K., Luo, H.H., and Shu, Z.Y. 1997. application of DIA (dot immunobinding assay) for rapid detection of *Xanthomonas axonopodis* pv. citri. *Journal of Southwest University,* 19:529-532.

Yamada, T., Palm, C., Brooks, B., and Kosuge, T. 1985. Nucleotide sequence of the *Pseudomonas savastanoi* indole acetic acid genes show homology with *Agrobacterium tumefaciens* T-DNA. *Proceedings of the National Academy of Sciences, USA*, 82:6522.

Yu, J., Wayadande, A.C., and Fletcher, J. 2000. *Spiroplasma citri* surface protein P89 implicated in adhesion to cells of vector *Circulifer tenellus*. *Phytopathology*, 90:716-722.

Zeng, X.M., Lai, W.J., and Xu, D. 1996. Serological specificity of leaf streak pathogen of rice. *Chinese Rice Research Newsletter*, 4(2):7.

Zhao, Y.F., Jones, W.T., Sutherland, P., Palmer, D.A., Mitchell, R.E., Reynolds, P.H.S., Damicone, J.P., and Bender, C.L. 2001. Detection of the phytotoxin coronatine by ELISA and localization in infected plant tissue. *Physiological and Molecular Plant Pathology*, 58:247-258.

Chapter 10

Abdalla, M.E. and Albrechtsen, S.E. 2001. Petri dish-ELISA, a simple and economic technique for detecting plant viruses. *Acta Phytopathologica et Entomologica, Hungarica*, 36:1-8.

Abraham, A. and Albrechtsen, S.E. 2000. Petri dish-agar dot immunoenzymatic assay (PADIA)—A new and inexpensive method for the detection and identification of plant viruses. *International Journal of Pest Management*, 46:161-164.

Abraham, A. and Albrechtsen, S.E. 2001. Comparison of penicillinase and alkaline phosphatase as labels in enzyme-linked immunosorbent assay (ELISA) for the detection of plant viruses. *Journal of Plant Diseases and Protection*, 108:49-57.

Acheche, H., Fattouch, S., M' Hirsi, S., Marzouki, N., and Marrakchi, M. 1999. Use of optimised PCR methods for the detection of GLRaV3: A closterovirus associated with grapevine in Tunisian grapevine plants. *Plant Molecular Biology Reporter*, 17:31-42.

Ahoonmanesh, A., Hajimorad, M.R., Ingham, B.L., and Francki, R.I.B. 1990. Indirect double antibody sandwich ELISA for detecting alfalfa mosaic virus in aphids after short probes on infected plants. *Journal of Virological Methods*, 30:271-282.

Almási, A., Apatini, D., Bóka, K., Böddi, B., and Gáborjányi, R. 2000. BSMV infection inhibits chlorophyll biosynthesis in barley plants. *Physiological and Molecular Plant Pathology*, 56:227-233.

Ammar, E.D., Jarlfors, U., and Pirone, T.P. 1994. Association of potyvirus helper component protein with virions and the cuticle lining the maxillary food canal and foregut of an aphid vector. *Phytopathology*, 84:1054-1060.

Antoniw, J.F., Ritter, C.E., Pierpoint, W.S., and Van Loon, L.C. 1980. Comparison of three pathogenesis-related proteins from plants of two cultivars of tobacco infected with TMV. *Journal of General Virology*, 30:375.

Arbatova, J., Lehto, K., Pehu, E., and Pehu, T. 1998. Localization of the P1 protein of potato associated with (1) cytoplasmic inclusion bodies and (2) cytoplasm of infected cells. *Journal of General Virology,* 79:2319-2323.

Bar-Joseph, M., Garnsey, S.M., Gonsalves, D., Moscovitz, M., Purcifull, D.E., Clark, M.F., and Loebenstein, G. 1979. The use of enzyme-linked immunosorbent assay for the detection of citrus tristeza virus. *Phytopathology,* 69:190-194.

Bass, H.W., Nagar, S., Hanley-Bowdoin, L., and Robertson, D. 2000. Chromosome condensation induced by geminivirus infection of mature plant cells. *Journal of Cell Science,* 113:1149-1160.

Berger, P.H., Thornbury, D.W., and Pirone, T.P. 1985. Detection of picogram quantities of potyviruses using a dot-blot immunobinding assay. *Journal of Virological Methods,* 12:31.

Bertozzi, T., Alberts, E., and Sedgley, M. 2002. Detection of *Prunus* necrotic ringspot virus in almond: Effect of sampling time on the efficiency of serological and biological indexing methodologies. *Australian Journal of Experimental Agriculture,* 42:207-210.

Boonham, N. and Barker, I. 1998. Strain-specific recombinant antibodies to potato virus Y potyvirus. *Journal of Virological Methods,* 74:193-199.

Bos, L. 1970. *Symptoms of Virus Diseases in Plants,* Oxford and IBH Publishing Co. Pvt. Co., New Delhi.

Boscia, D., Digiaro, M., Safi, M., Garau, R., Zhou, I., Minafra, A., Ghanem-Sabanadzonic, N., Bottalico, G., and Potere, O. 2001. Production of monoclonal antibodies to grapevine virus D and contribution to the study of its aetiological role in grapevine disease. *Vitis,* 40:69-74.

Cambra, M., Camarasa, E., Gorris, M.T., Roman, M.P., Asensio, M., Perez, E., Serra, J., and Cambra, M.A. 1994. Detection of structural proteins of virus by immunoprinting: ELISA and its use for diagnosis. *Investigacion Agraria, Produccion y Protection Vegetales, Fuera de Serie* No. 2:221-230.

Chamberlain, J.R., Culbreath, A.K., Todd, J.W., and Demski, J.W. 1993. Detection of tomato spotted wilt virus in tobacco thrips (Thysanoptera: Thripidae) overwintering in harvested peanut fields. *Journal of Economic Entomology,* 86: 40-45.

Chatenet, M., Delage, C., Ripolles, M., Irey, M., Lockhart, B.E.L., and Rott, P. 2001. Detection of sugarcane yellow leaf virus in quarantine and production of virus-free sugarcane by apical meristem culture. *Plant Disease,* 85:1177-1180.

Chen, C.C. and Chang, C.A. 1999. The improvement for the detection of tuberose mild mosaic potyvirus in tuberose bulbs by temperature treatment and direct tissue blotting. *Plant Pathology Bulletin,* 8:83-88.

Chen, C.C., Chang, C.A., and Dac, T. 1999. Variable distribution patterns of bean yellows mosaic potyvirus and cucumber mosaic cucumoviruses in gladiolus plants and their influence on virus detection. *Plant Pathology Bulletin,* 8:117-120.

Chen, T.H. and Lu, Y.T. 2000. Preparation of rabbit IgG-alkaline phosphatase conjugate by maleimide method and its application for detecting bamboo mosaic potexvirus. *Bulletin of National Pingtung University of Science and Technology,* 9:125-132.

Chitra, T.R., Prakash, H.S., Albrechtsen, S.E., Shetty, H.S., and Mathur, S.B. 1999. Infection of tomato and bell pepper by ToMV and TMV at different growth stages and establishment of virus in seeds. *Journal of Plant Pathology,* 81:123-126.

Cho, J.J., Mau, R.F.L., Hamsaki, R.T., and Gonsalves, D. 1988. Detection of tomato spotted wilt virus in individual thrips by enzyme-linked immunosorbent assay. *Phytopathology,* 78:1348-1352.

Chu, P.W.G., Waterhouse, P.M., Martin, R.R., and Gerlach, W.L. 1989. New approaches to the detection of microbial plant pathogens. *Biotechnology and Genetic Engineering Reviews,* 7:45-111.

Clark, M.F. and Adams, A.N. 1977. Characteristics of the microplate method of enzyme-linked immunosorbent assay for the detection of plant viruses. *Journal of General Virology,* 34:475-483.

Clark, M.F. and Bar-Joseph, M. 1984. Enzyme immunosorbent assays in plant virology. *Methods in Virology,* 7:51-85.

Clover, G.R.G., Ratti, C., Rubies-Autonell, C., and Henry, C.M. 2002. Detection of European isolates of oat mosaic virus. *European Journal of Plant Pathology,* 108:87-91.

Comas, J., Pons, X., Albajes, R., and Plumb, R.T. 1993. The role of maize in the epidemiology of barley yellow dwarf virus in northeast Spain. *Journal of Phytopathology,* 138:244-248.

Coutts, B.A. and Jones, R.A.C. 2000. Viruses infecting canola (*Brassica napus*) in south west Australia: Incidence, distribution, spread and infection reservoir in wild radish *(Raphanus raphinistrum). Australian Journal of Agricultural Research,* 51:925-936.

Darda, G. 1998. A new test-combination: DAS-ELISA with subsequent amplified ELISA on the same microtitre plate for detection of potato virus X in potato tubers. *Journal of Plant Diseases and Protection,* 105:105-113.

de Assis Filho, F.M., Paguio, O.R., Sherwood, J.L., and Deom, C.M. 2002. Symptom induction by cowpea chlorotic mottle virus on *Vigna unguiculata* is determined by amino acid residue 151 in the coat protein. *Journal of General Virology,* 83:879-883.

Dicenta, F., Martinez-Gómez, P., Bellanger, I., and Audergon, J.M. 2000. Localization of plum pox virus in stem and petiole tissues of apricot cultivars by immuno tissue printing. *Acta Virologica,* 44:323-328.

Dohi, K., Mori, M., Furusawa, I., Mise, K., and Okuno, T. 2001. Brome mosaic virus replicase proteins localize with the movement protein at infection specific cytoplasmic inclusions in infected barley leaf cells: A brief report. *Archives of Virology,* 146:1607-1615.

Dolores-Talens, A.C., Hill, J.H., and Durand, D.P. 1989. Application of enzyme-linked fluorescent assay (ELFA) to detection of lettuce mosaic virus in seeds. *Journal of Phytopathology*, 124:149-154.

D'Onghia, A.M., Djelouah, K., Frasheri, D., and Potere, O. 2001. Detection of citrus psorosis virus by direct tissue blot immunoassay. *Journal of Plant Pathology*, 83:139-142.

Dovas, C.L., Mamolos, A.P., and Katis, N.L. 2002. Fluctuations in concentrations of two potyviruses in garlic during the growing period and sampling conditions for reliable detection by ELISA. *Annals of Applied Biology*, 140:21-28.

Du, Z.Q., Li Li, Wang, X.X., and Zhou, G.H. 2000. Detection of PAV serum type of BYDV with ISEM. *Plant Protection*, 26:31.

Duarte, K.M.R., Gomes, L.A., Andrino, F.G., Leal Junior, G.A., da Silva, F.H.B., Pascholal, J.A.R., Giacomelli, A.M.B., and Tavares, F.C.A. 2002. Identification of tomato mosaic tobamovirus (ToMV) using monoclonal antibodies. *Scientia Agricola*, 59:107-112.

Erhardt, M., Stussi-Garaud, C., Guilley, H., Richards, K.E., Jonard, G., and Bouzoubaa, S. 1999. The first triple gene block proteins of peanut clump virus localizes to the plasmodesmata during virus infection. *Virology (New York)*, 264:220-229.

Erokhina, T.N., Zinovkin, R.A., Vitushkina, M.V., Jelkmann, W., and Agranovsky, A.A. 2000. Detection of beet yellows closterovirus methyl transferase-like and helicase-like proteins in vitro using monoclonal antibodies. *Journal of General Virology*, 81:597-603.

Eun, A.J.C. and Wong, S.M. 1999. Detection of cymbidium mosaic potexvirus and odontoglossum ringspot tobamovirus using immuno capillary zone electrophoresis. *Phytopathology*, 89:522-528.

Fischer, R., Schumann, D., Zimmermann, S., Drossard, J., Sack, M., and Schillberg, S. 1999. Expression and characterization of biospecific single-chain Fv-fragments produced in transgenic plants. *European Journal of Biochemistry*, 262: 810-816.

Fox, R.T.V. 1998. Plant disease diagnosis. In *The Epidemiology of Plant Diseases*, (ed.) D. Gareth Jones, pp. 14- 41. Kluwer Academic Publishers, Dordrecht, The Netherlands.

Franconi, R., Roggero, P., Pirazzi, P., Arias, F.J., Desiderio, A., Bitti, O., Pashkonlo, V.D., Mattei, B., Bracci, L., Masenga, V., et al. 1999. Functional expression in bacteria and plants of scFv antibody fragment against tospoviruses. *Immunotechnology*, 4:180-201.

Franz., A., Makkouk, K.M., and Vetten, H.J. 1998. Acquisition, retention and transmission of faba bean necrotic yellows virus by two of its aphid vectors, *Aphis craccivora* (Koch) and *Acyrthosiphon pisum* (Harris). *Journal of Phytopathology*, 146:347-355.

Garret, A., Kerlan, C., and Thomas, D. 1993. The intestine is a site of passage for potato leaf roll virus from the gut lumen into the hemocoel in the aphid vector *Myzus persicae* Sulz. *Archives of Virology*, 31:377-392.

Gera, A., Hsu, H.T., Cohen, J., Watad, A., Beckelman, E., and Hsu, Y.H. 2000. Effect of cucumber mosaic virus monoclonal antibodies on virus infectivity and transmission by *Myzus persicae*. *Journal of Plant Pathology*, 82:119-124.

Gillaspie, A.G., Jr., Pittman, R.N., Pinnow, D.L., and Cassidy, B.G. 2000. Sensitive method for testing peanut seed lots for peanut stripe and peanut mottle viruses by immuno capture-reverse transcription-polymerase chain reaction. *Plant Disease*, 84:559-561.

Giunchedi, L. and Pollini, C.P. 1992. Cytopathological negative staining and serological electron microscopy of clostero-like virus associated with pear vine yellows disease. *Journal of Phytopathology*, 134:329-335.

Gough, K.C., Cockburn, W., and Whitelam, G.C. 1999. Selection of phage-displayed peptides that bind to cucumber mosaic virus coat protein. *Journal of Virological Methods*, 79:169-180.

Griep, R.A., Prins, M., van Twisk, C., Keller, H.J. H.G., Kerschbaumer, R.J., Kormelink, R., Goldbach, R.W., and Schots, A. 2000. Application of phage display in selecting tomato spotted wilt virus-specific single-chain antibodies (scFvs) for sensitive diagnosis in ELISA. *Phytopathology*, 90:183-190.

Harper, K., Kerschbaumer, R.J., Ziegler, A., Macintosh, S.M., Cowan, G.H., Himmler, G., Mayo, M.A., and Torrance, L. 1997. A scFv-alkaline phosphatase fusion protein which detects potato leaf roll luteovirus in plant extracts by ELISA. *Journal of Virological Methods*, 63:237-242.

Harper, K., Toth, R.L., Mayo, M.A., and Torrance, L. 1999. Properties of a panel of single chain variable fragments against potato leaf roll virus obtained from two phage display libraries. *Journal of Virological Methods*, 81:159-168.

He, X., Harper, K., Grantham, G., Yang, C.H., and Creamer, R. 1998. Serological characterization of the 3'-proximal encoded proteins of beet yellows closterovirus. *Archives of Virology*, 143:1349-1363.

He, X.H., Liu, S.J., and Perry, K.C. 1998. Identification of epitopes in cucumber mosaic virus using a phage-displayed random peptide library. *Journal of General Virology*, 79:3145-3153.

Heinze, C., Rogger, O.P., Sohn, M., Vaira, A.M., Masenga, V., and Adam, G. 2000. Peptide-derived broad-reacting antisera against tospovirus NSs-protein. *Journal of Virological Methods*, 89:137-146.

Helguera, M., Bravo-Almonacid, F., Kobayashi, K., Rabinowicz, P.D., Conci, V., and Mentaberry, A. 1997. Immunological detection of a Gar V-type virus in Argentine garlic cultivars. *Plant Disease*, 81:1005-1010.

Hobbs, H.A., Reddy, D.V.R., Rajeshwari, R., and Reddy, A.S. 1987. Use of direct antigen coating and protein A coating-ELISA procedures for detection of three peanut viruses. *Plant Disease*, 71:747-749.

Hoffmann, K., Sackey, S.T., Maiss, E., Adomako, D., and Vetten, H.J. 1997. Immunocapture polymerase chain reaction for the detection and characterization of cacao swollen shoot virus 1A isolates. *Journal of Phytopathology*, 145:205-213.

Hoffmann, K., Uhde, C., Lessemann, D.E., Sackey, S.T., Adomako, D., and Vetten, H.J. 1999. Production, characterization and application of monoclonal antibodies to the cocao swollen shoot virus isolate 1A. *Journal of Phytopathology,* 147:725-735.

Horn, N.M., Reddy, S.V., and Reddy, D.V.R. 1994. Virus-vector relationships of chickpea chlorotic dwarf geminivirus and the leafhopper *Orosius orientalis* (Hemiptera: Cicadellidae). *Annals of Applied Biology,* 124:441-450

Hsu, H.T., Barzuna, L., Hsu, Y.H., Bliss, W., and Perry, K.L. 2000. Identification and subgrouping of cucumber mosaic virus with mouse monoclonal antibodies. *Phytopathology,* 90:615-620.

Hsu, H.T., Lawson, R.H., Lin, N.S., and Hsu, Y.H. 1995. Direct tissue blot immunoassay for analysis of plant pathogens. In *Molecular Methods in Plant Pathology,* (eds.) R.P. Singh and U.S. Singh, pp. 367-376. CRC Lewis Publishers, Boca Raton, Florida.

Hsu, H.T., Vongsasitron, D., and Lawson, R.H. 1992. An improved method for serological detection of cymbidium mosaic potexvirus infection in orchids. *Phytopathology,* 82:491-495.

Hu, J.S., Sether, D.M., Liu, X.P., Wang, M., Zee, F., and Ullman, D.F. 1997. Use of tissue blotting immunoassay to examine the distribution of pineapple closterovirus in Hawaii. *Plant Disease,* 81:1150-1154.

Hu, J.S., Sether, D.M., and Ullman, D.E. 1996. Detection of pineapple *Closterovirus* in pineapple plants and mealybugs using monoclonal antibodies. *Plant Pathology,* 45:829-836.

Iannelli, D., D'Apice, L., Cottone, C., Viscardi, M., Scala, F., Zoina, A., del Sorbo, G., Spigno, P., and Capparelli, R. 1997. Simultaneous detection of cucumber mosaic virus, tomato mosaic virus and potato virus Y by flow cytometry. *Journal of Virological Methods,* 69:137-145.

Jacob, T. and Usha, R. 2001. 3'-terminal sequence analysis of the RNA genome of the Indian isolate of cardamom mosaic virus: A new member of genus *Macluravirus* of Potyviridae. *Virus Genes,* 23:81-88.

Jadãq, A.S., Pavan, M.A., da Silva, N., and Zerbini, F.M. 2002. Seed transmission of lettuce mosaic virus (LMV) pathotypes II and IV in different lettuce genotypes. *Summa Phytopathologica,* 28:58-61.

James, D. and Mukerji, S. 1996. Comparison of ELISA and immunoblotting techniques for the detection of cherry mottle leaf virus. *Annals of Applied Biology,* 129:13-23.

Jianping, C., Swaby, A.G., Adams, M.J., and Yili, R. 1991. Barley mild mosaic virus inside its fungal vector, *Polymyxa graminis. Annals of Applied Biology,* 118:615-621.

Karanastasi, E., Vassilakos, N., Roberts, I.M., MacFarlane, S.A., and Brown, D.J.F. 2000. Immunogold localization of tobacco rattle virus particles within *Paratrichodorus anemones. Journal of Nematology,* 32:5-12.

Karyeija, R.F., Kreuze, J.F., Gibson, R.W., and Valkonen, J.P.T. 2000. Two sero-types of sweet potato feathery mottle virus in Uganda and their interaction with resistant sweet potato cultivars. *Phytopathology,* 90:1250-1255.

Kastirr, R. 1990. Problems in the detection of plant pathogenic viruses in aphids by means of ELISA and interpretation of the results. *Nachrichtenblatt Pflanzen-schutz,* 44:201-204.

Kawano, T. and Takahashi, Y. 1997. Simplified detection of plant viruses using high density latex. *Annals of Phytopathological Society, Japan,* 63:403-405.

Keller, K.E., Johansen, E., Martin, P.R., and Hampton, R.O. 1998. Potyvirus ge-nome-linked protein (VPg) determines pea seed-borne mosaic virus pathotype-specific virulence in *Pisum sativum. Molecular Plant-Microbe Interactions,* 11:124-130.

Khan, M.A. and Slack, S.A. 1978. Studies on the sensitivity of a latex agglutination test for the serological detection of potato virus S and potato virus X in Wiscon-sin. *American Potato Journal,* 55:627-637.

Kumar, P.L., Duncan, G.H., Roberts, I.M., Jones, A.T., and Reddy, D.V.R. 2002. Cytopathology of pigeonpea sterility mosaic virus in pigeonpea and *Nicotiana benthamiana:* Similarities with those of eriophyid mite-borne agents of uniden-tified aetiology. *Annals of Applied Biology,* 140:87-96.

Kumari, S.G., Makkouk, K.M., Katul, L., and Vetten, H.J. 2001. Polyclonal anti-bodies to the bacterially expressed coat protein of faba bean necrotic yellows vi-rus. *Journal of Phytopathology,* 149:543-550.

Kundu, A.K., Oshima, K., Sako, N., and Yaegashi, H. 2001. Cross-reactive and ma-jor virus specific epitopes are located at the N-terminal halves of the cylindrical inclusion proteins of turnip mosaic and zucchini yellow mosaic potyviruses. *Ar-chives of Virology,* 145:1437-1447.

Kuntze, L., Bauer, E., and Forough-Wehr, B. 2000. An improved method for inocu-lation and detection of Ba YMV-2 in winter barley. *Journal of Plant Diseases and Protection,* 107:310-317.

Lange, L. and Heide, M. 1986. Dot immunobinding (DIB) for detection of virus in seed. *Canadian Journal of Plant Pathology,* 8:373-379.

Lemmetty, A., Susi, P., Latvala, S., and Lehto, K. 1998. Detection of the putative causal agent of black currant reversion disease. *Acta Horticulturae,* No. 471, 93-98.

Leonardon, S.L., Gordan, D.T., and Gregory, R.E. 1993. Serological differentiation of maize dwarf potyvirus strains A, D, E and F by electroblot immunoassay. *Phytopathology,* 83:86-91.

Li, X.D., Fan, Z.F., Li, H.F., and Qiu, W.F. 2001. Accumulation and immuno-localization of maize dwarf mosaic virus HC. Pro in infected maize leaves. *Acta Phytopathologica Sinica,* 31:310-314.

Li, Y., Wei,C., Tien, P., Pan, N., and Chen, Z. 1996. Immunodetection of beet ne-crotic yellow vein RNA3-encoded protein in different host plants and tissues. *Acta Virologica,* 40:67-72.

Lin, J.H., Xie, L.H., Rundell, P.A., and Powell, C.A. 2000. Development of western blot procedure for using polyclonal antibodies to study the proteins of citrus tristeza virus. *Acta Phytopathologica Sinica,* 30:250-266.

Lin, R.F., Yu, J.B., Jin, D.D., and Ye, L. 1991. Studies on methods of detection of two solanaceous vegetable viruses. *Acta Agriculture Zheijiangensis,* 3:127-132.

Lin, Y.J., Rundell, P.A., and Powell, C.A. 2002. In situ immunoassay (ISIA) of field grapefruit trees inoculated with mild isolates of citrus tristeza virus indicates mixed infection with severe isolates. *Plant Disease,* 86:458-461.

Lin, Y.J., Rundell, P.A., Xie, L., and Powell, C.A. 2000. In situ immunoassay for detection of citrus tristeza virus. *Plant Disease,* 84:937-940.

Ling, K.S., Zhu, H.Y., Jiang, Z.Y., and Gonsalves, D. 2000. Effective application of DAS-ELISA for detection of grapevine leafroll-associated closterovirus-3 using polyclonal antiserum developed from recombinant coat protein. *Journal of Plant Pathology,* 106:301-309.

Liu, F., Schubert, J., and Ambrosova, S.M. 1999. Monoclonal antibodies against RNA-dependent RNA polymerase of potato Y potyvirus and their use in the detection of virus infection. *Journal of Plant Diseases and Protection,* 106:181-187.

Lu, X.B., Peng, B.Z., Zhou, G.Y., Jing, D.D., Chen, S.X., and Gong, Z.X. 1999. Localization of PS9 in rice ragged stunt *Oryzavirus* and its role in virus transmission by brown planthopper. *Acta Biochemica et Biophysica Sinica,* 31:180-184.

Lunello, P., Bravo-Almonacid, F., Kobayashi, K., Helguera, M., Nome, S.F., Mentaberry, A., and Conci, V.C. 2000. Distribution of garlic virus A in different garlic production regions of Argentina. *Journal of Plant Pathology,* 82:17-21.

Lunsgaard, T. 1992. N Protein of *Festuca* leaf streak virus (Rhabdoviridae) detected in cytoplasmic viroplasms by immunogold labelling. *Journal of Phytopathology,* 134:27-32.

Manickam, K., Sabitha Doraiswamy, Ganapathy, T., Mala, T.G., and Rabindran, R. 2001. Characterization and serological detection of banana bunchy top virus in India. *Journal of Plant Diseases and Protection,* 108:490-499.

Más, P., Sánchez-Pina, M.A., Balsalobre, J.M., and Pallas, V. 2000. Subcellular localization of cherry leafroll virus coat protein and genomic RNAs in tobacco leaves. *Plant Science (Limerick),* 153:113-124.

McCafferty, J., Griffiths, A.D., Winter, G., and Chiswell, D.J. 1990. Phage-antibodies: Filamentous phage displaying antibody variable domains. *Nature,* 348: 552-554.

Medeiros, R.B., Ullman, D.E., Sherwood, J.L., and German, T.L. 2000. Immunoprecipitation of a 50 kDa protein: A candidate receptor component for tomato spotted wilt tospovirus (Buny aviridae) in its main vector, *Frankliniella occidentalis. Virus Research,* 67:109-118.

Merits, A., Guo, D., and Saarma, M. 1998. VPg, coat protein and five nonstructural proteins of potato A potyvirus bind RNA in a sequence unspecific manner. *Journal of General Virology,* 79:3123-3127.

Milne, R.G. 1992. Immunoelectron microscopy of plant viruses and phytoplasmas. *Advances in Disease Vector Research,* 9:283-312.

Milne, R.G. 1993. Solid phase immunoelectron microscopy of virus preparations. In *Immune Electron Microscopy for Virus Diagnosis,* (eds.) A.D. Hyat and B.T. Eaton, pp. 27-70. CRC Press, Boca Raton, Florida.

Minafra, A., Casati, P., Elicio, V., Rowhani, A., Saldarelli, P., Savino, V., and Martelli, G.P. 2000. Serological detection of grapevine rupestris stem pitting-associated virus (GRSPa V) by a polyclonal antiserum to recombinant virus coat protein. *Vitis,* 39:115-118.

Monis, J. 2000. Development of monoclonal antibodies reactive to a new grapevine leafroll-associated closterovirus. *Plant Disease,* 84:858-862.

Moore, C.J., Sutherland, P.W., Foster, R.L.S., Gardner, R.C., and MacDiarmid, R.M. 2001. Dark green islands in plant virus infection are the result of post-transcriptional gene silencing. *Molecular Plant-Microbe Interactions,* 14:939-946.

Moury, B., Cardin, L., Onesto, J.P., Candresse, T., and Poupet, A. 2001. Survey of *Prunus* necrotic ringspot virus in rose and its variability in rose and *Prunus* spp. *Phytopathology,* 91:84-91.

Myrta, A., Potere, O., Crescenzi, A., Nuzzaci, M., and Boscia, D. 2000. Properties of two monoclonal antibodies specific to the cherry strain of plum pox virus. *Journal of Plant Pathology,* 82 95-101.

Nagata, T., Inou-Nagata, A.K., Smid, H.M., Goldbach, R., and Peters, D. 1999. Tissue tropism related to vector competence of *Frankliniella occidentalis* for tomato spotted wilt virus. *Journal of General Virology,* 80:507-515.

Narayanasamy, P. 2001. *Plant Pathogen Detection and Disease Diagnosis,* Second Edition. Marcel Dekker, Inc., New York.

Narayanasamy, P. and Jaganathan, T. 1975. Seed transmission of blackgram leaf crinckle virus. *Phytopathologische Zeitschrift,* 82:107-110.

Narayanasamy, P. and Palaniswami, A. 1973. Studies on yellow mosaic disease of soybean. I. Effect of virus infection on plant pigments. *Experientia,* 29:1166-1167.

Narayanasamy, P. and Ramakrishnan, K. 1965. Studies on the sterility mosaic disease of pigeonpea. II. Carbohydrate metabolism of infected plants. *Proceedings of Indian Academy of Sciences,* B. 62:130-139.

Ndowora, T.C.R. and Lockhart, B.E.L. 2000. Development of a serological assay for detecting serologically diverse banana streak virus isolates. *Acta Horticulturae,* No. 540, 377-388.

Neustroeva, N.P., Dzantier, B.B., Markaryan, A.N., Bobkova, A.F., Igorov, A.M., and Atabekov, I.G. 1989. Enzyme-immunoassay of potato virus X using antibodies labeled by β-galactosidase of *Escherichia coli. Brologiya,* 1:115-118.

Niimi, Y., Gondaira, T., Kutsuwada, Y., and Tsuji, H. 1999. Detection by ELISA and DIBA tests of lily symptomless virus (LSV), tulip breaking virus-lily (TBV-

L) and cucumber mosaic virus (CMV). *Journal of the Japanese Society for Horticultural Science,* 68:176-183.

Njukeng, A.P., Atiri, G.I., Hughes, J.d'A., Agindotan, B.O., Mignouna, H.D., and Thottappilly, G. 2002. A sensitive TAS-ELISA for the detection of some West African isolates of yam mosaic virus in *Dioscorea. Tropical Science,* 42:65-74.

Pal, N., Moon, J.S., Sandhu, J., Domier, L.L., and D'Arcy, C.J. 2000. Production of barley yellow dwarf virus antisera by DNA immunization. *Canadian Journal of Plant Pathology,* 22:410-415.

Peng Ritte, Han Cheng Gui, Yang Lili, Yu Jia-Lin, and Liu Yi, 1998. Cytological localization of beet necrotic yellow vein virus transmitted by *Polymyxa betae. Acta Phytopathologica Sinica,* 28:257-261.

Peralta, E.L., Diaz, C., Lima, H., and Martinez, Y. 1997. Diagnosis of citrus tristeza virus utilizing the fluorogenic ultramicro ELISA system. *Fitopatologia,* 32:112-115.

Pesic, Z., Hiruki, C., and Chen, M.H. 1988. Detection of viral antigen by immunogold cytochemistry in ovules, pollen and anthers of alfalfa infected with alfalfa mosaic virus. *Phytopathology,* 78:1027-1032

Petrzik, K., Mráz, I., and Kubeková, D. 2001. Preparation of recombinant coat protein of *Prunus* necrotic ringspot virus. *Acta Virologica,* 45:61-63.

Quillec, F.L., Plantegenest, M., Riault, G., and Dedryver, C.A. 2000. Analyzing and modeling temporal disease progress of barley yellow dwarf virus serotypes in barley fields. *Phytopathology,* 90:860-866.

Ramiah, M., Bhat, A.I., Jain, R.K., Pant, R.P, Ahlawat, Y.S., Prabhakar, K., and Varma, A. 2001. Partial characterization of an isometric virus causing sunflower necrosis disease. *Indian Phytopathology,* 54:246-250.

Riedel, D., Lesemann, D.E., and Marss, E. 1998. Ultrastructural localization of nonstructural and coat proteins of 19 potyviruses using antisera to bacterially expressed proteins of plum pox potyvirus. *Archives of Virology,* 143:2133-2158.

Roberts, I.M. and Harrison, B.D. 1979. Detection of potato leaf roll and potato moptop viruses by immunosorbent electron microscopy. *Annals of Applied Biology,* 93:289-297.

Rouis, S., Traincard, F., Gargouri, R., Dartevelle, S., Jeannequin, O., Mazié, J.C., and Ayadi, H. 2001. Inhibition of potato virus Y NIa activity: Preparation of monoclonal antibody directed against PVY NIa protein that inhibits cleavage of PVY polyprotein. *Archives of Virology,* 146:1297-1306.

Ryu, K.H., Kim, C.H., and Palukaitis, P. 1998. The coat protein of cucumber mosaic virus is a host range determinant for infection of maize. *Molecular Plant-Microbe Interactions,* 11:351-367.

Sáena, P., Quiot, L., Quiot, J.P., Candresse, T., and Garcia, J.A. 2001. Pathogenicity determinants in the complex virus population of a plum pox virus isolate. *Molecular Plant-Microbe Interactions,* 14:278-287.

Saito, Y., Yamanaka, K., Watanabe, Y., Takamatsu, N., Meshi, T., and Okada, Y. 1989. Mutational analysis of the coat protein gene of tobacco mosaic virus in relation to hypersensitive response in tobacco plants with the N' gene. *Virology*, 173:11-20.

Saldarelli, P. and Minafra, A. 2000. Immunodetection of the 20kDa protein encoded by ORF2 of grapevine virus B. *Journal of Plant Pathology*, 82:157-158.

Saldarelli, P., Minafra, A., Castellano, M.A., and Martelli, G.P. 2000. Immunodetection and subcellular localization of the proteins encoded by ORF3 of grapevine viruses A and B. *Archives of Virology*, 145:1535-1542.

Santa Cruz, S. and Baulcombe, D.C. 1993. Molecular analysis of potato virus X isolates in relation to the potato hypersensitivity gene *Nx*. *Molecular Plant-Microbe Interactions*, 6:707-714.

Seddas, A., Haidar, M.M., Greif, C., Jacquet, C., Cloquemin, G., and Walter, B. 2000. Establishment of a relationship between grapevine leafroll closteroviruses 1 and 3 by use of monoclonal antibodies. *Plant Pathology*, 49:80-85.

Sefc, K.M., Leonhardt, W., and Steinkellner, H. 2000. Partial sequence identification of grapevine leafroll-associated virus-1 and development of a highly sensitive IC-RT-PCR detection method. *Journal of Virological Methods*, 86:101-106.

Sether, D.M., Ullman, D.E., and Hu, J.S. 1998. Transmission of pineapple mealybug wilt-associated virus by two species of mealybug (*Dysmicoccus* spp.). *Phytopathology*, 88:1224-1230.

Shalitin, D. and Wolf, S. 2000. Interaction between phloem proteins and viral movement proteins. *Australian Journal of Plant Physiology*, 27:801-806.

Shepard, J.F. 1970. A radial-immunodifussion test for the simultaneous diagnosis of potato virus S and X. *Phytopathology*, 60:1669-1671.

Sherwood, J.L., Sanborn, M.R., and Keyser, G.C. 1987. Production of monoclonal antibodies to peanut mottle virus and their use in enzyme-linked immunosorbent assay and dot immunobinding assay. *Phytopathology*, 77:1158-1161.

Simmonds, D.H. and Cumming, B. 1979. Detection of lily symptomless viruses by immunodiffusion. *Phytopathology*, 69:1212-1215.

Slack, S.A. and Shepherd, R.J. 1975. Serological detection of seed-borne barley stripe mosaic virus by a simplified radical-diffusion technique. *Phytopathology*, 65:948-955.

Soler, S., Díez, M.J., Roselló, S., and Nuez, F. 1999. Movement and distribution of tomato spotted wilt virus in resistant and susceptible accessions of *Capsicum* spp. *Canadian Journal of Plant Pathology*, 21:317-325.

Spence, N.J. 1997. The molecular genetics of plant virus interactions. In *The Gene for Gene Relationship in Plant-Parasite Interactions* (eds.) I.R. Crute, E.B. Holub, and J.J. Burdon, pp. 347-359. CAB International, United Kingdom.

Stevens, M., Hull, R., and Smith, H.G. 1997. Comparison of ELISA and RT-PCR for the detection of beet yellows closterovirus in plants and aphids. *Journal of Virological Methods*, 68:9-16.

Stevens, M., Smith, H.G., and Hallsworth, P.A. 1995. Detection of the luteoviruses, beet mild yellowing virus and beet western yellow virus in aphids caught in sugar beet and oilseed rape crops. *Annals of Applied Biology*, 127:309-320.

Su, H.J. and Wu, R.Y. 1989. Characterization and monoclonal antibodies of the virus causing banana bunchy top. *Technical Bulletin,* No. 115:1-10. Food and Fertilizer Technology Center, Taipei, Taiwan.

Sudharsana, M.R. and Reddy, D.V.R. 1989. Penicillinase-based enzyme-linked immunosorbent assay for the detection of plant viruses. *Journal of Virological Methods,* 26:45-52.

Sugiyama, M., Saito, H., Karasawa, A., Hase, S., Takahashi, H., and Ehara, Y. 2000. Characterization of symptom determinants in two mutants of cucumber mosaic virus Y strain, causing distinct mild green mosaic symptoms in tobacco. *Physiological and Molecular Plant Pathology,* 56:85-90.

Sukhacheva, E., Novikov, V., Plaksin, D., Pavlova, I., and Ambrosova, S. 1996. Highly sensitive immunoassays for detection of barley stripe mosaic virus and beet necrotic yellow vein virus. *Journal of Virological Methods,* 56:199-207.

Terrada, E., Kerschbaumer, R.J., Giunta, G., Galeffi, P., Himmler, G., and Cambra, M. 2000. Fully "recombinant enzyme-linked immunosorbent assays" using genetically engineered single-chain-antibody fusion proteins for detection of citrus tristeza virus. *Phytopathology,* 90:1337-1344.

Thottappilly, G., Dahal, G., and Lockhart, B.E.L. 1998. Studies on a Nigerian isolate of banana streak badnavirus. I. Purification and enzyme-linked immunosorbent assay. *Annals of Applied Biology,* 132:253-261.

Tomassoli, L., Lumia, V., Cerato, C., and Ghedini, R. 1998. Occurrence of potato tuber necrotic ringspot disease (PTNRD) in Italy. *Plant Disease,* 82:350.

Tremaine, J.H. and Wright, N.S. 1967. Cross-reactive antibodies in antisera to two strains of southern bean mosaic virus. *Virology,* 31:481-488.

Uhde, K., Kerschbaumer, R.J., Koenig, R., Hirschl, S., Lemaire, O., Boonham, N., Roake, W., and Himmler, G. 2000. Improved detection of beet necrotic yellow vein virus in a DAS-ELISA by means of antibody single chain fragments (scFv) which were selected to protease-stable epitopes from phage display libraries. *Archives of Virology,* 145:179-185.

Ullman, D.E., German, T.L., Sherwood, J.L., Westcot, D.M., and Cantone, F.A. 1993. Tospovirus replication in insect vector cells: Immunochemical evidence that the nonstructural protein encoded by S RNA of tomato spotted wilt virus is present in thrips vector cells. *Phytopathology,* 83:456-463.

Vaira, A.M., Roggero, P., Luisoni, E., Masenga, Milne, R.G., and Lisa, V. 1993. Characterization of two tospoviruses in Italy: Tomato spotted wilt and impatiens necrotic spot. *Plant Pathology,* 42:530-542.

Vaira, A.M., Vecchiati, M., Masenga, V., and Accotto, G.P. 1996. A polyclonal antiserum against a recombinant viral protein combines specificity with versatility. *Journal of Virological Methods,* 48:209-219.

Van der Vlugt, R.A.A., Berendsen, M., and Koenraadt. 1997. Immunocapture reverse transcriptase PCR for detection of lettuce mosaic virus. In *Seed Health Testing Progress Toward the Twenty-First Century,* (eds.) J.D. Hutchins and J.C. Reeves, pp. 185-198. CAB International, United Kingdom.

Van der Vlugt, R.A.A., Steffens, P., Cuperus, C., Barg, E., Lesemann, D.E., Bos, L., and Vetten, H.J. 1999. Further evidence that shallot yellow stripe virus (SYSV) is a distinct potyvirus and reidentification of Welsh onion yellow stripe virus as a SYSV strain. *Phytopathology,* 89:148-155.

Van Loon, L.C. 1999. Occurrence and properties of plant pathogenesis-related proteins. In *Pathogenesis-Related Proteins in Plants* (eds.) S.K. Datta and S. Muthukrishnan, pp. 1-19. CRC Press, Boca Raton, Florida.

Van Loon, L.C. and Van Kammen, A. 1970. Polyacrylamide disc electrophoresis of the soluble leaf proteins from *Nicotiana tabacum* var. Samsun and Samsun NN. II. Changes in protein constitution after infection with tobacco mosaic virus. *Virology,* 40:199.

Van Regenmortel, M.H.V. 1982. *Serology and Immunochemistry of Plant Viruses.* Academic Press, New York.

Van Regenmortel, M.H.V. and Dubs, M.C. 1993. Serological procedures. In *Diagnosis of Plant Virus Diseases,* (ed.) R.E.F. Matthews, pp. 160-214. CRC Press, Boca Raton, Florida.

Varveri, C. and Boutsika, K. 1998. Application of immunocapture PCR technique for plum pox potyvirus detection under field conditions in Greece and assays to simplify standard technique. *Acta Horticulturae,* No. 472:475-481.

Vejaratpimol, R., Channuntapipat, C., Pewnim, T., Ito, K., Iizuka, M., and Minamiura, N. 1999. Detection and serological relationships of cymbidium mosaic potexvirus isolates. *Journal of Bioscience and Bioengineering,* 87:161-168.

Verchot, J., Driskel, B.A., Zhu, Y., Hunger, R.M., and Littlefield, L.J. 2001. Evidence that soil-borne wheat mosaic virus moves along distance through the xylem in wheat. *Protoplasma,* 218:57-66.

Viswanathan, R. and Premachandran, M.N. 1998. Occurrence and distribution of sugarcane bacilliform virus in the sugarcane germplasm collection in India. *Sugar Cane,* No. 6:9-18.

Walkey, D.G.A., Lyons, N.F., and Taylor, J.D. 1992. An evaluation of a virobacterial agglutination test for the detection of plant viruses. *Plant Pathology,* 41:462-471.

Wang, S.H. and Gergerich, R.C. 1998. Immunofluorescent localization of tobacco ringspot nepovirus in the vector nematode *Xiphinema americanum. Phytopathology,* 88:885-889.

Wang, S.H., Gergerich, R.C., Wickizer, S.L., and Kim, K.S. 2002. Localization of transmissible and nontransmissible viruses in the vector nematode *Xiphinema americanum. Phytopathology,* 92:646-653.

Weilbach, A. and Sander, E. 2000. Quantitative detection of potato viruses X and Y (PVX, PVY) with antibodies raised in chicken egg yolk (IgY) by ELISA variants. *Journal of Plant Diseases and Protection,* 107:318-328.

Whitworth, J.L., Mosley, A.R., and Reed, G.L. 2000. Monitoring current season potato leafroll virus movement with an immunosorbent direct tissue blot assay. *American Journal of Potato Research,* 77:1- 9.

Wijkamp, I., Lent, J., Van Kormelin, K.R., Golbach, R., and Peters, D. 1993. Multiplication of tomato spotted wilt virus in its vector. *Frankliniella occidentalis. Journal of General Virology,* 74:341-349.

Wu, Y.Q., Chen, S.Y., Wang Wen Hui, and Wang Xiao Feng. 1998. Comparison of three ELISA methods for the detection of apple chlorotic leaf spot virus and apple stem grooving virus. *Acta Phytophylacia, Sinica,* 25:245-248.

Yang, Y., Kim, K.S., and Anderson, E.J. 1997. Seed transmission of cucumber mosaic virus in spinach. *Phytopathology,* 87:924-931.

Ye, R., Xu, L., Gao, Z.Z., Yang, J.P., Chen, J., Chen, J.P., Adams, M.J., and Yu, S.Q. 2000. Use of monoclonal antibodies for the serological differentiation of wheat and oat furoviruses. *Journal of Phytopathology,* 148:257-262.

Ye, R., Zheng, T., Xu, L., Lei, J.L., Chen, J.P., and Yu, S.Q. 2000. Coinfection of wheat yellow mosaic virus and rodshaped virus related to soil-borne wheat mosaic virus on winter wheat in Yantai district. *Chinese Journal of Virology,* 16:80-82.

Ziegler, A., Cowan, G.H., Torrance, L., Ross, H.A., and Davies, H.V. 2000. Facile assessment of cDNA constructs for expression of functional antibodies in plants using the potato virus X vector. *Molecular Breeding,* 6:327-335.

Ziegler, A., Mayo, M.A., and Torrance, L. 1998. Synthetic antigen from a peptide library can be an effective positive control in immunoassays for the detection and identification of two geminiviruses. *Phytopathology,* 88:1302-1305.

Zotto, A.D., Nome, S.F., di Rienzo, J.A., and Docampo, D.M. 1999. Fluctuations of *Prunus* necrotic ringspot virus (PNRSV) at various phenological stages in peach cultivars. *Plant Disease,* 83:1055-1057.

Chapter 11

Ahl, P. and Gianinazzi, S. 1982. *b*-Protein as a constitutive component in highly (TMV) resistant interspecific hybrids of *Nicotiana glutinosa* × *N. debneyi. Plant Science Letters,* 26:173-181.

Angenent, G.C., Van den Ouweland, J.M.W., and Bol, J.F. 1990. Susceptibility to virus infection of trangenic tobacco plants expressing structural and nonstructural genes of tobacco rattle virus. *Virology,* 175:191-198.

Anguelova-Merhar, V.S., van der Westhuizen, A.J., and Pretorius, Z.A. 1999. Intercellular proteins and β-1,3-glucanase activity associated with leaf rust resistance in wheat. *Physiologia Plantarum,* 106:393-401.

Anguelova-Merhar, V.S., van der Westhuizen, A.J., and Pretorius, Z.A. 2001. β-1,3-glucanase and chitinase activities and the resistance response of wheat to leaf rust. *Journal of Phytopathology,* 149:381-384.

Antoniw, J.F., Ritter, C.E., Pierpoint, W.S., and Van Loon, L.C. 1980. Comparison of three pathogenesis-related proteins from plants of two cultivars of tobacco infected with TMV. *Journal of General Virology,* 30:375.

Arce, P., Moreno, M., Gutierrez, M., Gebauer, M., Dell'orto, P., Torres, H., Acuña, L., Oliger, P., Venegas, A., Jordana, X., et al. 1999. Enhanced resistance to bacterial infection by *Erwinia carotovora* ssp. *atroseptica* in transgenic potato plants expressing the attacin or cecropin analogue *SB-37* genes. *American Journal of Potato Research,* 76:169-177.

Asíns, M.J., Mestre, P.F., Navarro, L., and Carbonell, E.A. 1999. Strategies to search for new citrus tristeza virus resistant genotypes in a germplasm bank. In *Genetics and Breeding for Crop Quality and Resistance* (eds.) G.T. Scarascia Mugnozza, E. Porceddu, and M.A. Pagnotta, pp. 251-256. Kluwer Academic Publishers, Dordrecht, Netherlands.

Astua-Monge, G., Minsavage, G.V., Stall, R.E., Davis, M.J., Bonas, U., and Jones, J.B. 2000. Resistance to tomato and pepper to T3 strains of *Xanthomonas campestris* pv. *vesicatoria* is specified by a plant inducible avirulence gene. *Molecular Plant-Microbe Interactions,* 13:911-912.

Bach, E.E. and Alba, A.P.C. 1993. Cross-reactive antigens between *Xanthomonas campestris* pv. *citri* pathotypes and *Citrus* species. *Journal of Phytopathology,* 138:84-88.

Baker, B., Zambryski, P., Staskawicz, B., and Dinesh-Kumar, S.P. 1997. Signaling in plant-microbe interactions. *Science,* 276:726-733.

Barker, H., Reavy, B., and McGeechy, K.D. 1998. High level of resistance in potato to potato mop-top virus induced by transformation with the coat protein gene. *European Journal of Plant Pathology,* 104:737-740.

Batchvarova, R., Nickolaeva, V., Slavov, S., Valkov, V., Atanassov, S., Guelemerov, S., Atanassov, A., and Anzai, H. 1998. Transgenic tobacco cultivars resistant to *Pseudomonas syringae* pv. *tabaci. Theoretical and Applied Genetics,* 97:981-989.

Beachy, R.N., Loesch-Fries, S., and Tumer, N.E. 1990. Coat protein-mediated resistance against virus infection. *Annual Review of Phytopathology,* 28:451-471.

Benhamou, N. 1996. Elicitor-induced plant defense pathways. *Trends in Plant Science,* 1:233-240.

Benhamou, N., Broglie, K., Broglie, R., and Chet, I. 1993. Antifungal effect of bean endochitinase on *Rhizoctonia solani*—Ultrastructural changes and cytochemical aspects of chitin breakdown. *Canadian Journal of Microbiology,* 39:318.

Bertioli, D.J., Cooper, J.I., Edwards, M.L., and Harves, W.S. 1992. Arabis mosaic nepovirus coat protein in transgenic tobacco lessens disease severity and virus replication. *Annals of Applied Biology,* 120:47-54.

Bolar, J.P., Norelli, J.L., Harman, G.E., Brown, S.K., and Aldwinckle, H.S. 2001. Synergistic activity of endochitinase from *Trichoderma atroviride (T. harzianum)* against pathogenic fungus *(Venturia inaequalis)* in transgenic apple plants. *Transgenic Research,* 10:533-543.

Bonman, H.G. and Hultmark, D. 1987. Cell free immunity in insects. *Annual Review of Microbiology,* 41:103-126.

Brault, V., Candresse, T., le Gall, O., Delbos, R.P., Lanneau, M., and Dunez, J. 1993. Genetically engineered resistance against grapevine chrome mosaic nepovirus. *Plant Molecular Biology,* 21:89-97.

Braun, C. and Hemenway, C. 1992. Expression of amino-terminal portions or full length viral replicase genes in transgenic plants confers resistance to potato virus X infection. *Plant Cell,* 4:735-744.

Broglie, K., Chet, I., Holliday, M., Cressman, R., Biddle, P., Knowlton, S., Mauvais, C.J., and Broglie, R. 1991. Transgenic plants with enhanced resistance to the fungal pathogen, *Rhizoctonia solani. Science,* 254:1194-1197.

Brunetti, A., Tavazza, M., Noris, E., Tavazza, R., Caciagle, P., Ancora, G., and Crespi, S. 1997. High expression of truncated viral rep-protein confers resistance to tomato yellow leafcurl virus in transgenic tomato plants. *Molecular Plant-Microbe Interactions,* 10:571-579.

Canady, M.A., Stevens, M.R., Barineau, M.S., and Scott, J.W. 2001. Tomato spotted wilt virus (TSWV) resistance in tomato derived from *Lycopersicon chilense* Dun L.A. 1938. *Euphytica,* 117:19-25.

Carvalho, A.O., Machado, O.L.T., da Cunha, M., Santos, I.S., and Gomes, V.M. 2001. Antimicrobial peptides and immunolocalization of a LTP in *Vigna unguiculata* seeds. *Plant Physiology and Biochemistry,* 39:137-146.

Carzaniga, R., Bowyer, P., and O'Connell, R.J. 2001. Production of extracellular matrices during development of infection structures by the downy mildew *Peronospora parasitica. New Phytologist,* 149:82-93.

Che, F.S., Nakajima, Y., Tanaka, N., Iwano, M., Yoshida, T., Takayama, S., Kadota, I., and Isogai, A. 2000. Flagellin from an incompatible strain of *Pseudomonas avenae* induces a resistance response in cultured rice cells. *Journal of Biological Chemistry,* 275:32347-32356.

Chen, Z.Y., Brown, R.L., Cleveland T.W., Damann, K.E., and Russin, J.S. 2001. Comparison of constitutive and inducible maize kernel proteins of genotypes resistant or susceptible to aflatoxin production. *Journal of Food Protection,* 64:1785-1792.

Christensen, A.B., Cho, B.H., Naesby, M., Gregersen, P.L., Brandt, J., Madriz-Ordeñana, K., Collinge, D.B., and Thordal-Christensen, H. 2002. The molecular characterization of two barley proteins establishes the novel PR-17 family of pathogenesis-related proteins. *Molecular Plant Pathology,* 3:135-144.

Clough, G.H. and Hamm, P.Q. 1995. Coat protein transgenic resistance to watermelon mosaic and zucchini yellow mosaic virus in squash and cantaloupe. *Plant Disease,* 79:1107-1109.

Coyne, C.J., Mehlenbacher, S.A., Hampton, R.O., Pinkerton, J.N., and Johnson, K.B. 1996. Use of ELISA to rapidly screen hazelnut for resistance to eastern filbert blight. *Plant Disease,* 80:1327-1330.

Croft, B.J. 2002. A method for rating sugarcane cultivars for resistance to ratoon stunting disease based on an enzyme-linked immunoassay. *Australian Journal of Plant Pathology,* 31:63-66.

Crosslin, J.M., Thomas, P.E., and Brown, C.R. 1999. Distribution of tobacco rattle virus in tubers of resistant and susceptible potatoes and systemic movement of virus into daughter plants. *American Journal of Potato Research,* 76:191-197.

Culver, J.N. 1996. Viral avirulence genes. In *Plant-Microbe Interactions* (eds.) G. Stacey and N.T. Keen, pp. 196-219. Chapman and Hall London, United Kingdom.

Culver, J.N. and Dawson, W.O. 1991. Tobacco mosaic virus elicitor coat protein genes produce a hypersensitive phenotype in transgenic *Nicotiana sylvestris* plants. *Molecular Plant-Microbe Interactions,* 4:458-463.

Culver, J.N., Sherwood, J.L., and Melonk, H.A. 1987. Resistance to peanut stripe virus in Arachis germplasm. *Plant Disease,* 71:1080-1082.

Cuozzo, M., O'Connell, K.M., Kaniewski, W., Fang, R.X., Chua, N.H., and Tumer, N.E. 1988. Viral protection in transgenic plants expressing the cucumber mosaic virus coat protein or its antisense RNA. *Bio/Technology,* 6:549-557.

Dardick, C.D. and Culver, J.N. 1997. Tobamovirus coat proteins: Elicitors of the hypersensitive response in *Solanum melongena* (eggplant). *Molecular Plant-Microbe Interactions,* 10:776-778.

Datta, K., Mutukrishnan, S., and Datta, S.K. 1999. Expression and function of PR-protein genes in trangenic plants in transgenic plants. In *Pathogenesis-related Proteins in Plants,* (eds.) S.K. Datta and S. Muthukrishnan, pp. 261-277. CRC Press, Boca Raton, Florida.

Datta, K., Valazhahan, R., Oliva, N., Ona, I., Mew, T., Khush, G.S., Muthukrishnan, S., and Datta, S.K. 1999. Over-expression of the cloned rice thaumatin-like protein (PR-5) gene in transgenic rice plants enhances environmental friendly resistance to *Rhizoctonia solani* causing sheath blight disease. *Theoretical and Applied Genetics,* 98:1138-1145.

Dempsey, D.A., Silva, H., and Klessig, D.F. 1998. Engineering disease and pest resistance in plants. *Trends in Microbiology,* 6:54-61.

Dohm, A., Ludwig, G., Schilling, D., and Debener, T. 2002. Transformation of roses with genes for antifungal proteins to reduce their susceptibility to fungal diseases. *Acta Horticulturae,* No. 572:105-111.

Doreste, V., Ramos, P.L., Enríquez, G.A., Rodríquez, R., Pearl, R., and Pujol, M. 2002. Transgenic potato plants expressing the potato virus X (PVX) coat protein gene developed resistance to the viral infection. *Phytoparasitica,* 30:177-185.

Duan, Y.P., Powell, C.A., Webb, S.E., Purcifull, D.E., and Hiebert, E. 1997. Geminivirus resistance in transgenic tobacco expressing mutated BC1 proteins. *Molecular Plant-Microbe Interactions,* 10:617-623.

El-Gendy, W., Brownleader, M.D., Ismail, H., Clarke, P.J., Gilbert, J., El-Bordiny, F., Trevan, M., Hopkins, J., Naldrett, M., and Jackson, P. 2001. Rapid deposition of wheat cell wall structural proteins in response to Fusarium-derived elicitors. *Journal of Experimental Botany,* 52:85-90.

Favaron, F. 2001. Gel detection of *Allium porrum* polygalacturonase-inhibitory protein reveals a high number of isoforms. *Physiological and Molecular Plant Pathology,* 58:239-245.

Favaron, F., Destro, T., and D'Ovidio, R. 2000. Transcript accumulation of polygalacturonase-inhibiting protein (PGIP) following pathogen infections in soybean. *Journal of Plant Pathology,* 82:103-109.

Fellers, J.P., Collins, G.B., and Hunt, A.G. 1998. The NIa-proteinase of different plant potyviruses provides specific resistance to viral infection. *Crop Science,* 38:1309-1319.

Ferreira, S.A., Pitz, K.Y., Manshardt, R., Zee, F., Fitch, M., and Gonsalves, D. 2002. Virus coat proteins transgenic papaya provides practical control of papaya ringspot virus in Hawaii. *Plant Disease,* 86:101-105.

Finnie, C., Andersen, C.H., Borch, J., Gjetting, S., Christensen, A.B., de Boer, A.H., Thordal-Christensen, H., and Collinge, D.B. 2002. Do 14-3-3- proteins and plasma membrane H+-ATPases interact in the barley epidermis in response to the barley powdery mildew fungus? *Plant Molecular Biology,* 49:137-147.

Fischer, R., Liao, Y.C., and Drossard, J. 1999. Affinity-purification of a TMV-specific recombinant full-size antibody from a transgenic tobacco suppression culture. *Journal of Immunological Methods,* 226:1-10.

Fischer, R., Schumann, D., Zimmermann, S., Drossard, J., Sack, M., and Schillberg, S. 1999. Expression and characterization of biospecific single-chain Fv-fragments produced in transgenic plants. *European Journal of Biochemistry,* 262: 810-816.

Fitch, M.M.M., Manshardt, R.M., Gonsalves, D., Slightom, J.L., and Sanford, C. 1992. Virus-resistant papaya plants derived from tissues bombarded with the coat protein gene of papaya ringspot virus. *Bio/Technology,* 10:1466-1472.

Franco-Lara, L. and Barker, H. 1999. Characterization of resistance to potato leafroll virus accumulation in *Solanum phureja. Euphytica,* 108:137-144.

Fraser, R.S.S. 1982. Are "pathogenesis-related" proteins involved in acquired systemic resistance of tobacco plants to tobacco mosaic virus? *Journal of General Virology,* 58:305.

Fuchs, M., Provvidenti, R., Slighton, J.L., and Gonsalves, D. 1996. Evaluation of transgenic tomato plants expressing the coat protein gene of cucumber mosaic virus strain WL under field conditions. *Plant Disease,* 80:270-275.

Gal-On, A., Wolf, D., Pilowsky, M., and Zelcer, A. 1999. A transgenic tomato F_1 hybrid harboring a defective viral replicase shows immunity to cucumber mosaic virus in field trials. *Acta Horticulturae,* No. 487, 329-333.

Gal-On, A., Wolf, D., Wang, Y., Faure, J.E., Pilowsky, M., and Zelcer, A. 1998. Transgenic resistance to cucumber mosaic virus in tomato: Blocking of long-distance movement of the virus in lines harboring a defective viral replicase gene. *Phytopathology,* 88:1101-1107.

Ghorbel, R., López, C., Fagoaga, C., Moreno, P., Navarro, L., Flores, R., and Peòa, L. Transgenic citrus plants expressing the citrus tristeza virus p23 protein exhibit virus-like symptoms. *Molecular Plant Pathology,* 2:27-36.

Gianinazzi, S., Pratt, H.M., Shewry, P.R., and Miflin, B.J. 1977. Partial purification and preliminary characterization of soluble proteins specific to virus-infected tobacco. *Journal of General Virology,* 34:345

Giglioti, E.A., Comstock, J.C., Davis, M.J., Matsuoka, S., and Tokeshi, H. 1999. Combining tissue blot enzyme immunoassay and staining by transpiration methods to evaluate colonization of sugarcane stalks by *Clavibacter xyli* ssp. *xyli* and its effects on the xylem functionality. *Summa Phytopathologica*, 25:125-132.

Golemboski, D.B., Lomonossoff, G.P., and Zaitlin, M. 1990. Plants transformed with a tobacco mosaic virus nonstructural gene sequence are resistant to the virus. *Proceedings of the National Academy of Sciences, USA*, 87:6311-6315.

Gonsalves, D., Chee, P., Provvidenti, R., Seem, R., and Slightom, J.L. 1992. Comparison of coat protein-mediated and genetically-derived resistance in cucumbers to infection by cucumber mosaic virus under field conditions with natural challenge inoculations by vectors. *Bio/Technology*, 10:1562-1570.

Grison, R., Grezes-Besset, B., Schneider, M., Lucanate, N., Olsen, L., Leguary, J.J., and Toppan, A. 1996. Field tolerance to fungal pathogens of *Brassica napus* constitutively expressing a chimeric chitinase gene. *Nature Biotechnology*, 14:643-646.

Gruden, K., Štruklej, B., Ravnikar, M., and Herzog-Velikonja, B. 2000. A putative resistance-connected protein isolated from potato cultivar Sante resistant to PVY [NTN] infection. *Phyton (Horn)*, 40:191-200.

Ham, K.S., Albersheim, P., and Darvill, A.G. 1995. Generation of β-glucan elicitors by plant enzymes and inhibition of the enzymes by a fungal protein. *Canadian Journal of Botany*, 73:1100.

Ham, K.S., Kauffmann, S., Albersheim, P., and Darvill, A.G. 1991. Host-pathogen interactions. XXXIX. A soybean pathogenesis-related protein with β-1,3-glucanase activity releases phytoalexin elicitor-active heat stable fragments from fungal cell walls. *Molecular Plant-Microbe Interactions*, 4:545.

Ham, K.S., Wu, S.C., Darvill, A.G., and Albersheim, P. 1997. Fungal pathogens secrete an inhibitor protein that distinguishes isoforms of plant pathogenesis-related endo-β-1,3-glucanases. *Plant Journal*, 11:169.

Hanselle, T. and Barz, W. 2001. Purification and characterization of the extracellular PR-2b, β-1,3-glucanase accumulating in different *Ascochyta rabiei*-infected chickpea (*Cicer arietinum* L.) cultivars. *Plant Science*, 161:773-781.

Hariri, D., Fouchard, M., and Pruïhomme, H. 2001. Incidence of soil-borne wheat mosaic virus in mixtures of susceptible and resistant cultivars. *European Journal of Plant Pathology*, 107:625-631.

Hasegawa, H., Miyazawa, J., and Iwata, M. 1999. Use of ELISA to quantify *Magnaporthe grisea* in rice leaf. *Annals of Phytopathological Society, Japan*, 65:312-314.

Hayakawa, T., Zhu, Y., Itoh, K., Kimura, Y., Izawa, T., Shimamoto, K., and Toriyama, S. 1992. Genetically engineered rice resistant to rice stripe virus and insect transmitted virus. *Proceedings of the National Academy of Sciences, USA*, 89:9865-9869.

Hemenway, C., Fang, R.X., Kaniewski, J., Chua, N.H., and Tumer, N.E. 1988. Analysis of the mechanism of protection in transgenic plants expressing the potato virus X coat protein or its antisense RNA. *EMBO Journal*, 7:1273-1280.

Herrero, S., Culbreath, A.K., Csinos, A.S., Pappu, H.R., Rufty, R.C., and Daub, M.E. 2000. Nucleocapsid gene-mediated transgenic resistance provides protection against tomato spotted wilt virus epidemics in the field. *Phytopathology,* 90:139-147.

Hilaire, E., Young, S.A., Willard, L.H., McGee, J.D., Sweat, T., Chittoor, J.M., Guikema, J.A., and Leach, J.E. 2001. Vascular defense responses in rice: Peroxidase accumulation in xylem parenchyma cells and xylem wall thickening. *Molecular Plant-Microbe Interactions,* 14:1411-1419.

Hill, K.K., Jarvis-Eagan, N., Halk, E.L., Krahn, K.J., Liao, L.W., Mathewson, R.S., Merlo, D.J., Nelson, S.E., Rashka, K.E., and Loesch-Fries, L.S. 1991. The development of virus-resistant alfalfa *Medicago sativa* L. *Bio/Technology,* 9:373-377.

Hoekema, A., Huisman, M.J., Molendijk, P., van den Elzen, P., and Cornelissen, B.J.C. 1989. The genetic engineering of two commercial potato cultivars for resistance to potato virus X. *Bio/Technology,* 7:273-278.

Hong, Y. and Stanley, J. 1996. Virus resistance in *Nicotiana benthamiana* conferred by African cassava mosaic virus replication-associated (AC1) transgene. *Molecular Plant-Microbe Interactions,* 9:219-225.

Huang, Y., Nordeen, R.O., Di, M., Owens, L.D., and Mc Beath, J.H. 1997. Expression of an engineered gene cassette in transgenic tobacco plants confers disease resistance to *Pseudomonas syringae* pv. *tabaci. Phytopathology,* 87:494-499.

Hutcheson, S.W. 1998. Current concepts of active defense in plants. *Annual Review of Phytopathology,* 36:59-90.

Huynh, O.K., Borgmeyer, J.R., and Zobel, J.F. 1992. Isolation and characterization of a 22 kDa protein with antifungal properties from maize seeds. *Biochemistry and Biophysics Research Communications,* 182:1-5.

Ioannidou, D., Lett, J.M., Pinel, A., Assigbetse, K., Brugidou, C., Ghesquiere, A., Nicole, M., and Fargette, D. 2000. Responses of *Oryza sativa japonica* subspecies to infection with rice yellow mottle virus. *Physiological and Molecular Plant Pathology,* 57:177-188.

Jach, G., Logemann, S., Wolf, G., Oppenheim, A., Chet, I., Schell, J., and Logemann, J. 1992. Expression of a bacterial chitinase leads to improved resistance of transgenic tobacco plants against fungal infection. *Biopractice,* 1:1.

Jenny, E., Amiet, J., Hecker, A., and Forrer, H.R. 1999. Control of *Septoria tritici* based on ELISA-thresholds. *Agrarforschung,* 6:329-332.

Jia, Y.L., Mc Adams, S.A., Bryan, G.T., Hershey, H.P., and Valent, B. 2000. Direct interaction of resistance gene and avirulence gene products confers rice blast resistance. *EMBO Journal,* 19:4004-4014.

Jones, D.A. and Jones, J.D.G. 1997. The role of leucine-rich repeat proteins in plant defenses. *Advances in Plant Pathology,* 24:89-167.

Jones, R.W. and Prusky, D. 2002. Expression of an antifungal peptide in *Saccharomyces:* A new approach for biological control of the postharvest disease caused by *Colletotrichum coccodes. Phytopathology,* 92:33-37.

Jongedijk, E., Tigelaar, H., van Roekel, J.S.C., Bres-Vloemans, S.A., Dekker, I., van den Elzen, P.J.M., Cornelissen, B.J.C., and Melchers, L.S. 1995. Synergistic activity of chitinases and β-1, 3-glucanases enhances resistance in transgenic tomato plants. *Euphytica,* 85:173.

Kaniewski, W., Ilardi, V. Tomassoli, L., Mitsky, T., Layton, J., and Barba, M. 1999. Extreme resistance to cucumber mosaic virus (CMV) in transgenic tomato expressing one or two viral coat proteins. *Molecular Breeding,* 5:111-119.

Karešová, R. and Paprštein, P. 1998. Occurrence of plum pseudopox (apple chlorotic leafspot virus) in plum germplasm. *Acta Horticulturae,* No. 478, 283-286.

Karpovich-Tate, N., Spanu, P., and Dewey, F.M. 1998. Use of monoclonal antibodies to determine biomass of *Cladosporium fulvum* in infected tomato leaves. *Molecular Plant-Microbe Interactions,* 11:710-716.

Kawchuck, L.M., Martin, R.R., and McPherson, J. 1990. Resistance in transgenic potato expressing the potato leafroll virus coat protein gene. *Molecular Plant-Microbe Interactions,* 3:301-307.

Keen, N.T. 1999. Plant disease resistance: Progress in basic understanding and practical application. *Advances in Botanical Research,* 30:291-328.

Kemp, G., Botha, A.M., Kloppers, F.J., and Pretorius, Z.A. 1999. Disease development and β-1,3-glucanase expression following leaf rust infection in resistant and susceptible near isogenic wheat seedlings. *Physiological and Molecular Plant Pathology,* 55:45-52.

Kommineni, K.V., Gillett, J.M., and Ramsdell, D.C. 1998. A study of tomato ringspot virus and prune brown line resistance in twenty-five root stock-scion combinations. *Hort Technology,* 8:349-353.

Kraft, J.M. and Boge, W.L. 1994. Development of an antiserum to quantify *Aphanomyces euteiches* in resistant pea lines. *Plant Disease,* 78:179-183.

Kraft, J.M. and Boge, W.L. 1996. Identification of characteristics associated with resistance to root rot caused by *Aphanomyces euteiches* in pea. *Plant Disease,* 80:1383-1386.

Krishnamurthy, K., Balconi, C., Sherwood, J.E., and Giroux, M.J. 2001. Wheat puroindolines enhance fungal disease resistance in transgenic rice. *Molecular Plant-Microbe Interactions,* 14:1255-1260.

Laugé, R., Goodwin, P.H., Wit, P.J.G.M. de and Joosten, M.H.A.J. 2000. Specific HR-associated recognition of secreted proteins from *Cladosporium fulvum* occurs in both host and nonhost plants. *Plant Journal,* 23:735-745.

Lawson, C., Kaniewski, W., Haley, L., Rozman, R., Newell, C., Sanders, P., and Tumer, N.E. 1990. Engineering resistance to mixed virus infection on a commercial potato cultivar: Resistance to potato virus X and potato virus Y in transgenic Russet Bulbank. *Bio/Technology,* 8:127-134.

Leclerc, D. and Abou Haider, M.G. 1995. Transgenic tobacco plants expressing a truncated form of the PAMV capsid protein (CP) gene show CP-mediated resistance to potato cucuba mosaic virus. *Molecular Plant-Microbe Interactions,* 8:58-65.

Lee, Y.K., Hippe-Sanwald, S., Jung, H.W., Hong, J.K., Hause, B., and Hwang, B.K. 2000. In situ localization of chitinase mRNA and protein in compatible and incompatible interactions of pepper stems with *Phytophthora capsici. Physiological and Molecular Plant Pathology,* 57:111-121.

Li, W.B., Zarka, K.A., Doucher, D.S., Coombs, J.J., Pett, W.L., and Grafius, E.J. 1999. Coexpression of potato PVY^O coat protein and *cryv-Bt* genes in potato. *Journal of American Society of Horticultural Sciences,* 124:218-223.

Li, W.L., Faris, J.D., Muthukrishnan, S., Liu, D.J., Chen, P.D., and Gill, B.S. 2001. Isolation and characterization of novel cDNA clones of acidic chitinases and β-1,3-glucanases from wheat spikes infected by *Fusarium graminearum. Theoretical and Applied Genetics,* 102:353-362.

Liljeroth, E., Santén, K., and Bryngelsson, T. 2001. PR-protein accumulation in seminal roots of barley and wheat in response to fungal infection—The importance of cortex senescence. *Journal of Phytopathology,* 149:447-456.

Lin, W., Anuratha, C.S., Datta, K., Potrykus, I., Muthukrishnan, S., and Datta, S.K. 1995. Genetic engineering of rice for resistance to sheath blight. *Bio/Technology,* 13:686.

Lin, Y.J., Rundell, P.A., and Powell, C.A. 2002. In situ immunoassay (ISIA) of field grapefruit trees inoculated with mild isolates of citrus tristeza virus indicates mixed infection with severe isolates. *Plant Disease,* 86:458-461.

Linthorst, H.J.M., Meuwissen, R.L.J., Kauffmann, S., and Bol, J.F. 1989. Constitutive expression of pathogenesis-related proteins PR-1, GRP and PR-5 in tobacco has no effect on virus infection. *Plant Cell,* 1:285-291.

Longstaff, M., Brigneti, G., Boccard, F., Chapman, S., and Baulcombe, D. 1993. Extreme resistance to potato virus X infection in plants expressing a modified component of the putative viral replicase. *EMBO Journal,* 12:379-386.

Ludwig, A.A. and Tenhaken, R. 2001. A new cell wall located N-rich protein is strongly induced during the hypersensitive response in *Glycine max L. European Journal of Plant Pathology,* 107:323-336.

MacFarlane, S.A. and Davies, J.W. 1992. Plants transformed with a region of the 201-kilodalton replicase gene from pea early browning virus RNA1 are resistant to virus infection. *Proceedings of National Academy of Sciences, USA,* 89:5829-5833.

Machinandiarena, M.F., Olivieri, F.P., Daleo, G.R., and Oliva, C.R. 2001. Isolation and characterization of a polygalacturonase-inhibiting protein from potato leaves. Accumulation in response to salicylic acid, wounding and infection. *Plant Physiology and Biochemistry,* 39:129-136.

Mackenzie, D.J. and Ellis, P.J. 1992. Resistance to tomato spotted wilt virus infection in transgenic tobacco expressing the viral nucleocapsid gene. *Molecular Plant-Microbe Interactions,* 5:34-40.

Madhu, D., Dharmesh, M.S., Arun Chandrashekar, Shetty, H.S., and Prakash, H.S. 2001. Role of H$^+$-ATPase in pearl millet downy mildew resistance. *Plant Science,* 161:799-806.

Maiti, I.B., Lanken, C., von Hong, Y.L., Dey, N., and Hunt, A.G. 1999. Expression of multiple virus-derived resistance determinants in transgenic plants does not lead to additive resistance properties. *Journal of Biochemistry and Biotechnology,* 8:67-73.

Mäki-Valkama, T., Pehu, T., Santala, A., Valkonen, J.P.T., Koivu, K., Lehto, K., and Pehu, E. 2000. High level of resistance to potato virus Y by expressing P1 sequence in antisense orientation in transgenic potato. *Molecular Breeding,* 6:95-104.

Mäki-Valkama, T., Valkonen, J.P.T., Kreuze, J.F., and Pehu, E. 2000. Transgenic resistance to PVY associated with post-transcriptional silencing of P_1 transgene is overcome by PVY^N strains that carry homologous P_1 sequences and recover transgene expression at infection. *Molecular Plant-Microbe Interactions,* 13:95-104.

Mäki-Valkama, T., Valkonen, J.P.T., Lehtinen, A., and Pehu, E. 2001. Protection against potato virus Y (PVY) in the field in potatoes transformed with the PVY-P1 gene. *American Journal of Potato Research,* 78:209-214.

Mauch, F., Mauch-Mani, B., and Boller, T. 1988. Antifungal hydrolases in pea tissue. II. Inhibition of fungal growth by combinations of chitinase and β-1,3-glucanase. *Plant Physiology,* 88:936.

McGarvey, J.A., Denny, T.P., and Schell, M.A. 1999. Spatial-temporal and quantitative analysis of growth and EPSI production by *Ralstonia solanacearum* in resistant and susceptible tomato cultivars. *Phytopathology,* 89:1233-1239.

Melander, M., Lee, M., and Sandgren, M. 2001. Reduction of potato mop top virus accumulation and incidence in tubers of potato transformed with a modified triple gene block gene of PMTV. *Molecular Breeding,* 8:197-206.

Métraux, J.P., Ahl-Goy, P., Staub, T., Speich, J., Steinemann, A., Ryals, J., and Ward, E. 1991. Induced resistance in cucumber in response to 2,6-dichloro-isonicotinic acid and pathogens. In *Advances in Molecular Genetics of Plant-Microbe Interactions,* Volume 1, (eds.) H. Hennecke and D.P.S. Verma, pp. 432-439. Kluwer Academic Publishers, Dordrecht, The Netherlands.

Meyer, U.M., Spotts, R.A., and Dewey, F.M. 2000. Detection and quantification of *Botrytis cinerea* by ELISA in pear stems during cold storage. *Plant Disease,* 84:1099-1103.

Mora, A.A. and Earle, E.D. 2001. Resistance to *Alternaria brassicola* in transgenic broccoli expressing a *Trichoderma harzianum* endochitinase gene. *Molecular Breeding,* 8:1-9.

Murillo, I., Cavallarin, L., and San Segundo, B. 1999. Cytology of infection of maize seedlings by *Fusarium moniliforme* and immunolocalization of the pathogenesis-related PRms protein. *Phytopathology,* 89:737-747.

Murphy, J.F., Sikora, E.J., Sammons, B., and Kaniewski, W.K. 1998. Performance of transgenic tomatoes expressing cucumber mosaic virus CP gene under epidemic conditions. *HortScience,* 33:1032-1035.

Narayanasamy, P. 2002. *Microbial Plant Pathogens and Crop Disease Management.* Science Publishers, Enfield, New Hampshire.

Ndjiondjop, M.N., Albar, L., Fargette, D., Fauquet, C., and Ghesquiere, A. 1999. The genetic basis of high resistance to rice yellow mottle virus (RYMV) in cultivars of two cultivated rice species. *Plant Disease,* 83:931-935.

Nelson, R.S., McCormick, S.M., Delannay, X., Debé, P., Layton, J., Anderson, E.J., Kaniewski, M., Proksch, R.K., Horsch, R.B., Rogers, S.G., et al. 1988. Virus tolerance, plant growth and field performance of transgenic tomato plants expressing coat protein from tobacco mosaic virus. *Bio/Technology*, 6:403-409.

Ogawa, T., Moniyama, T., Fujita, C., and Uesugi, T. 2000. Immunochemical characterization of soybean allergen, Gly m Bd 30k as a syringolide receptor. *Soy Protein Research, Japan*, 3:67-72.

Okada, Y., Saito, A., Nishiguchi, M., Kimura, T., Mori, M., Hanada, K., Sakai, J., Miyazaki, C., Matsuda, Y., and Murata, T. 2001. Virus resistance in transgenic sweet potato (*Ipomoea batatus* L. Lam) expressing the coat protein gene of sweet potato feathery mottle virus. *Theoretical and Applied Genetics*, 103:743-751.

Padgett, H.S. and Beachy, R.N. 1993. Analysis of a tobacco mosaic virus strain capable of overcoming N-gene-mediated resistance. *Plant Cell*, 5:577-586.

Pang, S.Z., Nagpala, P., Wang, M., Slighton, J.L., and Gonsalves, D. 1992. Resistance to heterologous isolates of tomato spotted wilt virus in transgenic tobacco expressing its nucleocapsid protein gene. *Phytopathology*, 82:1223-1229.

Porat, R., Vinokur, V., Holland, D., McCollum, T.G., and Droby, S. 2001. Isolation of a citrus chitinase cDNA and characterization of its expression in response to elicitation of fruit pathogen resistance. *Journal of Plant Physiology*, 158:1585-1590.

Poscai, E. and Muranyi, I. 2000. Resistance to barley yellow dwarf virus in Hungarian breeding materials of winter barley. *Rasteniev "dni Nauki,"* 37:788-790.

Powell, P.A., Nelson, R.S., De, B., Hoffman, N., Rogers, S.G., Fraley, R.T., and Beachy, R.N. 1986. Delay of disease development in transgenic plants that express the tobacco mosaic virus coat protein gene. *Science*, 232:738-743.

Powell, P.A., Sanders, P.R., Tumer, N., Fraley, R.T., and Beachy, R.N. 1990. Protection against tobacco mosaic virus infection in transgenic plants requires accumulation of coat protein rather than coat protein RNA sequences. *Virology*, 175:124-130.

Powell, P.A., Stark, D.M., Sanders, P.R., and Beachy, R.N. 1989. Protection against tobacco mosaic virus in transgenic plants that express tobacco mosaic virus antisense RNA. *Proceedings of National Academy of Sciences, USA*, 86:6949-6952.

Prins, M., de Haan, P., Luyten, R., van Veller, M., and van Grinsven, M.Q.J.M. 1995. Broad resistance to tospovirus in transgenic tobacco plants expressing three nucleoprotein gene sequences. *Molecular Plant-Microbe Interactions*, 8:85-91.

Prins, M., De Oliveira Resende, R., Anker, C., van Sahepen, A., De Haan, P., and Goldbach, R. 1996. Engineered RNA-mediated resistance to tomato spotted wilt virus is sequence specific. *Molecular Plant-Microbe Interactions*, 9:416-418.

Priou, S., Gutarra, L., and Aley, P. 1999. Highly sensitive detection of *Ralstonia solanacearum* in latently infected potato tubers by post enrichment enzyme-linked immunosorbent assay on nitrocellulose membrane. *Bulletin OEPP*, 29: 117-125.

Priou, S., Salas, C., de Mendiburu, F., Aley, P., and Gutarra, L. 2001. Assessment of latent infection frequency in progeny tubers of advanced potato clones resistant to bacterial wilt: A new selection criterion. *Program Report* (1999-2000), pp. 105-116. International Potato Center, Lima, Peru.

Pritsch, C., Muehlbauer, G.J., Bushnell, W.R., Somers, D.A., and Vance, C.P. 2000. Fungal development and induction of defense response genes during early infection of wheat spikes by *Fusarium graminearum*. *Molecular Plant-Microbe Interactions*, 13:159-169.

Rabinowicz, P.D., Bravo-Almonacid, F.F., Lampasona, S., Rodriguez, F., Gracia, O., and Mentaberry, A.N. 1998. Resistance against pepper severe mosaic potyvirus in transgenic tobacco plants. *Journal of Phytopathology*, 146:315-319.

Racman, D.S., Mc Geachy, K., Reavy, B., Štruklej, B., Zel, J., and Barker, H. 2001. Strong resistance to potato tuber necrotic ringspot disease in potato induced by transformation with coat protein gene sequences from an NTN isolate of potato virus Y. *Annals of Applied Biology*, 139:269-275.

Ravelonandro, M., Scorza, R., Bachelier, J.C., Laborne, G., Levy, L., Damsteegt, V., Callahan, A.M., and Dunez, J. 1997. Resistance of transgenic *Prunus domestica* to plum pox virus infection. *Plant Disease*, 81:1231-1235.

Renault, A.S., Deloire, A., Letinois, I., Kraeva, E., Tesniere, C., Ageorges, A., Redon, C., and Bierne, J. 2000. β-1, 3-glucanase gene expression in grapevine leaves as a response to infection with *Botrytis cinerea*. *American Journal of Enology and Viticulture*, 51:81-87.

Reynoird, J.P., Mourgues, F., Chevreau, E., and Brisset, M.N. 1999. First evidence for differences in fire blight resistance among transgenic pear clones expressing Attacin E gene. *Acta Horticulturae*, No. 489, 245-246.

Robert, N., Ferran, J., Breda, C., Coutos-Thévenot, P., Boulay, M., Buffard, D., and Esnault, R. 2001. Molecular characterization of the incompatible interaction of *Vitis vinifera* leaves with *Pseudomonas syringae* pv. *pisi:* Expression of genes coding for stilbene synthase and class 10 PR-protein. *European Journal of Plant Pathology*, 107:249-261.

Ronald, P.C. 1997. The molecular basis of disease resistance in rice. *Plant Molecular Biology*, 35:179-186.

Rubino, L., Lopo, R., and Russo, M. 1993. Resistance to cymbidium ringspot virus in transgenic *Nicotiana benthamiana* plants expressing full-length viral replicase gene. *Molecular Plant-Microbe Interactions*, 6:729-734.

Saiga, T., Fujiwara, M., Saitoh, H., Ohki, S.T., and Osaki, T. 1998. Comparative analysis for replication and movement of cucumber mosaic virus in *Cucumis figarei* and *C. melo*. *Annals of Phytopathological Society, Japan*, 64:255-263.

Sanders, P.R., Sammons, B., Kaniewski, W., Haley, L., Layton, J., La Vallee, B.J., Delannay, X., and Tumer, N.E. 1992. Field resistance of transgenic tomatoes expressing the tobacco mosaic virus or tomato mosaic virus coat protein genes. *Phytopathology*, 82:683-690.

Sanford, J.C. and Johnson, S.A. 1985. The concept of parasite-derived resistance: Deriving resistance genes from the parasites' own genome. *Journal of Theoretical Biology,* 115:395-405.

Schillberg, S., Zimmermann, S., Findlay, K., and Fischer, R. 2000. Plasma membrane display of antiviral single chain Fv fragments confers resistance to tobacco mosaic virus. *Molecular Breeding,* 6:317-326.

Selvarajan, R., Gupta, M.D., and Misra, R.L. 1998. Screening of gladiolus cultivars against bean yellow mosaic virus. *Journal of Ornamental Horticulture (New Series),* 1:69-71.

Serrano, C., Arce-Johnson, P., Torres, H., Gebauer, M., Gutierrez, M., Moreno, M., Jordana, X., Venegas, A., Kalazich, J., and Holuigue, L. 2000. Expression of the chicken lysozyme gene in potato enhances resistance to infection by *Erwinia carotovora* ssp. *atroseptica. American Journal of Potato Research,* 77:191-199.

Sessa, G. and Martin, G.B. 2000. Protein kinases in the plant defense response. *Advances in Botanical Research,* 32:379-404.

Sether, D.M., Karasev, A.V., Okumura, C., Arakawa, C., Zee, F., Kislan, M.M., Busto, J.L., and Hu, J.S. 2001. Differentiation, distribution and elimination of two different pineapple mealybug wilt-associated viruses found in pineapple. *Plant Disease,* 85:856-864.

Shan, Z.H., Liao, B.S., Tan, Y.J., Li, D., Lei, Y., and Shen, M.Z. 1997. ELISA technique used to detect latent infection of groundnut by bacterial wilt *(Pseudomonas solanacearum). Oilcrops of China,* 19:45-47.

Shankar, M., Gregory, A., Kalkhoven, M.J., Cowling, W.A., and Sweetingham, M.W. 1998. A competitive ELISA for detecting resistance to latent stem infection by Diaporthe toxica in narrow-leaved lupins. *Australasian Plant Pathology,* 27:251-258.

Sharon, A., Amsellem, Z., and Gressel, J. 1993. Quantification of infection by *Alternaria cassiae* using leaf immuno-autoradiography and radioimmunosorbent assays. *Journal of Phytopathology,* 138:233-243.

Simmons, C.R., Grant, S., Altier, D.J., Dowd, P.F., Crasta, O., Folkerts, O., and Yalpani, N. 2001. Maize *rhm1* resistance to *Bipolaris maydis* is associated with few differences in pathogenesis-related differences in pathogenesis-related and global mRNA profiles. *Molecular Plant-Microbe Interactions,* 14:947-954.

Sivamani, E., Brey, C.N., Talbert, L.E., Young, M.A., Dyer, W.E., Kaniewski, W.K., and Qu, R. 2002. Resistance to wheat streak mosaic virus in transgenic wheat engineered with the viral coat protein gene. *Transgenic Resistance,* 11:31-41.

Soler, S., Díez, M.J., Roselló, S., and Nuez, F. 1999. Movement and distribution of tomato spotted wilt virus in resistant and susceptible accessions of *Capsicum* spp. *Canadian Journal of Plant Pathology,* 21:317-325.

Somssich, I.E. and Hahlbrock, K. 1998. Pathogen defense in plants—A paradigm of biological complexity. *Trends in Plant Science,* 3:86-90.

Spielmann, A., Douet-Orhant, V., Gugerli, P., and Krastanova, S. 2000. Resistance to nepoviruses in grapevine and *Nicotiana benthamiana:* Expression of several putative resistance genes in transgenic plants. *Acta Horticulturae*, No. 528:373-378.

Spielmann, A., Krastanova, S., Douet-Orhant, V., and Gugerli, P. 2000. Analysis of transgenic grapevine *(Vitis rupestris)* and *Nicotiana benthamiana* plants expressing an arabis mosaic virus coat protein gene. *Plant Science (Limerick)*, 156:235-244.

Sta Cruz, F.C., Koganezawa, H., and Hibino, H. 1993. Comparative cytology of rice tungro viruses in selected rice cultivars. *Journal of Phytopathology*, 138:274-282.

Stark, D.M. and Beachy, R.N. 1989. Protection against potyvirus infection in transgenic plants: Evidence for broad-spectrum resistance. *Bio/Technology*, 7:1257-1262.

Steinlage, T.A., Hill, J.H., and Nutter, F.W., Jr. 2002. Temporal and spatial spread of soybean mosaic virus in soybeans transformed with coat protein gene of soybean mosaic virus. *Phytopathology*, 92:478-486.

Sticher, L., Mauch-Mani, B., and Métraux, J.P. 1997. Systemic acquired resistance. *Annual Review of Phytopathology*, 35:235-270.

Suo, Y. and Leung, D.W.M. 2002. Accumulation of extracellular pathogenesis-related proteins in rose leaves following inoculation of in vitro shoots with *Diplocarpon rosae*. *Scientia Horticulturae*, 93:167-178.

Tabei, Y., Kitade, S., Nishizawa, Y., Kikuchi, N., Kayano, T., Hibi, T., and Akutsu, K.E. 1997. Transgenic cucumber plants harboring a rice chitinase gene exhibit enhanced resistance to grey mold *(Botrytis cinerea)*. *Plant Cell Reports*, 17:159-164.

Takács, P.A., Horváth, J., Kazinczi, G., and Fribek, D. 2000. Reaction of different tobacco cultivars to infection with NTN and O strains of potato Y potyvirus (PVY[NTN], PVY[O]). *Növénytermelés*, 49:347-351.

Tavladoraki, P., Benvenuto, E., Trinca, S., De Martinis, D., Cattaneo A., and Galeffi, P. 1993. Transgenic plants expressing a functional single chain Fv antibody are specifically protected from virus attack. *Nature*, 366:469-472.

Thomas, P.E., Kaniewski, W.K., and Lawson, E.C. 1997. Reduced field spread of potato leaf roll virus in potatoes transformed with potato leaf roll virus coat protein gene. *Plant Disease*, 81:1447-1453.

Thomas, P.E., Lawson, E.C., Zalewski, J.C., Reed, G.L., and Kaniewski, W.K. 2000. Extreme resistance to potato leafroll virus in potato cv. Russet Burbank mediated by viral replicase gene. *Virus Research*, 71:49-62.

Tomassoli, L., Ilardi, V., Barba, M., and Kaniewski, W. 1999. Resistance of transgenic tomato to cucumber mosaic cucumovirus under field conditions. *Molecular Breeding*, 5:121-130.

Truve, E., Aaspôllu, A., Honkanen, J., Puska, R., Mehto, M., Hassi, A., Teeri, T.H., Kelve, M., Seppänen, P., and Saarma, M. 1993. Transgenic potato plants expressing mammalian 2'-5'-oligoadenylate synthetase are protected from potato virus X infection under field conditions. *Bio/Technology*, 11:1048-1052.

Tu, J., Ona, I., Zhang, Q., Mew, T.W., Khush, G.S., and Datta, S.K. 1998. Transgenic rice variety IR 72 with Xa21 is resistant to bacterial blight. *Theoretical and Applied Genetics,* 97:31-36.

Tumer, N.E., O'Connell, K.M., Nelson, R.S., Sanders, P.R., and Beachy, R.N. 1987. Expression of alfalfa mosaic virus coat protein gene confers cross-protection in transgenic tobacco and tomato plants. *EMBO Journal,* 6:1181-1188.

Vaira, A.M., Berio, T., Accotto, G.P., Vecchiati, M., and Allavena, A. 2000. Evaluation of resistance in *Osteospermum ecklonis* (DC) Nort. Plants transgenic for the N protein gene of tomato spotted wilt virus. *Plant Cell Reports,* 19:983-988.

van de Wetering, F., Posthuma, K. Goldbach, R., and Peters, D. 1999. Assessing the susceptibility of chrysanthemum cultivars to tomato spotted wilt virus. *Plant Pathology,* 48:693-699.

Van Dun, C.M. and Bol, J.F. 1988. Transgenic plants accumulating tobacco rattle virus coat protein resist infection with tobacco rattle virus and pea early browning virus. *Virology,* 167:649-652.

Van Dun, C.M., Bol, J.F., and Van-Vloten-Doting, L. 1987. Expression of alfalfa mosaic virus and tobacco rattle virus protein genes in transgenic tobacco. *Virology,* 159:299-305.

Van Dun, C.M., Overdurin, B., Van Vloten-Doting L., and Bol. J.F. 1988. Transgenic tobacco expressing tobacco streak virus or mutated alfalfa mosaic virus coat protein does not cross-protect against alfalfa mosaic virus infection. *Virology,* 164:383-389.

Van Loon, L.C. 1999. Occurrence and properties of plant pathogenesis-related proteins. In *Pathogenesis-Related Proteins in Plants,* (eds.) S.K. Datta and S. Muthukrishnan, pp. 1- 19. CRC Press, Boca Raton, Florida.

Van Loon, L.C. and Antoniw, J.F. 1982. Comparison of the effects of salicylic acid and etephon with virus-induced hypersensitivity and acquired resistance in tobacco. *Netherlands Journal of Plant Pathology,* 88:237.

Van Loon, L.C. and Van Kammen, A. 1970. Polyacrylamide disc electrophoresis of the soluble leaf proteins from *Nicotiana tabacum* var. Samsun and Samsun NN. II. Changes in protein constitution after infection with tobacco mosaic virus. *Virology,* 40:199.

Van Pelt-Heerschap, H. and Smit-Bakker, O. 1999. Analysis of defense-related proteins in stem tissue of carnation inoculated with a virulent and avirulent race of *Fusarium oxysporum* f. sp. *dianthi. European Journal of Plant Pathology,* 105: 681-691.

Vardi, E., Sela, I., Edelbaum, O., Livneh, O., Kuznetsova, L., and Strain, Y. 1993. Plants transformed with a cistron of a potato virus Y protease (NIa) are resistant to virus infection. *Proceedings of National Academy of Sciences, USA,* 90:7513-7517.

Veit, S., Wörle, J.M., Nürenberger, T., Koch, W., and Seitz, H.V. 2001. A novel protein elicitor (PaNie) from *Pythium aphanidermatum* induces multiple defense responses in carrot, *Arabidopsis* and tobacco. *Plant Physiology,* 127:832-841.

Viswanathan, R. 1997. Detection of phytoplasmas associated with grassy shoot disease of sugarcane by ELISA techniques. *Journal of Plant Diseases and Protection,* 104:9-16.

Viswanathan, R., Padmanaban, P., Mohanraj, D., and Jothi, R. 2000. Indirect-ELISA technique for the detection of the red rot pathogen in sugarcane (*Saccharum* spp. Hybrid) and resistance screening. *Indian Journal of Agricultural Sciences,* 70:308-311.

Wang, H., Li, J., Bostock, R.M., and Gilchrist, D.G. 1996. Apoptosis: A functional paradigm for programmed cell death induced by a host-selective phytotoxin and invoked during development. *Plant Cell,* 8:375-391.

Wang, M.B., Abbott, D.C., and Waterhouse, P.M. 2001. A single copy of a virus-derived transgene encoding hairpin RNA gene gives immunity to barley yellow dwarf virus. *Molecular Plant Pathology,* 1:347-356.

Wang, Y.Y., Gardner, R.C., and Pearson, M.N. 1997. Resistance to vanilla necrosis potyvirus in transgenic *Nicotiana benthamiana* plants containing the virus coat protein gene. *Journal of Phytopathology,* 145:7-15.

Ward, E., Uknes, S., and Ryals, J. 1994. Molecular biology and genetic engineering to improve plant disease resistance. In *Molecular Biology in Crop Protection,* (eds.) G. Marshall and D. Walters, pp. 121-145. Chapman and Hall, London.

Ward, E.R., Uknes, S.J., Williams, S.C., Dincher, S.S., Wiederhold, D.L., Alexander, A., Ahl-Goy, P., Métraux, J.P., and Ryals, J.A. 1991. Coordinate gene activity in response to agents that induce systemic acquired resistance. *Plant Cell,* 3:1085-1094.

Weber, H., Schultze, S., and Pfitzner, A.J.P. 1994. Two amino acid substitutions in the tomato mosaic virus 30-kDa movement protein counter the ability to overcome the Tm-2^2 resistance gene in tomato. *Journal of Virology,* 67:6432-6438.

Xiao, X.W., Chu, P.W.G., Frenkel, M.J., Tabe, L.M., Shukla, D.D., Hanna, P.J., Higgins, T.J.V., Müller, W.J., and Ward, C.W. 2000. Antibody-mediated improved resistance to ClYVV and PVY infections in transgenic tobacco plants expressing a single-chain variable region antibody. *Molecular Breeding,* 6:421-431.

Xue, B., Ling, K.S., Reid, C.L., Krastanova, S., Sekiya, M., Momol, E.A., Süle, S., Mozsar, J., Gonsalves, D., and Burr, T.J. 1999. Transformation of five grape-stocks with plant virus genes and a *virE2* gene from *Agrobacterium tumefaciens. In Vitro Cellular and Developmental Biology-Plant,* 35:226-231.

Yalpani, N., Silverman, P.T., Michael, A.W., Kleier, D.A., and Rastin, I. 1991. Salicylic acid is a systemic signal and an inducer of pathogenesis-related proteins in virus-infected tobacco. *Plant Cell,* 3:809-818.

Yeh, S.D., Bau, H.J., Cheng, Y.H., Yu, T.A., and Yang, J.S. 1998. Greenhouse and field evaluations of coat protein transgenic papaya resistant to papaya ring spot virus. *Acta Horticulturae,* 461:321-328.

Yoshikawa, N., Oogake, S., Terada, M., Miyabayashi, S., Ikeda, Y., Takahashi, T., and Ogawa, K. 1999. Apple chlorotic leaf spot virus 50 kDa protein is targeted to plasmodesmata and accumulates in sieve elements in transgenic plant leaves. *Archives of Virology,* 144:2475-2483.

Yuan, H., Ming, X., Wang, L., Hu, P., An, C., and Chen, Z. 2002. Expression of a gene encoding trichosanthin in transgenic rice plants enhances resistance to fungus blast disease. *Plant Cell Reports,* 20:992-998.

Zhang, H.L., Li, T.R., Hsai, A., Er, D., Zhang, T., Sun, Y., and Pang, R.J. 1995. Resistance of transgenic potato cultivars to potato leaf roll. *Chinese Journal of Virology,* 11:342-350.

Zhu, C.X., Song, Y.Z., Zhang, S., Guo, X.Q., and Wen, F.J. 2001. Production of transgenic Chinese cabbage by transformation with the CP gene of turnip mosaic virus. *Acta Phytopathologica Sinica,* 31:257-264.

Chapter 12

Ahl, P. and Gianinazzi, S. 1982. *b*-Protein as a constitutive component in highly (TMV) resistant interspecific hybrids of *Nicotiana glutinosa* × *N. debneyi. Plant Science Letters,* 26: 173-181.

Alström, S. 1991. Induction of disease resistance in common bean susceptible to haloblight bacterial pathogen after seed bacterization with rhizosophere *Pseudomonas. Journal of General and Applied Microbiology,* 37:495-501.

Anfoka, G. and Buchenauer, H. 1997. Systemic acquired resistance in tomato against *Phytophthora infestans* by preinoculation with tobacco necrosis virus. *Physiological and Molecular Plant Pathology,* 50:85-101.

Ayabe, M. and Sumi, S. 2001. A novel and efficient tissue culture method—"stem-disc dome culture"—for producing virus-free garlic (*Allium sativum* L.). *Plant Cell Reports,* 20:503-507.

Benhamou, N., Lafontaine, P.J., and Nicole, M. 1994. Induction of systemic resistance to *Fusarium* crown and root rot in tomato plants by seed treatment with chitosan. *Phytopathology,* 84:1432-1444.

Buysens, S., Heungens, K., Poppe, J., and Höfte, M. 1996. Involvement of pyochelin and pyoverdin in suppression of Pythium-induced damping off of tomato by *Pseudomonas aeruginosa* 7NSK2. *Applied Environmental Microbiology,* 62:865-871.

Chakraborty, M.K. and Ahlawat, Y.S. 2001. Indexing of cross-protected and non-cross-protected citrus cultivars/species for citrus tristeza virus by enzyme-linked immunosorbent assay. *Plant Disease Research,* 16:10-16.

Chatenet, M., Delage, C., Ripolles, M., Irey, M., Lockhart, B.E.L., and Rott, P. 2001. Detection of sugarcane yellow leaf virus in quarantine and production of virus-free sugarcane by apical meristem culture. *Plant Disease,* 85:1177-1180.

Chen, C.C. and Chang, C.A. 1999. The improvement for the detection of tuberose mild mosaic potyvirus in tuberose bulbs by temperature treatment and direct tissue blotting. *Plant Pathology Bulletin,* 8:83-88.

Chen, W.P., Chen, P.D., Liu, D.J., Kynast, R., Friebe, B., Velazhahan, R., Muthukrishnan, S., and Gill, B.S. 1999. Development of wheat scab symptoms is delayed in transgenic wheat plants that constitutively express a rice thaumatin-like protein gene. *Theoretical and Applied Genetics,* 99:755-760.

Chen, Z.C., White, R.F., Antoniw, J.F., and Lin, Q. 1991. Effect of pokeweed antiviral protein (PAP) on the infection of plant viruses. *Plant Pathology,* 40:612-620.

Chester, K.S. 1933. The problem of acquired physiological immunity in plants. *Quarterly Review of Biology,* 8:275-324.

Chiari, A. and Bridgen, M.P. 2002. Meristem culture and virus eradication in *Alstroemeria. Plant Cell, Tissue and Organ Culture,* 68:49-55.

Christ, U. and Mösinger, E. 1989. Pathogenesis-related proteins of tomato: I. Induction by *Phytophthora infestans* and other biotic inducers and correlations with resistance. *Physiological and Molecular Plant Pathology,* 35:53-65.

Cook, R.J. and Baker, K.F. 1983. *The Nature and Practice of Biological Control of Plant Pathogens.* The American Phytopathological Society, St. Paul, Minnesota.

Dann, E.K., Meuwly, P., Métraux, J.P., and Deverall, B.J. 1996. The effect of pathogen inoculation or chemical treatment on activities of chitinase and β-1,3-glucananse and accumulation of salicylic acid in leaves of green bean, *Phaseolus vulgaris* L. *Physiological and Molecular Plant Pathology,* 49:307-319.

Datta, K., Mutukrishnan, S., and Datta, S.K. 1999. Expression and function of PR-protein genes in trangenic plants in transgenic plants. In *Pathogenesis-Related Proteins in Plants,* (eds.) S.K. Datta and S. Muthukrishnan, pp. 261-277. CRC Press, Boca Raton, Florida.

De Cal, A., Pascual, S., and Melgarejo, P. 1997. Involvement of resistance induction by *Penicillium oxalicum* in the biocontrol of tomato wilt. *Plant Pathology,* 46:72-79.

Dempsey, D.A., Wobbe, K.K., and Klessig, D.F. 1993. Resistance and susceptible responses of *Arabidopsis thaliana* to turnip crinkle virus. *Phytopathology,* 83: 1021-1029.

El-Ghaouth, A., Arul, J., Grenier, J., Benhamou, N., Asselin, A., and Bélanger, R. 1994. Effect of chitosan on cucumber plant: Suppression of *Pythium aphanidermatum* and induction of defense reactions. *Phytopathology,* 84:313-320.

Enkerli, J., Gist, Y., and Mösinger, E. 1993. Systemic acquired resistance to *Phytophthora infestans* in tomato and the role of pathogenesis-related proteins. *Physiological and Molecular Plant Pathology,* 43:161-171.

Fitch, M.M.M., Lehrer, A.T., Konor, E., and Moore, P.H. 2001. Elimination of sugarcane yellow leaf virus from infected sugarcane plants by meristem tip culture visualized by tissue blot immunoassay. *Plant Pathology,* 50:676-680.

Gal- On, A. 2000. A point mutation in the FRNK motif of the potyvirus helper component protease gene alters symptom expression in cucurbits and elicits protection against the severe homologous virus. *Phytopathology,* 90:467-473.

Gianinazzi, S. 1983. Genetic and molecular aspects of resistance induced by infection and chemicals. In *Plant-Microbe Interactions: Molecular and Genetic Perspectives,* (eds.) E.W. Wester and T. Kosuge, pp. 321-342. MacMillan, New York.

Gottstein, H.D. and Kuæ, J. 1989. Induction of systemic resistance to anthracnose in cucumber by phosphates. *Phytopathology,* 79:176-179.

Hernandez, P.R., Prado, F.O., Pichardo, M.T., and Noa, C.J.C. 1995. Use of alternation of temperature for sanitation against potato virus X. *Centro Agricola,* 22: 68-71.

Hoffland, E., Hakulinen, J., and Pett, J.A. 1996. Comparison of systemic resistance induced by avirulent and nonpathogenic *Pseudomonas* species. *Phytopathology,* 86:757-762.

Jadãq, A.S., Pavan, M.A., da Silva, N., and Zerbini, F.M. 2002. Seed transmission of lettuce mosaic virus (LMV) pathotypes II and IV in different lettuce genotypes. *Summa Phytopathologica,* 28:58-61.

Kandan, A., Radja Commare, R., Nandakumar, R., Ramiah, M., Raguchander, T., and Samiyappan, R. 2002. Induction of phenylpropanoid metabolism by *Pseudomonas fluroescens* against tomato spotted wilt virus in tomato. *Folia Microbiologia,* 47:121-129.

Kim, D.H. and Lee, J.M. 2000. Seed treatment for cucumber green mottle mosaic virus (CGMMV) in gourd *Lagenaria siceraria* seeds and its detection. *Journal of Korean Society for Horticultural Science,* 41:1-6.

Kim, H.S., Jeon, J.H., Choi, K.H., Young, Y.H., Park, S.W., and Joung, H. 1996. Eradication of PVS (potato virus S) by thermo-and chemotherapy in potato tissue. *Journal of Korean Society of Horticultural Science,* 37:533-536.

Kuæ, J. 1987. Plant immunization and its applicability for disease control. In *Innovative Approaches to Plant Disease Control,* (ed.) L. Chet, pp. 255-274, John Wiley, New York.

Kuæ, J. 1990. Immunization for the control of plant disease. In *Biological Control of Soil-borne Pathogens,* (ed.) D. Hornby, pp. 355-373. CAB International, Wallingford, United Kingdom

Kuæ, J. 1991. Plant immunization—A non-pesticide control of plant disease. *Petria,* 1:79-83.

Latunde-Dada, A.O. and Lucas, J.A. 2001. The plant defence activator acibenzolar-S-methyl primes cowpea [*Vigna unguiculata* (L.) Walp] seedlings for rapid induction of resistance. *Physiological and Molecular Plant Pathology,* 58:199-208.

Leeman, M., Van Pelt, J.A., den Ouden, F.M., Heinsbroek, M., Bakker, P.A.H.M., and Schippers, B. 1995. Induction of systemic resistance against *Fusarium* wilt of radish by lipopolysaccharides of *Pseudomonas fluorescens. Phytopathology,* 85:1021-1027.

Lim, S.T., Wong, S.M., Yeong, C.Y., Lee, S.C., and Goh, C.J. 1993. Rapid dectection of Cymbidium mosaic virus by the polymerase chain reaction (PCR). *Journal of Virological Methods,* 41:37-46.

Lin, Y.J., Rundell, P.A., and Powell, C.A. 2002. In situ immunoassay (ISIA) of field grapefruit trees inoculated with mild isolates of citrus tristeza virus indicates mixed infection with severe isolates. *Plant Disease,* 86:458-461.

Liu, L., Kloepper, J.W., and Tuzun, S. 1995a. Induction of systemic resistance in cucumber against *Fusarium* wilt by plant growth-promoting rhizobacteria. *Phytopathology,* 75:695-698.

Liu, L., Kloepper, J.W., and Tuzun, S. 1995b. Induction of systemic resistance in cucumber by plant growth promoting rhizobacteria: Duration of protection and effect of host resistance on protection and root colonization. *Phytopathology,* 85:1064-1068.

Lodge, J.K., Kaniewski, W.K., and Turner, N.E. 1993. Broad spectrum virus resistance in transgenic plants expressing pokeweed antiviral proteins. *Proceedings of National Academy of Sciences, USA,* 90:7089-7093.

Maekawa, A., Umenito, Y., Yamashita, H., and Yamashita, H. 1995. Elimination of viruses by meristem culture. 3. Elimination of grapevine fan leaf virus. *Research Bulletin of Plant Protection Service, Japan,* 31:113-116.

Mauch-Mani, B. and Slusarenko, A.J. 1996. Production of salicylic acid precursors is a major function of phenyl alanine-ammonia lyase in the resistance of Arabidopsis to *Peronospora parasitica. Plant Cell,* 8:203-213.

Maurhofer, M., Hase, C., Meuwly, P., Métraux, J.P., and Défago, G. 1994. Induction of systemic resistance in tobacco to tobacco necrosis virus by root-colonizing *Pseudomonas fluorescens* CHAO: Influence of the *gacA* gene and of pyoverdine production. *Phytopathology,* 84:139-146.

Métraux, J.P., Ahl-Goy, P., Staub, T., Speich, J., Steinemann, A., Ryals, J., and Ward, E. 1991. Induced resistance in cucumber in response to 2,6-dichloroisonicotinic acid and pathogens. In *Advances in Molecular Genetics of Plant-Microbe Interactions,* (eds.) H. Hennecke and D.P.S. Verma, Vol.1, pp. 432-439. Kluwer Academic Publishers, Dordrecht, The Netherlands.

Métraux, J.P., Strait, L., and Staub, Th. 1988. A pathogenesis-related protein in cucumber is a chitinase. *Physiological and Molecular Plant Pathology,* 33:1-9.

Meyer, G.D. and Höfte, M. 1997. Salicylic acid produced by the rhizobacterium *Pseudomonas aeruginosa* 7NSK2 induces resistance to leaf infection by *Botrytis cinerea* on bean. *Phytopathology,* 87:588-593.

Muthulakshmi, P. 1997. Studies on rice tungro disease with special reference to molecular basis of induced resistance. Doctoral thesis. Tamil Nadu Agricultural University, Coimbatore, India.

Muthulakshmi, P. and Narayanasamy, P. 1996. Serological evidence for induction of resistance to rice tungro viruses in rice, using antiviral principles. *International Rice Research Notes,* 21:(2-3):77.

Muthulakshmi, P. and Narayanasamy, P. 2000. Induction of PR-proteins in Co43 rice plants following antiviral principles application and inoculation with rice tungro virus. *Madras Agricultural Journal,* 87:398-400.

Narayanasamy, P. 1990. Antiviral principles for virus disease management. In *Basic Research for Crop Disease Management,* (ed.) P. Vidhyasekaran, pp. 139-150. Daya Publishing House, New Delhi, India.

Narayanasamy, P. 1995. Induction of disease resistance in rice by studying the molecular biology of diseased plants. Final Report to Department of Biotechnology, Government of India, New Delhi, India.

Narayanasamy, P. 2002. *Microbial Plant Pathogens and Crop Disease Management,* Science Publishers, Enfield, New Hampshire.

Narayanasamy, P. and Ganapathy, T. 1986. Characteristics of antiviral principles effective against tomato spotted wilt virus on groundnut. *Proceedings of Fourteenth International Congress of Microbiology.* Manchester, England.

Narayanasamy, P. and Ramiah, M. 1983. Characterization of antiviral principle from sorghum leaf. In *Management of Diseases of Oilseed Crops,* (ed.) P. Narayanasamy, pp. 15-17. Tamil Nadu Agricultural University, Coimbatore, India.

Naylor, M., Murphy, A.M., Berry, J.O., and Carr, J.P. 1998. Salicylic acid can induce resistance to plant virus movement. *Molecular Plant-Microbe Interactions,* 11:860-868.

Niimi, Y., Han, D.S., and Fujisaki, M. 2001. Production of virus-free plantlets by anther culture of *Lilium* × 'Enchantment'. *Scientia Horticulturae,* 90:325-334.

Oh, S.K., Choi, D., and Yu, S.H. 1998. Development of integrated pest management techniques using biomass for organic farming. I. Suppression of late blight and *Fusarium* wilt of tomato by chitosan involving both antifungal and plant activating activities. *Korean Journal of Plant Pathology,* 14:278-285

Park, S.W., Jeon, J.H., Kim, H.S., and Young, H. 1994. Effects of antiviral chemical on eradication of potato virus S in potato (*Solanum tuberosum* L.) shoot tip culture. *Journal of Korean Society of Horticulture Science,* 35:32-35.

Pennazio, S. and Roggero, P. 1991. Systemic acquired resistance to virus infection and ethylene biosynthesis in asparagus bean. *Journal of Phytopathology,* 131: 177-183.

Pieterse, C.M.J., van Wees, S.C.M., Hoffland, E., Van Pelt, J.A., and Van Loon, L.C. 1996. Systemic resistance in *Arabidopsis* induced by biocontrol bacteria independent of salicylic acid accumulation and pathogenesis-related gene expression. *Plant Cell,* 8:1225-1237.

Porat, R., McCollum, T.G., Vinokur, V., and Droby, S. 2002. Effects of various elicitors on the transcription of a β-1,3-glucanase gene in citrus fruit. *Journal of Phytopathology,* 150:70-75.

Powell, C.A., Pelosi, R.R., Rundell, P.A., Stover, E., and Cohen, M. 1999. Cross-protection of grapefruit from decline-inducing isolates of citrus tristeza virus. *Plant Disease,* 83:989-991.

Price, W.C. 1936. Virus concentration in relation to acquired immunity from tobacco ringspot. *Phytopathology,* 26:503-529.

Quintanilla, P. and Brishammar, S. 1998. Systemic induced resistance to late blight in potato by treatment with salicylic acid and *Phytophthora cryptogea. Potato Research,* 41:135-142.

Raggi, V. 1998. Hydroxy proline-rich glycoprotein accumulation in TMV-infected tobacco showing systemic acquired resistance to powdery mildew. *Journal of Phytopathology,* 146:321-325.

Ramamoorthy, V., Raguchander, T., and Samiyappan, R. 2002. Induction of defense-related proteins in tomato roots treated with *Pseudomonas fluorescens* Pf1 and *Fusarium oxysporum* f. sp. *lycopersici. Plant and Soil,* 239:55-68.

Ramamoorthy, V., Viswanathan, R., Raguchander, T., Prakasam, V., and Samiyappan, R. 2001. Induction of systemic resistance by plant growth promoting rhizobacteria in crop plants against pests and diseases. *Crop Protection,* 20:1-11.

Raupach, G.S. and Kloepper, J.W. 1998. Mixtures of plant growth promoting rhizobacteria enhance biological control of multiple cucumber pathogens. *Phytopathology,* 88:1158-1164.

Raupach, G.S., Liu, L., Murphy, J.F., Tuzun, S., and Kloepper, J.W. 1996. Induced systemic resistance in cucumber and tomato against cucumber mosaic cucumovirus using plant growth-promoting rhizobacteria (PGPR). *Plant Disease,* 80: 891-894.

Reiss, E. and Bryngelsson, T. 1996. Pathogenesis-related proteins in barley leaves, induced by infection with *Drechslera teres* (Sacc.) Shoem and by treatment with other biotic agents. *Physiological and Molecular Plant Pathology,* 49:331-341.

Repka, 2001. Elicitor-stimulated induction of defense mechanism and defense gene activation in grapevine cell suspension cultures. *Biologia Plantarum,* 44:555-565.

Reuveni, M., Agapov, V., and Reuveni, R. 1997. A foliar spray of micronutrient solution induces local and systemic protection against powdery mildew *(Sphaerotheca fuliginea)* in cucumber plants. *European Journal of Plant Pathology,* 103:581-588.

Reuveni, R., Agapov, V., and Reuveni, M. 1994. Foliar spray of phosphates induces growth increase and systemic resistance to *Puccinia sorghi* in maize. *Plant Pathology,* 43:245-250.

Ross, A.F. 1961a. Localized acquired resistance to plant virus infection in hypersensitive hosts. *Virology,* 14:329-339.

Ross, A.F. 1961b. Systemic acquired resistance induced by localized virus infections in plants. *Virology,* 14:340-358.

Ross, A.F. 1966. Systemic effects of local lesion formation. In *Viruses of Plants,* (eds.) A.B.R. Beemster and J. Dijkstra, pp. 127-150. North-Holland, Amsterdam.

Senula, A., Keller, E.R.J., and Leseman, D.E. 2000. Elimination of viruses through meristem culture and chemotherapy for the establishment of an in vitro collection of garlic *(Allium sativum)*. *Acta Horticulturae,* No. 530:121-128.

Sether, D.M., Karasev, A.V., Okumura, C., Arakawa, C., Zee, F., Kislan, M.M., Busto, J.L., and Hu, J.S. 2001. Differentiation, distribution and elimination of two different pineapple mealybug wilt-associated viruses found in pineapple. *Plant Disease,* 85:856-864.

Shanmugam, V., Raguchander, T., Balasubramanian, P., and Samiyappan, R. 2001. Inactivation of *Rhizoctonia solani* toxin by a putative β-glucosidase from coconut leaves for control of sheath blight disease in rice. *World Journal of Microbiology and Biotechnology,* 17:545-553.

Shanmugam, V., Sriram, S., Babu, S., Nandakumar, R., Raguchander, T., Balasubramanian, P., and Samiyappan, R. 2001. Purification and characterization of an extracellular α-glucosidase protein from *Trichoderma viride* which degrades a phytotoxin associated with sheath blight in rice. *Journal of Applied Microbiology,* 90:320-329.

Shelly Parveen, Savarni Tripathi, and Varma, A. 2001. Isolation and characterization of an inducer protein (Crip-31) from *Clerodendrum inerme* leaves responsible for induction of systemic resistance against viruses. *Plant Science,* 161:453-459.

Srivastava, A.K., Tanuja Singh, Jana, T.K., and Arora, D.K. 2001. Induced resistance and control of charcoal rot in *Cicer arietinum* (chickpea) by *Pseudomonas fluorescens. Canadian Journal of Botany,* 79:789-795.

Sticher, L., Mauch-Mani, B., and Métraux, J.P. 1997. Systemic acquired resistance. *Annual Review of Phytopathology,* 35:235-270.

Sundar, A.R., Viswanathan, R., and Padmanabhan, P. 2001. Induction of systemic resistance to *Colletotrichum falcatum* in sugarcane by a synthetic signal molecule, acibenzolar-*S*-methyl (CGA-245704). *Proceedings of Annual Convention of Sugar Technologists Association, India,* pp. 112-134, Jaipur, India.

Suo, Y. and Leung, D.W.M. 2002a. Accumulation of extracellular pathogenesis-related proteins in rose leaves following inoculation of in vitro shoots with *Diplocarpon rosae. Scientia Horticulturae,* 93:167-178.

Suo, Y. and Leung, D.W.M. 2002b. BTH-induced accumulation of extracellular proteins and black spot disease in rose. *Biologia Plantarum,* 45:273-279.

Takács, A. and Dolej, S. 1998. The effect of resistance activator "BION" on the relationship of *Fusarium oxysporum* f. sp. *lycopersici* with tomato. *Nönvényvédelem,* 34:257-259.

Thrane, C., Nielsen, M.N., Sørensen, J., and Olsson, S. 2001. *Pseudomonas fluorescens* DR 54 reduces sclerotia formation, biomass development and disease incidence of *Rhizoctonia solani* causing damping-off in sugarbeet. *Microbial Ecology,* 42:438-445.

Tosi, L., Luigetti, R., and Zazzerini, A. 2000. Occurrence of PR-proteins in sunflower *(Helianthus annuus* L.) treated with some plant activators and inoculated with *Plasmopara helianthi* Novot. *Atti Giornate fitopatologiche, Perugia,* 2: 283-290.

Triolo, E., Panattoni, A., and Mainardi, M. 1999. Antiviral chemotherapy in vitro: Effects of some molecules on Mr. S. 2/5 infected with PNRSV. *Italus Hortus,* 6:98-99.

Tuzun, S. and Kuæ, J. 1991. Plant immunization: An alternative to pesticides for control of plant diseases in greenhouse and field. In *The Biological Control of Plant Diseases,* (ed.) J. Bay Peterson, pp. 30-40. Food and Fertilizer Technology Center, Taipei, Taiwan.

Velazhahan, R., Chen-Cole, K., Anuratha, C.S., and Muthukrishan, S. 1998. Induction of thaumatin-like proteins (TLPs) in *Rhizoctonia solani*-infected rice and characterization of two new cDNA clones. *Physiologia Plantarum,* 102:21-28.

Verma, H.N., Srivastava, S., Varsha, N., and Kumar, D. 1996. Induction of systemic resistance in plants against viruses by a basic protein from *Clerodendrum aculeatum* leaves. *Phytopathology,* 86:485-492.

Vidhyasekaran, P. and Muthamilan, M. 1999. Evaluation of powder formulation of *Pseudomonas fluorescens* Pf1 for control of rice sheath blight. *Biocontrol Science and Technology,* 9:67-74.

Vidhyasekaran, P., Rabindran, R., Muthamilan, M., Nayar, R., Rajappan, K., Subramanian, N., and Vasumathi, K. 1997. Development of powder formulation of *Pseudomonas fluorescens* for control of rice blast. *Plant Pathology,* 46:291-297.

Viswanathan, R. and Samiyappan, R. 2001. Role of chitinases in *Pseudomonas* spp.-induced systemic resistance against *Colletotrichum falcatum* in sugarcane. *Indian Phytopathology,* 54:418-423.

Viswanathan, R. and Samiyappan, R. 2002. Induced systemic resistance by fluorescent pseudomonas against red rot disease of sugarcane caused by *Colletotrichum falcatum. Crop Protection,* 21:1-10.

Walters, D.R. and Murray, D.C. 1992. Induction of systemic resistance to rust in *Vicia faba* by phosphate and EDTA: Effects of calcium. *Plant Pathology,* 41:444-448.

Wang, H.L. and Gonsalves, D. 1999. Utilization of monoclonal and polyclonal antibodies to monitor the protecting and challenging strains of zucchini yellow mosaic virus in cross-protection. *Plant Pathology Bulletin,* 8:111-116.

Wang, S.C., Xu, L.L., Li, G.J., Chen, P.Y., Xia, K., and Zhou, X. 2002. An ELISA for the determination of salicylic acid in plants using a monoclonal antibody. *Plant Science,* 162:529-535.

Ward, E.R., Uknes, S.J., Williams, S.C., Dincher, S.S., Wiederhold, D.L., Alexander, A., Ahl-Goy, P., Métraux, J.P., and Ryals, J.A. 1991. Coordinate gene activity in response to agents that induce systemic acquired resistance. *Plant Cell,* 3:1085-1094.

Wei, L., Kloepper, J.W., and Tuzun, S. 1996. Induced systemic resistance to cucumber diseases and increased plant growth by plant growth-promoting rhizobacteria under field conditions. *Phytopathology,* 86:221-224.

Xie, C. and Kuæ, J. 1997. Induction of resistance to *Peronospora tabacina* in tobacco leaf disks by leaf disks with induced resistance. *Physiological and Molecular Plant Pathology,* 51:279-286.

Xue, L., Charest, P.M., and Jabaji-Hare, S.H. 1998. Systemic induction of peroxidases, 1,3-β-glucanases, chitinases and resistance in bean plants by binucleate *Rhizoctonia* species. *Phytopathology,* 88:359-365.

Yalpani, N., Shulaev, V., and Raskin, I. 1993. Endogenous salicylic acid levels correlate with the accumulation of pathogenesis-related proteins and virus resistance in tobacco. *Phytophathology,* 83:702-708.

Ye, X.S., Pan, S.Q., and Kuæ, J. 1990. Association of pathogenesis-related proteins and activities of peroxidase, β -1,3-glucanase and chitinase with systemic induced resistance to blue mold, but not to systemic tobacco mosaic virus. *Physiological and Molecular Plant Pathology,* 36:523-531.

Zhang, S., Reddy, M.S., Kokalis-Burelle, N., Wells, L.W., Nightingale, S.P., and Kloepper, J.W. 2001. Lack of induced systemic resistance in peanut to late leaf spot disease by plant growth-promoting rhizobacteria and chemical elicitors. *Plant Disease,* 85:879-884.

Zhang, Y.M., Tian, Y.T., and Luo, X.F. 1998. Microapical culture of grapevine and detection of grapevine fanleaf virus by ELISA and probe. *Journal of Beijing Forestry University,* 20:54-58.

Chapter 13

Agarwal, V.K. and Sinclair, J.B. 1996. *Principles of Seed Pathology,* Lewis Publishers, Boca Raton, Florida.

Anderson, H.W., Nehring, E.W., and Wisher, W.R. 1975. Aflatoxin contamination of corn in the field. *Journal of Agricultural and Food Chemistry,* 23:775.

Anonymous. 1996. Food and Agriculture Organization of the United Nations-Basic facts of the world cereal situation. *Food Outlook* 5/6.

Ansari, A.A. and Shrivastava, A.K. 1990. Natural occurrence of *Alternaria* mycotoxins in sorghum and ragi from North Bihar, India. *Food Additives and Contaminants,* 7:815.

Ashworth, Jr. L. J., McMeans, J.L., and Brown, M. 1969. Infection of cotton by *Aspergillus flavus:* Time of infection and influence of fiber moisture. *Phytopathology,* 59:383.

Azcona-Olivera, J.I., Abouzeid, M.M, and Pestka, J.J. 1990. Detection of zearalenone by tandem immuno-affinity-enzyme-linked immunosorbent assay and its application to milk. *Journal of Food Protection,* 53:577.

Azcona-Olivera, J.I., Abouzeid, M.M., Plattner, R.D., Norred, W.P., and Pestka, J.J. 1992. Generation of antibodies reactive with fumonisins B_1, B_2 and B_3 by using chlolera toxin as the carrier adjuvant. *Applied Environmental Microbiology,* 58:169.

Azcona-Olivera, J.I., Abouzeid, M.M., Plattner, R.D., and Pestka, J.J. 1992. Production of monoclonal antibodies to the mycotoxins fumonisins B_1, B_2 and B_3. *Journal of Agricultural and Food Chemistry,* 40:531-534.

Bai, G.H., Desjardins, A.E., and Plattner, R.D. 2002. Deoxynivalenol-non-producing *Fusarium graminearum* causes initial infection but does not cause disease spread in wheat spikes. *Mycopathologia,* 153:91-98.

Barna-Vetró, I., Szabó, E., Fazekas, B., and Solti, L. 1999. Developing a sensitive ELISA for determination of fumonisin B_1 in cereals. *Növényvédelem,* 35:609-615.

Barna-Vetró, I. Szabó, E., Fazekas, B., and Solti, L. 2000. Development of a sensitive ELISA for the determination of fumonisin B_1 in cereals. *Journal of Agricultural and Food Chemistry,* 48:2821-2825.

Bassa, S., Mestres, C., Champiat, D., Hell, K., Vernier, P., and Cardwell, K. 2001. First report of aflatoxin in dried yarm chips in Benin. *Plant Disease,* 85:1032.

Bennett, G.A., Nelsen, T.C., and Miller, B.M. 1994. Enzyme-linked immunosorbent assay for detection of zearalenone in corn, wheat and pig feed: Collaborative study. *Journal of Association of Official Analytical Chemists, International,* 77:1500.

Bhat, R.V. 1991. Aflatoxins: Successes and failures of three decades of research. In *Fungi and Mycotoxins in Stored Products,* (eds.) B.R. Chemp, E. Highley, A.D. Hocking, and J.I. Pitt. Proceedings of International Conference, ACIAR Proceedings No. 36, p.170, Bangkok.

Bhattacharya, D., Bhattacharya, R., and Dhar, T.K. 1999. A novel signal amplification technology for ELISA based on catalyzed reporter deposition: Demonstration of its applicability for measuring aflatoxin B_1. *Journal of Immunological Methods,* 230:71-86.

Bilgrami, K.S. and Choudhary, A.K. 1998. Mycotoxins in preharvest contamination of agricultural crops. In *Mycotoxins in Agriculture and Food Safety,* (eds.) K.K. Sinha and D. Bhatnagar, pp. 1-43. Marcel Dekker, New York.

Bilgrami, K.S., Choudhary, A.K., and Ranjan, K.S. 1992. Aflatoxin contamination in field mustard *(Brassica juncea)* cultivars. *Mycotoxin Research,* 8:21.

Busby, W.F. and Wogan, G.N. 1984. Aflatoxins. In *Chemical Carcinogens,* (ed.) C.E. Searle, p. 945. *American Chemical Society,* Washington, DC.

Candlish, A.A.G., Aidoo, K.E., Smith, J.E., and Pearson, S.M. 2000. A limited survey of aflatoxins and fumonisins in retail maize-based products in UK using immunoassay detection. *Mycotoxin Research,* 16:2-8.

Casale, W.L., Pestka, J.J., and Hart, P. 1988. Enzyme-linked immunosorbent assay employing monoclonal antibody specific for deoxinivalenol (vomitoxin) and several analogues. *Journal of Agricultural and Food Chemistry,* 36:663.

Chen, L.M. and Chen, Y.X. 1998a. Direct competitive ELISA screening method for aflatoxin B_1. *Journal of Nanjing Agricultural University,* 21:62-65.

Chen, L.M. and Chen, Y.X. 1998b. Indirect competitive ELISA for quantitative analysis of aflatoxin B_1 and its application. *Journal of Nanjing Agricultural University,* 21:65-72.

Christensen, H.R., Yu, F.Y., and Chiu, F.S. 2000. Development of a polyclonal antibody-based sensitive enzyme-linked immunosorbent assay for fumonisinB$_4$. *Journal of Agricultural and Food Chemistry*, 48:1977-1984.

Chu, F.S. 1988. Immunoassay for mycotoxins. In *Modern Methods in the Analysis and Structural Elucidation of Mycotoxins,* (ed.) R.J. Cole, pp. 207-237. Academic Press, New York.

Chu, F.S. 1991. Current immunochemical methods for mycotoxin analysis. In *Immunoassays for Trace Chemical Analysis,* (eds.) M. Vanderlaan, L.H. Stankers, B.E. Watkins, and D.W. Roberts. pp. 140-157. American Chemical Society, Washington, DC.

Chu, F.S. 1992. Recent progress on analytical techniques for mycotoxins in feed stuffs. *Journal of Animal Science*, 70:3950.

Chu, F.S. 1994. Development of antibodies against aflatoxins. In *Toxicology of Aflatoxins: Human Health, Veterinary and Agricultural Significance,* (eds.) D.L. Eaton and J.D. Groopman, p. 451. Academic Press, San Diego, California.

Ciegler, A., Lillehoj, E.B., Peterson, R.E., and Hall, H.H. 1966. Microbial detoxification of aflatoxin. *Applied Microbiology*, 14:934.

Coker, R.D., Jones, B.D., and Nagler, M.J. 1984. *Mycotoxins Training Manual.* Tropical Development Research Institute, London.

Cole, R.J., Kirksey, J.W., and Blankenship, B.R. 1972. Conversion of aflatoxin B$_1$ to isomeric hydroxy compounds by *Rhizopus* spp. *Journal of Agricultural and Food Chemistry*, 20:1100.

de Côrtes, N.A., Cassetari Neto, D., and Correa, B. 2000. Occurrence of aflatoxins in maize produced by traditional cultivation methods in agricultural communities in the state of Marto Grosso. *Higiene Alimentar*, 14:16-26.

de Saeger, S. and van Peteghem, C. 1999. Flow-through membrane-based enzyme immunoassay for rapid detection of ochratoxin A in wheat. *Journal of Food Protection*, 62:65-69.

Diener, U.L., Cole, R.J., Sanders, T.H., Payne, G.A., Lee, L.S., and Klich, M.A. 1987. Epidemiology of aflatoxin formation by *Aspergillus flavus. Annual Review of Phytopathology*, 25:249.

Doyle, M.P., Applebaum, R.S., Bracket, R.E., and Marth, E.H. 1982. Physical, chemical and biological degradation of mycotoxins in foods and agricultural commodities. *Journal of Food Protection*, 45:964.

Dunlap, J.R. 1995. Indentifying heat tolerant sources of corn germplasm with reduced susceptibility to aflatoxin contamination. *Proceedings of the USDA—ARS Aflatoxin Elimination Workshop*, p. 12. Atlanta, Georgia.

El-Nakib, O., Pestka, J.J., and Chu, S.S. 1981. Determination of aflatoxin B$_1$ in corn, wheat and peanut butter by enzyme-linked immunosorbent assay and solid phase radioimmunoassay. *Journal of Association of Analytical Chemists*, 64: 1077.

Fajardo, J.E., Dexter, J.E., Roscoe, M.M., and Norwiki, T.W. 1995. Retention of ergot alkaloids in wheat during processing. *Cereal Chemistry*, 72:291-298.

Fremy, J.M. and Chu, F.S. 1989. Immunochemical methods of analysis for aflatoxin M₁. In *Mycotoxin in Dairy Products,* (ed.) H.P. van Egmond, p. 97. Elsevier, London.

Fukal, L., Prosek, J., Rauch, P., Sova, Z., and Kas, J. 1987. Selection of the separation step in the radioimmunoassay for aflatoxin B₁ using ¹²⁵I as marker. *Journal of Radioanalytical and Nuclear Chemistry Articles,* 109:383.

Fuller, J.C. 1968. *The Day of St. Anthony's Fire.* MacMillan, New York.

Gelderblom, W.C.A., Jasiewicz, K., Marasas, W.F.O., Thiel, P.G., Horak, R.M, Vleggaar, R., and Kriek, N.P.J. 1988. Fumonisins—Novel mycotoxins with cancer-promoting activity produced by *Fusarium moniliforme. Applied Environmental Microbiology,* 45:1806.

Gendloff, E.H., Pestka, J.J., Swanson, S.P., and Hart, L.P. 1984. Detection of T-2 toxin in *Fusarium sporotrichioides*-infested corn by enzyme-linked immunosorbent assay. *Applied and Environmental Microbiology,* 47:1161.

Hart, L.P., Casper, H., Schabenberger, O., and Ng, P. 1998. Comparison of gas chromatography-electron capture and enzyme-linked immunosorbent assay for deoxynivalenol in milled fractions of naturally contaminated wheat. *Journal of Food Protection,* 61:1695-1697.

Holbrook, C.C., Wilson, D.M., and Matheron, M.E. 1995. An update on breeding peanut for resistance to preharvest aflatoxin contamination. *Proceedings of the USDA—ARS Aflatoxins Elimination Workshop,* p. 3. Atlanta, GA.

Hu, W.J., Woychik, N., and Chu, F.S. 1983. ELISA of picogram quantities of aflatoxin M₁ in urine and milk. *Journal of Food Protection,* 47:126.

Ichinoe, M. and Kurata, H. 1983. Trichothecene-producing fungi. In *Developments in Food Science,* Volume IV, *Trichothecenes: Chemical, Biological and Toxicological Aspects,* (ed.) Y. Ueno, p. 73. Elsevier, Amsterdam.

Jewers, K. and John, A.E. 1990. *Alternaria* mycotoxin-possible contaminants of few sorghum varieties. *Tropical Science,* 30:397.

Kang, Z. and Buchenauer, H. 1999. Immunocytochemical localization of fusarium toxins in infected wheat spikes by *Fusarium culmorum. Physiological and Molecular Plant Pathology,* 55:275-288.

Kang, Z. and Buchenauer, H. 2000a. Ultrastructural and immunocytochemical investigation of pathogen development and host responses in resistant and susceptible wheat spikes by *Fusarium culmorum. Physiological and Molecular Plant Pathology,* 57:255-268.

Kang, Z. and Buchenauer, H. 2000b. Ultrastructural and cytochemical studies on cellulose-xylan and pectin degradation in wheat spikes infected by *Fusarium culmorum. Journal of Phytopathology,* 148:263-275.

Kang, Z.S., Huang, L.L., Krieg, U., Mauler-Machink, A., and Buchenauer, H. 2001. Effects of tebuconazole on morphology, structure, cell wall components and trichothecene production by *Fusarium culmorum* in vitro. *Pest Management Science,* 57:491-500.

Kemp, H.A., Mills, E.N.C., and Morgan, M.R.A. 1986. Enzyme-linked immuno-sorbent assay of 3-acetyldeoxynivalenol applied to rice. *Journal of Food and Agricultural Science,* 37:888.

Krogh, P. 1987. *Mycotoxins in Food.* Academic Press, New York.

Kulisek, E.S. and Hazebroek, J.P. 2000. Comparison of extraction buffers for the detection of fumonisinB$_1$ in corn by immunoassay and high performance liquid chromatography. *Journal of Agricultural and Food Chemistry,* 48:65-69.

Landsteiner, K. 1945. *The Specificity of Serological Reactions.* Harvard University Press, Boston.

Lee, L.S., Lee, Jr. L.V., and Russel, M. 1986. Aflatoxin in Arizona cotton seed: Field inoculation of bolls by *Aspergillus flavus* spores in wind-driven soil. *Journal of American Oil Chemical Society,* 63:530.

Lee, S. and Chu, F.S. 1981a. Enzyme-linked immunosorbent assay of ochratoxin A in wheat. *Journal of Association of Official Analytical Chemists,* 67:45.

Lee, S. and Chu, F.S. 1981b. Radioimmunoassay of T-2 toxin in corn and wheat. *Journal of Association of Official Analytical Chemists,* 64:156.

Lew, H., Alder, A., and Edinger, W. 1991. Moniliformin and the European corn borer *(Ostrinia nubilalis). Mycotoxin Research,* 7:71.

Li, F.Q., Luo, X.Y., and Yoshizawa, T. 1999. Mycotoxins (trichothecenes, zearalenone and fumonisins) in cereals associated with human red mold intoxications stored since 1989 and 1991 in China. *Natural Toxins,* 7:93-97.

Li, F.Q., Yoshizawa, T., Kawamura, O., Luo, X.Y., and Li, Y.W. 2001. Aflatoxins and fuminotoxins in corn from the high incidence area of human hepatocellular carcinoma in Guangxi, China. *Journal of Agriculture and Food Chemistry,* 49:4122-4126.

Lindsey, D.L. 1970. Effect of *Aspergillus flavus* on peanuts grown under gnotobiotic conditions. *Phytopathology,* 60:208.

Lopéz-García, R. and Park, D.L. 1998. Effectiveness of postharvest procedures in management of mycotoxin hazards. In *Mycotoxins in Agriculture and Food Safety,* (eds.) K.K. Sinha and D. Bhatnagar, pp. 407-430. Marcel Dekker, New York.

Makarananda, K. and Neal, G.E. 1992. Competitive ELISA. In *Immunochemical Protocols,* (ed.) M.M. Manson, pp. 267-272. Humana Press, Totowa, New Jersey.

Maragos, C.M. and Thompson, V.S. 1999. Fibre-optic immunosensor for mycotoxins. *Natural Toxins,* 7:371-376.

Marquardt, R. and Frohlich, A. 1990. Ochratoxin A: An important western Canadian storage mycotoxin. *Canadian Journal of Physiology and Pharmacology,* 68:991.

Mehan, V.K., Nageswara Rao, R.C., McDonald, D., and Williams, J.H. 1988. Management of drought stress to improve field screening of peanuts for resistance to *Aspergillus flavus. Phytopathology,* 78:659-663.

Morgan, M.R.A. 1989. Mycotoxin immunoassay with special reference to ELISA. *Tetrahedron,* 45:2237.

Morgan, M.R.A., McNerney, R., and Chan, H.W.S. 1983. Enzyme-linked immunosorbent assay of ochratoxin A in barley. *Journal of Association of Official Analytical Chemists,* 66:472.

Ngoko, Z., Marasas, W.F.O., Rheeder, J.P., Shephard, G.S., Wingfield, M.J., and Cardwell, K.F. 2001. Fungal infection and mycotoxin contamination of maize in the humid forest and western highlands of Cameroon. *Phytoparasitica,* 29:352-360.

Ono, E.Y.S., Kawamura, O., Ono, M.A., Ueno, Y., and Hiroka, E.Y. 2000. A comparative study of indirect competitive ELISA and HPLC for fumonisins detection in corn of the state of Parana, Brazil. *Food and Agricultural Immunology,* 12:5-14.

Ono, E.Y.S., Ono, M.A., Funo, F.Y., Medina, A.E., Oliveira, T.R.C.M., Kawamura, O., Ueno, Y., and Hirooka, E.Y. 2001. Evalution of fuminosin-aflatoxin co-occurrence in Brazilian corn hybrids by ELISA. *Food Additives and Contaminants,* 18:719-729.

Pascale, M., Visconti, A., Avantaggiato, G., Prończuk, M., and Chelkowski, J. 1999. Mycotoxin contamination of maize hybrids after infection with *Fusarium proliferatum. Journal of the Science of Food and Agriculture,* 79:2094-2098.

Perkowski, J., Meidaner, T., Geiger, H.H., Muller, H.M., and Chelkowski, J. 1995. Occurrence of deoxynivalenol (DON), 3-acetyl-DON. Zearalenone and ergosterol in winter rye inoculated with *Fusarium culmorum. Cereal Chemistry,* 72:205.

Pestka, J.J., Azcona-Olivera, J.I., Plattner, R.D., Minervini, F., Doko, M.B., and Visconti, A. 1994. Comparative assessment of fumonisin in grain-based foods by ELISA, GC-MS and HPLC. *Journal of Food Protection,* 57:109.

Pestka, J.J., Lee, S.S., Lau, H.P., and Chu, F.S. 1981. Enzyme-immunosorbent assay for T-2 toxin. *Journal of American Oil Chemists Society,* 58:940 A.

Pichler, H., Krska, R., Székács, A., and Grasserbauer, M. 1998. An enzyme-immunoassay for the detection of the mycotoxin zearalenone by use of yolk antibodies. *Freseniuœ Journal of Analytical Chemistry,* 362:176-177.

Pitt, J.I. 1991. After 30 years: Real mycotoxins, real mycotoxicoses. *Australian Mycotoxin Newsletter,* 2:1.

Pitt, J.I. 1992. After 30 years, real mycotoxins of lesser importance. *Australian Mycotoxin Newsletter,* 3:2

Plattner, R.D., Neffed, W.P., Bacon, C.W., Voss, K.A., Peterson, R., Shackelford, D.D., and Weisleder, D. 1990. A method of detection of fumonisins in corn sample associated with field cases of equine leukoencephalomalacia. *Mycologia,* 82:898.

Proctor, R.H., Desjardins, A.E., Plattner, R.D., and Hohn, T.M. 1999. A polyketide synthase gene required for biosynthesis of fumonisin mycotoxins in *Gibberella fujikuroi* mating population A. *Fungal Genetics and Biology,* 27:100-112.

Putnam, M.L. and Binkerd, K.A. 1992. Comparison of a commercial ELISA kit and TLC for detection of deoxynivalenol in wheat. *Plant Disease,* 76:1078.

Radová, H., Hajšlová, J., Karálová, J., Popušková, L., and Sýkorová, S. 2001. Analysis of zearalenone in wheat using high performance liquid chromatography with fluorescence detection and/or enzyme-linked immunosorbent assay. *Cereal Research Communications,* 29:435-442.

Rauch, P., Fukal, L., Prosek, J., Bresina, P., and Kos, J. 1987. Radioimmunoassay of aflatoxin M_1. *Journal of Radioanalytical and Nuclear Chemistry Letters,* 117: 163.

Reddy, S.V., Mayi, D.R., Reddy, M.V., Thirumala Devi, R., and Reddy, D.V.R. 2001. Aflatoxin B_1 in different grades of chillies *(Capsicum annuum)* in India as determined by indirect competitive ELISA. *Food Additives and Contaminants,* 18:553-558.

Rousseau, D.M., Slegers, G.A., and van Peteghem, C.H. 1985. Radioimmunoassay of ochratoxin A in barley. *Applied and Environmental Microbiology,* 50:529.

Ruprich, J., Piskac, A., and Mala, J. 1988. Using the radioimmunological assays of the ochratoxin A for screening control of cereal type foods and feeds of plant origin. *Veterinary Medicine, Praha,* 33:165.

Sauer, D.B., Seitz, L.M., and Burrough, R., et al. 1978. Toxicity of *Alternaria* metabolites found in weathered sorghum grain at harvest. *Journal of Agriculture and Food Chemistry,* 26:1380.

Savard, M.E., Sinha, R.C., Seaman, W.L., and Fedak, G. 2000. Sequential distribution of the mycotoxin deoxynivalenol in wheat spikes after inoculation with *Fusarium graminearum. Canadian Journal of Plant Pathology,* 22:280-285.

Schurbring, S.L. and Chu, F.S. 1987. An indirect enzyme-linked immunosorbent assay for the detection of diacetoxyscirpenol in wheat and corn. *Mycotoxin Research,* 3:97.

Scott, P.M. 1989. The natural occurrence of tricothecenes. In *Trichothecene Mycotoxicoses, Pathophysiological Effects,* Volume I, (ed.) V.R. Beasley. CRC Press, Boca Raton, Florida.

Scott, P.M., Lombaert, G.A., Pellaers, P., Bacler, S., and Lappi, J. 1992. Ergot alkaloids in grain foods sold in Canada. *Journal of AOAC International,* 75:773-779.

Seitz, L.M., Sauer, D.B., Mohr, H.E., and Burrough, R. 1975. Weathered grain sorghum: Natural occurrence of alternariols and storability of the grain. *Phytopathology,* 65:1259.

Shelby, R.A. 1999. Toxicology of ergot alkaloids in agriculture. In *Ergot—The Genus* Claviceps, (eds.) V. Køen and L. Cvak, pp. 469-477. Harwood Academic Publishers, Amsterdam, The Netherlands.

Shelby, R.A., Rottinghaus, G.E., and Minor, H.C. 1994. Comparison of thin-layer chromatography and competitive immunoassay methods for detection of fumonisinin maize. *Journal of Agriculture and Food Chemistry,* 42:2064.

Shivendra Kumar and Roy, A.K. 2000. Occurrence of ochratoxin in maize grains— A report. *National Academy Science Letters,* 23:101-103.

Sibanda, L., de Saeger, S., van Peteghem, C., Grabarkiewicz-Szczesna, J., and Tomczak, M. 2000. Detection of T-2 toxin in different cereals by flow-through enzyme immunoassay with a simultaneous internal reference. *Journal of Agricultural and Food Chemistry*, 48:5864-5867.

Sinha, A.K. 2000. Testing methods for aflatoxins in foods. *Food and Nutrition Bulletin*, 21:458-464.

Sutton, J.C., Balico, W., and Funnel, H.S. 1980. Relation of weather variables to incidence of zearalenone in corn in southern Ontario. *Canadian Journal of Plant Sciences*, 60:149.

Sydenham, E.W., Shephard, G.S., Thiel, R.G., Bird, C., and Miller, B.M. 1996. Determination of fumonisins in corn: Evaluation of competitive immunoassay and HPLC techniques. *Journal of Agriculture and Food Chemistry*, 44:159.

Sydenham, E.W., Stockenström, S., Thiel P.G., Rheeder, J.P., Doko, M.B., Bird, C., and Miller, B. 1996. Polyclonal antibody-based ELISA and HPLC methods for the determination of fumonisins in corn: Comparative study. *Journal of Food Protection*, 59:893.

Sydenham, E.W., Thiel, P.G., Marasas, W.F.O., Shepherd, G.S., Van Schalkwyk, D.J., and Koch, K.R. 1990. Natural occurrence of some *Fusarium* mycotoxins in corn from low and high esophageal cancer prevalence areas of the Transkei, South Africa. *Journal of Agriculture and Food Chemistry*, 38:1900.

Teshima, R., Hirai, K., Sato, M., Ikebuchi, H., Ichinoe, M., and Terao, T. 1990. Radioimmunoassay of nivalenol in barley. *Applied and Environmental Microbiology*, 56:764.

Thirumala Devi, K., Mayo, M.A., Gopal Reddy, Emmanuel, R.E., Larondelle, Y., and Reddy, D.V.R. 2001. Occurrence of ochratoxin A in black pepper, coriander, ginger and turmeric in India. *Food Additives and Contaminants*, 18:830-835.

Thuvander, A., Möller, T., Barbieri, H.E., Jansson, A., Salomonsson, A.C., and Olsem, M. 2001. Dietary intake of some important mycotoxins by Swedish population. *Food Additives and Contaminants*, 18:696-706.

Trucksess, M.W., Stack, M.E., Allen, S., and Barrion, N. 1995. Immunoaffinity column coupled with liquid chromatography for determination of fumonisinB_1 in canned and frozen sweet corn. *Journal of Association of Analytical Chemists International*, 78:705.

Trucksess, M.W., Stack, M.E., Nesheim, S. et al. 1991. Immunoaffinity column coupled with solution fluorometry or liquid chromatography post-column derivatization for determination of aflatoxins in corn, peanuts and peanut butter: Collaborative study. *Journal of Association of Official Analytical Chemists*, 74:81.

Trucksess, M.W., Stack, M.E., Nesheim, S., Park, D.L., and Pohland, A.E. 1989. Enzyme-linked immunosorbent assay of aflatoxins B_1, B_2 and G_1 in corn, cottonseed, peanut butter and poultry feed: Collaborative study. *Journal of Association of Analytical Chemists*, 72:957.

Tsai, G.J. and Yu, S.C. 1999. Detecting *Aspergillus parasiticus* in cereals by an enzyme-linked immunosorbent assay. *International Journal of Food Microbiology,* 50:181-189.

Tseng, T.C. 1995. Immunochemical methods for detection of toxins and pesticides. In *Molecular Methods in Plant Pathology,* (eds.) R.P. Singh and U.S. Singh, pp. 445-460. CRC Press, Boca Raton, Florida.

Tseng, T.C., Lee, K.L., Deng, C.Y., Liu, C.Y., and Huang, J.W. 1995. Production of fumonisins by *Fusarium* species of Taiwan. *Mycopathologia,* 130:117.

Usleber, E., Dietrich, R., Schneider, E., and Märtlbauer, E. 2001. Immunochemical method for ochratoxin A. In *Mycotoxin Protocols,* (eds.) M.W. Trucksess and A.E. Pohland, pp. 81-94. Humana Press, Totowa, New Jersey.

Usleber, E., Märtlbauer, E., Dietrich, R., and Terplan, G. 1991. Direct enzyme-linked immunosorbent assays for the detection of the 8-ketotrichothecene mycotoxins deoxynivalenol, 3-acetyl deoxynivalenol and 15-acetyl-deoxynivalenol in buffer solutions. *Journal of Agricultural and Food Chemistry,* 39:2091.

Usleber, E., Straka, M., and Terplan, G. 1994. Enzyme immunoassay for fumonisin B_1 applied to corn-based food. *Journal of Agricultural and Food Chemistry,* 42:1392.

Vrabcheva, T.M. 2000. Mycotoxins in spices. *Voprosy Pitaniya,* 69:40-43.

Vrabcheva, T., Usleber, E., Dietrich, R., and Märtlabauer, E. 2000. Cooccurrence of ochratoxin A and citrinin in cereals from Bulgarian villages with a history of Balkan endemic nephropathy. *Journal of Agricultural and Food Chemistry,* 48:2483-2488.

Warden, B.A., Allam, K., Sentissi, A., Cecchini, D.J., and Giese, R.W. 1987. Repetitive hit-and-run fluoroimmunoassay for T-2 toxin. *Analytical Biochemistry,* 162:363.

Warner, R., Ram, B.P., Hart, H.P., and Pestka, J.J. 1986. Screening for zearalenone in corn by competitive direct enzyme-linked immunosorbent assay. *Journal of Agricultural and Food Chemistry,* 34:714.

Widstrom, N.W., McMillan, W.W., and Wilson, D. 1987. Segregation for resistance to aflatoxin contamination among seeds on an ear of hybrid maize. *Crop Science,* 27:961-963.

Wilson, D.M., Sydenham, E.W., Lombaert, G.A., Trucksess, M.W., Abramson, D., and Bennett, G.A. 1998. Mycotoxin analytical techniques. In *Mycotoxins in Agriculture and Food Safety,* (eds.) K.K. Sinha and D. Bhatnagar, pp. 135-181. Marcel Dekker, New York.

Wogan, G.N. 1991. Aflatoxins as risk factors for primary hepatocellular carcinoma in human. In *Mycotoxins, Cancer and Health,* (eds.) G.A. Bray and D.H. Ryan, p.18. Louisiana University Press, Baton Rouge, Louisiana.

Wood, G.E. and Trucksess, M.W. 1998. Regulatory control programs for mycotoxins-contaminated food. In *Mycotoxins in Agriculture and Food Safety,* (eds.) K.K. Sinha and D. Bhatnagar, pp. 459-481. Marcel Dekker Inc., New York.

Xu, Y.C., Zhang, G.S., and Chu, F.S. 1988. Enzyme-linked immunosorbent assay for deoxynivalenol in corn and wheat. *Journal of Association of Official Analytical Chemists,* 71:945.

Yates, I.E., Meredith, F., Smart, W., Bacon, C.W., and Jaworski, A.J. 1999. *Trichoderma viridae* suppresses fumonisinB$_1$ production by *Fusarium moniliforme. Journal of Food Protection,* 62:1326-1332.

Yong, R.K. and Cousin, M.A. 2001. Detection of moulds producing aflatoxin in maize and peanuts by an immunoassay. *International Journal of Food Microbiology,* 65:27-38.

Yuan, Q.P., Hu, W.Q., Pestka, J.J., He, S.Y., and Hart, L.P. 2000. Expression of a functional antizearalenone single-chain Fv antibody in transgenic *Arabidopsis* plants. *Applied and Environmental Microbiology,* 66:3499-3505.

Chapter 14

Abad, A., Manclús, J.J., Moreno, M.J., and Montoya, A. 2001. Determination of thiabendazole in fruit juices by a new monoclonal immunoassay. *Journal of Association of Official Analytical Chemists, International,* 84:156-161.

Abad, A., Moreno, M.J., and Montoya, A. 1997. A monoclonal immunoassay for carbofuran and its application to the analysis of fruit juices. *Analytica Chemica Acta,* 347:103-110.

Abad, A., Moreno, M.J., and Montoya, A. 1999. Development of monoclonal antibody-based immunoassays to the *N*-methylcarbamate pesticide carbofuran. *Journal of Agricultural and Food Chemistry,* 47:2475-2485.

Abad, A., Moreno, M.J., Pelegrí, R., Martínez, M.J., Saez, A., Gamon, M., and Montoya, A. 2001. Monoclonal enzyme immunoassay for the analysis of carbaryl in fruits and vegetables without sample cleanup. *Journal of Agricultural and Food Chemistry,* 49:1707-1712.

Agusti, N., Aramburu, J., and Gabarra, R. 1999. Immunological detection of *Helicoverpa armigera* (Lepidoptera: Noctuidae) ingested by heteropteran predators: Time-related decay and effect of meal size on detection period. *Annals of Entomological Society of America,* 92:56-62.

Bauer, E.R.S., Bitsch, N., Brunn, H., Sauerwein, H., and Meyer, H.H.D. 2002. Development of an immuno-immobilized androgen receptor assay (IRA) and its application for the characterization of the receptor binding affinity of different pesticides. *Chemosphere,* 45:1107-1115.

Brent, K.J. 2003. Fungicides, an overview. In *Encyclopedia of Agrochemicals,* Volume two, (Eds.) J.R. Plimmer, D.W. Gammon, and N.N. Ragsdale, pp. 528-531, Wiley-Interscience, UK.

Bushway, R.J. and Fan, Z.H. 1998. Determination of chlorpyrifos in fruits and vegetables by ELISA and confirmation by GC-AED. *Food and Agricultural Immunology,* 10:215-221.

Danks, C., Chaudhry, M.Q., Parker, L., Barker, I., and Banks, J.N. 2001. Development and validation of an immunoassay for the determination of tebuconazole residues in cereal crops. *Food and Agricultural Immunology,* 13:151-159.

Dreher, R.M. and Podratzki, B. 1988. Development of an enzyme immunoassay for endosulfan and its degradation products. *Journal of Agricultural and Food Chemistry,* 36:1072.

Ercegovich, C.D., Vallejo, R.D., Gettig, R.D., Woods, L., Bogus, E.R., and Mumnia, R.O. 1981. Development of a radioimmunoassay for parathion. *Journal of Agricultural and Food Chemistry,* 29:559.

Forlani, F. and Pagani, S. 1998. Strategies of ELISA development of pesticide detection. *Recent Research Developments in Agricultural and Food Chemistry,* 2:155-164.

Gueguen, F., Boisdé, F., Queffalec, A.L., Haelters, J.P., Thouvenot, D., Corbel, B., and Nodet, P. 2000. Hapten synthesis for the development of a competitive inhibition enzyme-immunoassay for thiram. *Journal of Agricultural and Food Chemistry,* 48:4492-4499.

Hagler, J.R. 1998. Variation in the efficacy of several predator gut content immunoassays. *Biological Control,* 12:25-32.

Hagler, J.R., Naranjo, S.E., Erickson, M.L., Matchley, S.A., and Wright, S.F. 1997. Immunological examinations of species variability in predator gut immunoassays: Effect of predator-prey protein ratio on immunosensitivity. *Biological Control,* 9:120-128.

Huang, L.L., Kang, Z., Heppner, C., and Buchenauer, H. 2001. Ultrastructural and immunocytochemical studies on effects of the fungicide Mon 65500 (Latitude ®) on colonization of roots of wheat seedlings by *Gaeumannomyces graminis* var. *tritici. Journal of Plant Diseases and Protection,* 108:188-203.

Jahn, C. and Schwack, W. 2001. Determination of cutin-bound residues of chlorothalonil by immunoassay. *Journal of Agricultural and Food Chemistry,* 49: 1233-1238.

Jahn, C., Zorn, H., Peterson, A., and Schwack, W. 1999. Structure-specific detection of plant cuticle bound residues of chlorothalonil by ELISA. *Pesticide Science,* 55:1167-1176.

Karanth, N.G.K., Pasha, A., Amita Rani, B.E., Asha, M.B., Udayakumari, C.G., and Vijayashankar, Y.N. 1998. Developing immunoassays in a developing nation: Challenges and success in India. *ACIAR Proceedings Series,* No. 85: 263-269.

Kendall, J.J., Hollomon, D.W., and Selley, A. 1998. Immunodiagnosis as an aid to timing of fungicide sprays for the control of *Mycosphaerella graminicola* on winter wheat in the U.K. *Brighton Crop Protection Conference: Pests and Diseases,* 2:701-706.

Le, H.M., Hegedüs, G., and Székács, J. 1998. Differential detection of N-heterocyclic compounds and their N-methylated derivatives of myclobutanil by immunoanalysis. *Acta Biologica Hungarica,* 49:455-462.

Lim, U.T. and Lee, J. 1999. Enzyme-linked immunosorbent assay used to analyze predation of *Nilaparvata lugens* (Homoptera: Delphacidae) by *Pirata subpiraticus* (Arneae: Lycosidae). *Environmental Entomology*, 28:1177-1182.

Mercader, J.V. and Montoya, A. 1997. A monoclonal antibody-based ELISA for the analysis of azinphos-methyl in fruit juices. *Analytica Chimica Acta*, 347:95-101.

Meulenberg, E.P., de Vree, L.G., and Dogterom, J. 1999. Investigation of indicative methods in The Netherlands: Validation of several commercial ELISAs for pesticides. *Analytica Chemica Acta*, 399:143-149.

Moreno, M.J., Abad, A., Pelegrí, R., Martínez, M.I., Saez, A., Gamon, M., and Montoya, A. 2001. Validtion of a monoclonal enzyme immunoassay for the determination of carbofuran in fruits and vegetables. *Journal of Agricultural and Food Chemistry*, 49:1713-1719.

Nakata, M., Fukushima, A., and Ohkawa, H. 2001. A monoclonal antibody-based ELISA for the analysis of the insecticide flucythrinate in environmental and crop samples. *Pest Management Science*, 57:269-277.

Naranjo, S.E. and Hagler, J.R. 2001. Toward quantification of predation with predator gut immunoassays: A new approach integrating functional response behavior. *Biological Control*, 20:175-187.

Newsome, W.H. and Collins, P.G. 1987. Enzyme-linked immunosorbent assay of benomyl and thiabendazole in foods. *Journal of Association of Official Analytical Chemists*, 70:1025.

Newsome, W.H. and Shields, J.B. 1981. A radioimmunoassay for benomyl and methyl 2-benzimidazole carbamate on food crops. *Journal of Agricultural and Food Chemistry*, 29:220-222.

Newson, W.H. 1985. An enzyme-linked immunosorbent assay for metalaxyl in foods. *Journal of Agricultural and Food Chemistry*, 33:528.

Newson, W.H. 1986. Development of an enzyme-linked immunosorbent assay for triadimefon in foods. *Bulletin of Environmental Contamination and Toxicology*, 36:9.

Nunes, G.S., Marco, M.P., Farré, M., and Barcalo, D. 1999. Direct application of an enzyme-linked immunosorbent assay method for carbaryl determination in fruits and vegetables. Comparison with liquid chromatograph-post column reaction fluorescence detection method. *Analytica Chemica Acta*, 387:245-253.

Oxley, S.J.P. 1999. Optimizing input and performance of triazole fungicides in winter wheat by monitoring residual activity. *HGCA Project Report*, No. 204: 66 pp.

Queffelec, A.L., Boisdé, F., Larue, J.P., Haelters, J.P., Corbel, B., Thouvenot, D., and Nodet, P. 2001. Development of an immunoassay (ELISA) for the quantification of thiram in lettuce. *Journal of Agricultural and Food Chemistry*, 49: 1675-1680.

Rosso, I., Giraudi, G., Gamberini, R., Baggiani, C., and Vanni, A. 2000. Application of an ELISA to the determination of benalaxyl in red wines. *Journal of Agricultural and Food Chemistry*, 48:33-36.

Schmaedick, M.A., Ling, K.S., Gonsalves, D., and Shelton, A.N. 2001. Development and evaluation of an enzyme-linked immunosorbent assay to detect *Pieris rapae* remains in guts of arthrapod predators. *Entomologia Experimentalis et Applicata*, 99:1-12.

Symondson, W.O.C., Gasull, T., and Liddell, J.E. 1999. Rapid identification of adult whiteflies in plant consignments using monoclonal antibodies. *Annals of Applied Biology*, 134:271-276.

Takahashi, Y., Odanaka, Y., Wada, Y., Minakawa, Y., and Fukita, T. 1999. Pesticide run off and mass balance in field model tests analyzed by commercially available immunoassay kits. *Journal of Pesticide Science*, 24:255-261.

Tseng, T.C. 1995. Immunochemical methods for detection of toxins and pesticides. In *Molecular Methods in Plant Pathology*, (eds.) R.P. Singh and U.S. Singh, pp. 445-460. CRC Press, Boca Raton, Florida.

Zeng, F., Ramaswamy, S.B., and Pruett, S. 1998. Monoclonal antibodies specific to tobacco budworm and bollworm eggs. *Annals of Entomological Society of America*, 91:677-684.

Zhang, G.R., Zhang, W.Q., and Gu, D.S. 1999. Quantifying predation by *Ummeliata insecticeps* Boes et Str. (Araneae: Linyphiidae) on rice planthoppers using ELISA. *Entomologia Sinica*, 6:77-82.

Zhang, G.R., Zhang, W.Q., and Gu, D.X. 1997. Application of ELISA method for determining control effects of predatory arthropods on rice planthoppers in rice field. *Acta Entomologica Sinica*, 40:171-176.

Index

Page numbers followed by the letter "f" indicate figures; those followed by the letter "t" indicate tables.

Order a copy of this book with this form or online at:
http://www.haworthpress.com/store/product.asp?sku=5131

IMMUNOLOGY IN PLANT HEALTH AND ITS IMPACT ON FOOD SAFETY

_____in hardbound at $79.95 (ISBN: 1-56022-286-7)

_____in softbound at $49.95 (ISBN: 1-56022-287-5)

Or order online and use special offer code HEC25 in the shopping cart.

COST OF BOOKS_____	☐ **BILL ME LATER:** (Bill-me option is good on US/Canada/Mexico orders only; not good to jobbers, wholesalers, or subscription agencies.)
	☐ Check here if billing address is different from shipping address and attach purchase order and billing address information.
POSTAGE & HANDLING_____ *(US: $4.00 for first book & $1.50 for each additional book)* *(Outside US: $5.00 for first book & $2.00 for each additional book)*	
	Signature_____
SUBTOTAL_____	☐ **PAYMENT ENCLOSED: $**_____
IN CANADA: ADD 7% GST_____	☐ **PLEASE CHARGE TO MY CREDIT CARD.**
STATE TAX_____ *(NJ, NY, OH, MN, CA, IL, IN, & SD residents, add appropriate local sales tax)*	☐ Visa ☐ MasterCard ☐ AmEx ☐ Discover ☐ Diner's Club ☐ Eurocard ☐ JCB
	Account # _____
FINAL TOTAL_____ *(If paying in Canadian funds, convert using the current exchange rate, UNESCO coupons welcome)*	Exp. Date_____
	Signature_____

Prices in US dollars and subject to change without notice.

NAME_____

INSTITUTION_____

ADDRESS_____

CITY_____

STATE/ZIP_____

COUNTRY_____ COUNTY (NY residents only)_____

TEL_____ FAX_____

E-MAIL_____

May we use your e-mail address for confirmations and other types of information? ☐ Yes ☐ No We appreciate receiving your e-mail address and fax number. Haworth would like to e-mail or fax special discount offers to you, as a preferred customer. **We will never share, rent, or exchange your e-mail address or fax number.** We regard such actions as an invasion of your privacy.

Order From Your Local Bookstore or Directly From
The Haworth Press, Inc.
10 Alice Street, Binghamton, New York 13904-1580 • USA
TELEPHONE: 1-800-HAWORTH (1-800-429-6784) / Outside US/Canada: (607) 722-5857
FAX: 1-800-895-0582 / Outside US/Canada: (607) 771-0012
E-mailto: orders@haworthpress.com

For orders outside US and Canada, you may wish to order through your local
sales representative, distributor, or bookseller.
For information, see http://haworthpress.com/distributors

(Discounts are available for individual orders in US and Canada only, not booksellers/distributors.)
PLEASE PHOTOCOPY THIS FORM FOR YOUR PERSONAL USE.
http://www.HaworthPress.com   BOF04